Daniel Härtl

Scattering Amplitudes in String Theory

Daniel Härtl

Scattering Amplitudes in String Theory

Correlators of Ramond-Neveu-Schwarz Fields

Südwestdeutscher Verlag für Hochschulschriften

Impressum/Imprint (nur für Deutschland/only for Germany)
Bibliografische Information der Deutschen Nationalbibliothek: Die Deutsche Nationalbibliothek verzeichnet diese Publikation in der Deutschen Nationalbibliografie; detaillierte bibliografische Daten sind im Internet über http://dnb.d-nb.de abrufbar.
Alle in diesem Buch genannten Marken und Produktnamen unterliegen warenzeichen-, marken- oder patentrechtlichem Schutz bzw. sind Warenzeichen oder eingetragene Warenzeichen der jeweiligen Inhaber. Die Wiedergabe von Marken, Produktnamen, Gebrauchsnamen, Handelsnamen, Warenbezeichnungen u.s.w. in diesem Werk berechtigt auch ohne besondere Kennzeichnung nicht zu der Annahme, dass solche Namen im Sinne der Warenzeichen- und Markenschutzgesetzgebung als frei zu betrachten wären und daher von jedermann benutzt werden dürften.

Verlag: Südwestdeutscher Verlag für Hochschulschriften GmbH & Co. KG
Heinrich-Böcking-Str. 6-8, 66121 Saarbrücken, Deutschland
Telefon +49 681 37 20 271-1, Telefax +49 681 37 20 271-0
Email: info@svh-verlag.de

Approved by: München, LMU, Diss., 2011

Herstellung in Deutschland:
Schaltungsdienst Lange o.H.G., Berlin
Books on Demand GmbH, Norderstedt
Reha GmbH, Saarbrücken
Amazon Distribution GmbH, Leipzig
ISBN: 978-3-8381-3044-6

Imprint (only for USA, GB)
Bibliographic information published by the Deutsche Nationalbibliothek: The Deutsche Nationalbibliothek lists this publication in the Deutsche Nationalbibliografie; detailed bibliographic data are available in the Internet at http://dnb.d-nb.de.
Any brand names and product names mentioned in this book are subject to trademark, brand or patent protection and are trademarks or registered trademarks of their respective holders. The use of brand names, product names, common names, trade names, product descriptions etc. even without a particular marking in this works is in no way to be construed to mean that such names may be regarded as unrestricted in respect of trademark and brand protection legislation and could thus be used by anyone.

Publisher: Südwestdeutscher Verlag für Hochschulschriften GmbH & Co. KG
Heinrich-Böcking-Str. 6-8, 66121 Saarbrücken, Germany
Phone +49 681 37 20 271-1, Fax +49 681 37 20 271-0
Email: info@svh-verlag.de

Printed in the U.S.A.
Printed in the U.K. by (see last page)
ISBN: 978-3-8381-3044-6

Copyright © 2012 by the author and Südwestdeutscher Verlag für Hochschulschriften GmbH & Co. KG and licensors
All rights reserved. Saarbrücken 2012

This doctoral thesis is based on the author's work conducted from May 2008 until April 2011 at the Max-Planck-Institut für Physik (Werner-Heisenberg-Institut), München, under the supervision of Dr. Stephan Stieberger and Prof. Dr. Dieter Lüst. The work has been published in [1–4].

Abstract

In this thesis we provide calculational tools in order to calculate scattering amplitudes in string theory at tree- and loop-level. In particular, we discuss the calculation of correlation functions consisting of Ramond–Neveu–Schwarz fields in four, six, eight and ten space-time dimensions and calculate the amplitude involving two gauge fields and four gauginos at tree-level.

Multi-parton superstring amplitudes are of considerable theoretical interest in the framework of a full-fledged superstring theory and of phenomenological interest in describing corrections to four-dimensional scattering processes. The Neveu–Schwarz fermions and Ramond spin fields enter the scattering amplitudes through vertex operators of bosonic and fermionic string states and determine the Lorentz structure of the total amplitude. Due to their interacting nature their correlators cannot be evaluated using Wick's theorem but must be calculated from first principles.

At tree-level such correlation functions can be determined by analyzing their Lorentz and singularity structure. In four space-time dimensions we show how to calculate Ramond–Neveu–Schwarz correlators with any number of fields. This method is based on factorizing the expressions into correlators involving only left- or right-handed spin fields and calculating these functions. This factorization property does not hold in higher dimensions. Nevertheless, we are able to calculate certain classes of correlators with arbitrary many fields. Additionally, in eight dimensions we can profit from $SO(8)$ triality to derive further tree-level correlation functions.

Ramond–Neveu–Schwarz correlators at loop-level can be evaluated by re-expressing the fermions and spin fields in terms of $SO(2)$ spin system operators. Using this method we present expressions for all correlators up to six-point level and show in addition results for certain classes of correlators with any number of fields. Our findings hold for string scattering at arbitrary loop order.

To complement the discussion we calculate the tree-level amplitude of two gauge fields and four gauginos for string compactifications to four dimensions and give its field theory limit. This open string amplitude is of particular interest because it can be related to an open-closed amplitude involving gauge fields and bulk moduli. In this way the mapping between the open and the open-closed sector can be studied in great detail and brane-bulk couplings can be determined in terms of open string couplings.

Acknowledgments

I am deeply indebted to my supervisor Stephan Stieberger for his excellent support. I could greatly profit from his knowledge and insight into this subject and enjoyed the conversations about our recent mountaineering trips. Likewise I want to thank Dieter Lüst for giving me the opportunity to work in the Munich String Theory group as well as constantly supporting and encouraging me during my doctoral studies. I also want to thank Johanna Erdmenger, who agreed to act as second referee for this thesis.

I am very grateful to Oliver Schlotterer for the fruitful collaboration and many interesting and helpful discussions. Additionally I would like to thank Martin Ammon, Ralph Blumenhagen, Stefan Groot Nibbelink, Sebastian Halter, Johannes Held, Benjamin Jurke, Patrick Kerner, Sebastian Moster, Hai Ngo Than, Erik Plauschinn, Thorsten Rahn, Maximilian Schmidt-Sommerfeld and Dimitrios Tsimpis for sharing their thoughts and ideas. Special thanks go to Johannes Held, Benjamin Jurke, Ananda Landwehr, Oliver Schlotterer and Martin Spinrath for comments on the manuscript.

I am happy to thank my fellow PhD students and colleagues at the MPI as well as the head of our IMPRS, Frank Daniel Steffen for contributing to a memorable time. I am also very grateful to my parents, my sister and my friends Markus Gick, Klaus Mühlbauer and Christopher Rose.

Most importantly, I want to thank Tamara for her love, patience and encouragement.

Contents

Abstract	ii
Acknowledgments	v

1 Introduction — 1
   1.1 Particle Physics and Gravity 1
   1.2 String Theory 4
   1.3 Scattering Amplitudes 7
   1.4 Motivation & Outline 10

2 Scattering in String Theory — 13
   2.1 Conformal Field Theory on the World-Sheet 13
       2.1.1 Matter Fields 14
       2.1.2 The Spin Field 17
       2.1.3 Ghost Fields 19
       2.1.4 Open strings 22
       2.1.5 Vertex Operators 24
   2.2 String Scattering Amplitudes 27
       2.2.1 Tree-level Amplitudes 29
       2.2.2 Loop-Level Amplitudes 33

3 Ramond–Neveu–Schwarz Correlators at Tree-Level — 35
   3.1 Prerequisites 35
       3.1.1 Spinors in Higher Dimensions 35
       3.1.2 The Underlying Conformal Field Theory 37
   3.2 The Evaluation of Correlators 38
       3.2.1 The Iterative Procedure 38
       3.2.2 From Fermions to Spin Fields 41
       3.2.3 Alternative Methods 43
   3.3 The Index Terms 45
       3.3.1 The Group Theory Behind the Correlators 45

	3.3.2 Relations between Index Terms	46
	3.3.3 Manipulation of Index Terms	49
3.4	Techniques in Four Dimensions	51
	3.4.1 The Number of Index Terms	52
	3.4.2 Replacing Fermions with Spin Fields	54
	3.4.3 Pure Spin Field Correlators	56
3.5	Techniques in Six Dimensions	59
3.6	Techniques in Eight Dimensions	60
	3.6.1 $SO(8)$ Triality	60
	3.6.2 Examples	62
3.7	Techniques in Ten Dimensions	66

4 Results in Four Space-Time Dimensions — 67

4.1	Review of Known Results	68
4.2	Five-Point Functions	69
4.3	Six-Point Functions	69
4.4	Seven-Point Functions	72
4.5	Eight-Point Functions	74
4.6	General Results	79

5 Ramond–Neveu–Schwarz Correlators at Loop-Level — 83

5.1	Prerequisites	83
	5.1.1 Generalized Θ Functions	83
	5.1.2 $SO(2)$ Spin Systems	85
5.2	Loop Correlators	88
	5.2.1 Correlators of $SO(2)$ Spin Systems	88
	5.2.2 Results in Lorentz Covariant Form	89
5.3	Results of RNS Loop Correlators	93
	5.3.1 Results for $D = 4$	94
	5.3.2 Results for $D = 6$	94
	5.3.3 Results for $D = 8$	97
	5.3.4 Results for $D = 10$	102
	5.3.5 Pure Spin Field Correlators	107
	5.3.6 General Results	108

6 A Full Amplitude — 113

6.1	Open vs. Open-Closed Amplitudes	113
6.2	Prerequisites	115

6.3	The Separate Correlators	116
6.4	A First Result	121
6.5	Gauge Invariance	122
6.6	Spinor Products	123
6.7	The Field Theory Limit	125

7 Conclusion   127
    7.1 Summary   127
    7.2 Outlook   128

A Gamma Matrices in D Dimensions   131
    A.1 Notation and conventions   131
    A.2 Symmetry properties   133
    A.3 Fierz identities   135
    A.4 A Concrete Representation   136

B Relations between Index Terms   139
    B.1 Relations for $D = 4$ 140
    B.2 Relations for $D = 6$ 146
    B.3 Relations for $D = 8$ 148
    B.4 Relations for $D = 10$ 150

C Generalized Θ Functions   153
    C.1 Periodicity Properties   153
    C.2 Fay's Trisecant Identity   154

D Details of the Amplitude Calculation   157
    D.1 The Kinematical Structure   157
    D.2 Gauge Invariance   159
    D.3 Results ins Spinor Product Notation   161

E Spinor Helicity Formalism   163
    E.1 Clifford Algebra   163
    E.2 Momentum Spinors   164
    E.3 Spinor Products   166
    E.4 Polarization Vectors   167

Bibliography   169

List of Figures

1.1	Intersecting D6-brane model	8
1.2	Stringy corrections to dijet production	9
2.1	Summing over world-sheet topologies	28
2.2	Mapping the string world-sheet onto the unit disk	29
2.3	Mapping the unit disk onto the upper half plane	31
2.4	Chan–Paton factors in a four-string interaction.	32
2.5	String contact interaction and exchange of Regge resonances	33
2.6	One-cycles α_I and β_I of a Riemann surface	34
3.1	Dynkin diagram of $SO(8)$	61
4.1	The web of limits for the correlators Ω_n and ω_n	81

List of Tables

3.1	Number of independent index terms for $\langle S_{\alpha_1} \ldots S_{\alpha_{2M}} \rangle$	42
3.2	Number of independent index terms for $\langle S_{\alpha_1} S^{\dot\beta_1} \ldots S_{\alpha_M} S^{\dot\beta_M} \rangle$	43
3.3	Number of independent index terms for correlators in various dimensions	47
3.4	Symmetry properties of index terms	51
3.5	Values of the coefficients $q(i,N)$	54
4.1	Number of index terms of the correlators Ω_n and ω_n	82
6.1	Kinematic quantities s_{ij} expressed through s_k and t_k	118

CHAPTER 1

Introduction

Particle physics is entering a very exciting time. In 2009 the Large Hadron Collider (LHC) at CERN began its operation after many years of construction. Experimental and theoretical physicists alike are excited what new insights can be gained in the coming years from the proton-proton collisions taking place in the detectors. With the same eagerness physicists gaze at other earth-bound experiments and satellite missions waiting for new results. Future findings are going to challenge our current understanding of the universe on small and large scales and will point the way to correct theories of particle physics and gravity.

1.1 Particle Physics and Gravity

Our current knowledge about the interactions of the smallest constituents of matter is manifested in the standard model of elementary particle physics (SM) [5–7]. It successfully describes the strong and electroweak interactions down to distances of at least 10^{-16} cm and has been tested to high accuracy. Its underlying mathematical concept is that of local quantum field theory [8], which is renormalizable. In detail, the gauge group of the SM is given by $SU(3)_c \times SU(2)_L \times U(1)_Y$ and the matter sector consists of three generations of leptons and quarks. Mass terms for the fermions and gauge bosons in the SM are not gauge invariant and therefore must be generated dynamically. This is achieved via spontaneous symmetry breaking [9–11]. One scalar $SU(2)$ doublet, the Higgs field, breaks the SM gauge group down to $SU(3)_c \times U(1)_{\text{em}}$. At the time of writing the Higgs particle is the only constituent in the SM yet to be discovered. In the broken theory the gluons represent the gauge bosons of the strong interaction, while the photon and the massive W and Z bosons depict the quanta of the electromagnetic and weak interactions.

Despite its great achievements the SM does not resolve all issues from a conceptual point of view[1]. The most prominent example is the hierarchy problem [13–15] which questions why the

---
[1] In discussing the problems of the SM we follow [12].

two fundamental scales in physics, namely the electroweak scale $M_{\text{ew}} \approx 10^2$ GeV and the reduced Planck scale $M_{\text{Pl}} \approx 10^{18}$ GeV differ by sixteen orders of magnitude. Although this discrepancy does not pose a threat to the SM itself it has far-reaching consequences for the Higgs mass. Through loop diagrams the square of the bare Higgs mass receives quantum corrections, which grow quadratically with a cut-off scale introduced in order to regulate the momentum integral. The natural scale for the physical Higgs mass would then also be this cut-off scale. If the SM is taken to be valid up to energies where a theory including quantum gravity takes over, this cut-off scale would be the Planck scale, whereas for a Grand Unified Theory (GUT) replacing the SM at higher energies this scale would be still be as high as $M_{\text{GUT}} \approx 10^{16}$ GeV. However, searches at the Large Electron-Positron Collider (LEP) and indirect constraints from electroweak precision measurements favor a Higgs mass between 114.4 GeV and 186 GeV [16]. There must be an enormously contrived fine-tuning over roughly 30 decimals such that the bare value and the radiative corrections cancel and yield such a low Higgs mass.

Another fine-tuning problem within the SM is the strong CP problem [17]. Gauge symmetry does not prohibit to introduce a new CP-violating term θ_{QCD} to the Lagrangian of Quantum Chromodynamics (QCD). Such a term induces an electric dipole moment for the neutron, which is however heavily constrained from experiment. The SM fails to explain why θ_{QCD} is so small compared to the observed CP violation in the electroweak sector.

Next to these naturalness problems there are further issues in the SM worth mentioning. The SM cannot explain why the gauge group has its peculiar structure and more important it fails to illustrate why charges are quantized. All leptons and bosons come with electromagnetic charges that are multiples of $e/3$, which is crucial for the neutrality of atoms. In addition, we do not know why there are exactly three families of leptons and quarks, although under ordinary terrestrial conditions all matter is only built out of the first generation. The masses of the fermions is another mystery. Going from the top quark to the electron they vary over five orders of magnitude. The masses arise from the Yukawa couplings, but the SM does not predict why these couplings have such a hierarchical structure. In the same way there is no indication why the other parameters of the SM, like mixing parameters of quarks and leptons and further parameters in the neutrino sector, have their precise values. Additionally, recent data, e.g. from cosmic microwave background measurements, shows that a large fraction of more than 80% of the matter in our universe is not accounted for by SM particles, but is existent in the unknown form of dark matter [18].

In view of all these shortcomings an extensions of the SM is highly desirable. A widely-used approach to address some of the open questions is guided by the idea to incorporate new symmetries. As the symmetry principle has been a major point in the construction of the SM itself such advances are with good prospects. An elegant solution to the strong CP problem in this sense is given by the Peccei–Quinn mechanism [19, 20]. Here, the QCD angle is promoted

1.1 Particle Physics and Gravity

to a dynamical field charged under an additional $U(1)$ symmetry, which then is dynamically broken and gives rise to a massless Goldstone boson, namely the axion. Such a particle has not yet been discovered, but various experiments are looking for possible signals.

The most promising extension of the SM with additional symmetries is the idea of low energy supersymmetry (SUSY) [21] which adds fermionic generators to the symmetry algebra [22] and thus evades the Coleman-Mandula theorem [23]. From a phenomenological point of view a supersymmetric extension like the Minimal Supersymmetric Standard Model (MSSM) [24] is very attractive as it keeps the radiative corrections to the Higgs mass under control and can provide a viable dark matter candidate in form of the lightest SUSY particle. However, SUSY cannot be an exact symmetry. The SUSY algebra implies that particles and their superpartners have the same mass but a superpatner of the electron with a mass of 511 keV has not been discovered. SUSY must therefore be broken so that the superpartners are heavy and have not yet shown up in experiments. This breaking introduces many new parameters, whose values are not addressed by the theory as well as the Yukawa matrices and neutrino parameters. In this sense a SUSY extension of the SM is not a fully satisfactory solution.

A very promising feature in SUSY theories though is gauge coupling unification. The renormalization group evolution of the gauge couplings is changed in such a way that they perfectly meet at an energy scale $M_{\text{GUT}} \approx 10^{16}$ GeV. This unification provides strong evidence that the SM gauge groups are replaced at such high energies by a single group like $SU(5)$ [25] or $SO(10)$ [26]. In such GUTs it is also possible to address the origin and nature of quark and lepton masses and their mixing because these particles appear in the same multiplets. Charge quantization can also be explained in such a framework. However, common problems in the construction of GUT models are to suppress proton decay to an acceptable level and further (little) hierarchy problems.

However, extending the SM in such ways fails to provide an answer for the cosmological constant problem. Dark matter and SM model particles alone cannot account for the current phase of accelerated expansion of our universe. Therefore another form of energy driving this expansion has to be introduced. This dark energy must amount to nearly 73% of the total energy content of our universe [18]. In the standard model of cosmology, the ΛCDM model, dark energy is incorporated by a cosmological constant in Einstein's field equations of general relativity (GR) [27–29]. Trying to explain dark energy naively as the vacuum energy of some quantum field leads to a discrepancy of 118 orders of magnitude between the theoretical and experimental value [30]. In SUSY this disagreement is reduced to 60 orders of magnitude, which is still tremendous. This defect might be deeply intertwined with our current lore of gravity in the form of GR.

The concept used in GR is that of curved space-time, i.e. a four-dimensional manifold with Minkowski signature. Every form of energy causes space-time to curve which is encoded in the metric. The motion of an object due to a gravitational field is explained as the object moving along geodesics on the manifold. Compared to the other fundamental interactions in the SM the gravitational force is by far the weakest and can safely be neglected when studying particle collisions at "low" energies up the electroweak scale. However, gravity is dominating on large distances and therefore GR constitutes a good framework to describe macroscopic motion within the universe and the extension of the universe itself. Einstein's theory of GR is a classical theory and it might be just this fact which makes our current predictions of the cosmological constant and its actual value gape so tremendously. Furthermore in the present formulation of gravity we cannot describe physical effects at energies around the Planck scale where gravity is expected to become as dominant as the other fundamental forces. Although these energies might never be accessible in a laboratory experiment, this situation was existent shortly after the big bang.

The theory seems to be incomplete in another fashion. The validity of general relativity ends if the Schwarzschild radius of an object exceeds its size. The object collapses into a black hole, which is mathematically described as a point-like singularity. Transitions, like a massive star collapsing into a black hole, from an initially well-defined setting to a singularity of space-time are troublesome and signal the break-down of the theory. A possible solution would be that such singularities get smeared by quantum effects. Yet further issues arise in the discussion of black holes if thermodynamics is taken into account [31, 32]. A black hole must carry a vast amount of entropy which is proportional to the size of its event horizon. Generally, the number of microstates leading to a thermodynamical configuration accounts for the entropy of the system. General relativity, however, does not have the necessary degrees of freedom to explain the enormous entropy of a black hole and fails at providing a microscopic explanation of the Bekenstein-Hawking formula.

These defects of general relativity propose to construct a quantum theory of gravity. Doing quantum field theory based on the action of general relativity yields, however, a non-renormalizable theory. Infinitely many parameters have to be introduced in order to render the theory finite which makes the theory un-predictive. Thus new revolutionary approaches are necessary.

1.2 String Theory

The most promising candidate for a theory of quantum gravity is string theory [33–36]. It was originally discovered as a by-product in the late 1960s in order to explain hadronic resonances appearing in the CERN accelerators and elsewhere. The discovered resonance peaks exhibit Regge behavior $j = j_0 + \alpha' M^2$, a relation between their mass M and spin j. Physicists were able to construct an S-matrix with these properties [37], which was later discovered to arise from the scattering of bosonic strings. Although QCD turned out to be the correct theory to

1.2 String Theory

describe strong interactions and asymptotic freedom the interest in string theory did not fade. It was discovered that string theory yields a massless spin-2 state which can be associated with the graviton, the exchange boson of gravity. In contrast to loop quantum gravity [38] the big advantage of string theory is that it also incorporates gauge interactions like in the SM and thus provides a unified framework to describe all fundamental interactions of nature.

At the moment only a perturbative formulation of weakly coupled string theory is available, by which the motion of quantized strings in a given space-time with certain background charges can be described. Five different string theories (type I, type IIA, type IIB, heterotic $E_8 \times E_8$ and heterotic $SO(32)$) are known which are related by different dualities. Although the formulation in this form is sufficient for the discussion of many phenomena the final goal is still to obtain a fully quantized version of string theory which might be provided by a so-far poorly understood theory in eleven dimensions called M-theory [39].

The integral parts of string theory are one-dimensional objects, namely strings, sweeping out a two-dimensional surface in space-time, the string world-sheet. The action is mathematically described by a non-linear σ-model embedded into a higher-dimensional target-space. The case of one-dimensional objects is singled out because in this case the symmetry algebra of conformal transformations is infinite dimensional. The strings can oscillate in different modes due to their extended nature which results in a discrete mass spectrum. This is also special to the case of strings and is lost if the quantization of higher-dimensional membranes is considered [40]. The mass gap in the spectrum is characterized by the single free parameter of the theory, the string length $l_s = 2\pi\sqrt{\alpha'}$, where α' is the Regge slope. Promoting the world-sheet action to a supersymmetric theory solves two problems of bosonic string theory. Firstly, it gives rise to fermionic states in the spectrum. Secondly, the troublesome tachyonic ground state is projected out via the GSO projection [41]. This results in a theory that exhibits target space SUSY. In order for the theory to be free of a superconformal anomaly the target space must be ten-dimensional space-time [42].

Strings come in two topologies, they can be either closed or open. Closed strings can propagate in all ten space-time dimensions. Upon quantization these yield at the lowest mass-level the spin-2 graviton. The situation is different for open strings. The end-points of the latter must satisfy either Neumann or Dirichlet boundary conditions. The latter imply in type II string theories that the endpoints of open strings are confined to higher-dimensional D-branes [43]. The spectrum of open strings gives rise at the lowest mass-level to gauge fields which endow the D-branes with a super-Yang-Mills (SYM) theory living on their world-volume. If D-branes intersect, open strings can stretch from one brane to the other in the vicinity of the intersection. In the presence of two-form flux these new string states depict massless chiral fermions. In such a way it is possible to construct string models which contain many of the phenomenological features of the SM [44].

With string theory at hand it is possible to confront the problems arising in general relativity. A big success has been the explanation of the Bekenstein-Hawking formula for extremal five-dimensional black holes. [45]. String theory as a fundamental theory is furthermore a distinguished candidate to discuss physics at high energy scales. Active research is going on in studying string cosmology [46], building string GUT models from F-theory [47] or heterotic orbifolds [48]. Many new insights into geometry and gauge theory in general have been gained, as well as the existence of non-commutative and non-associative structures in string theory [49–51]. The most significant achievement of string theory within the last years has been the discovery of gauge/gravity duality. This correspondence relates the partition function of a gravity theory on the one hand with the generating functional of correlation functions in a conformal field theory (CFT) on the other hand. In its original form the correspondence links type IIB string theory on $AdS_5 \times S^5$ with $\mathcal{N} = 4$ SYM in four dimensions [52–54]. The correspondence is often studied in the limit of large t'Hooft coupling and a large number of colors. Type IIB string theory then reduces to type IIB supergravity (SUGRA), which is well understood. Furthermore, the relation between gauge and gravity theory becomes a weak/strong duality. This opens up to the possibility to investigate either theory at strong coupling via its weakly coupled dual theory. This is a very promising attempt to study strongly coupled effects in physics as they occur e.g. in QCD [55] or certain condensed matter systems [56–59].

The problem of the cosmological constant in string theory is deeply related to the vacuum problem. Obviously there is a mismatch between the ten dimensions of string theory and the four dimensions in which the SM is set. The standard approach to resolve this issue is compactification, where six of the ten spatial dimensions in string theory do not extend to infinity but form a compact manifold, while the remaining four dimensions constitute four-dimensional Minkowski space. The claim to obtain a phenomenological attractive theory in four dimensions with $\mathcal{N} = 1$ SUSY constricts the compactification manifolds to Calabi-Yau (CY) manifolds [60], where in type II theories orientifold projections must be included. Today also further mechanisms to break SUSY in the compactification procedure are available. Carrying out the compactification without further ingredients leads to a vast number of massless scalar fields in four dimensions characterizing the shape and volume of the internal manifold. These moduli fields would mediate long-range fifth-force-like interactions which are phenomenologically not acceptable. A proposed solution is to turn on background fluxes that result in a potential for the moduli and thus renders them massive [61]. The vacuum energy of the (local) minimum of the potential then corresponds to the cosmological constant in string theory which can accomplished to be small and positive [62, 63]. However, there is no distinguished minimum. String theory loses a lot of its uniqueness by going down from ten to four dimensions in the sense that there does not exist a unique six-dimensional compactification manifold and the fluxes have to be quantized but apart from that are arbitrary. This leads to an incredible large number of different string vacua, a commonly quoted estimate is 10^{500} [64]. So there might not be just one vacuum that can explain the observed four-dimensional physics, although it has turned out hard so far to construct four-dimensional string models, which satisfy all consistency conditions of the compactification and yield the gauge and matter content of the SM or MSSM in a generic way.

1.3 Scattering Amplitudes

A different approach in connecting string theory with low energy particle physics rests on the calculation of quantities that do not depend on the exact details of the compactification. In particular scattering amplitudes in weakly coupled string theory involving gluons and at most two fermions are such quantities which are insensitive to the respective compactification model. If in addition the mass scale of string theory is as low as the TeV scale, such a scenario gives rise to fascinating phenomenology of physics beyond the SM [65–68] that could be detected in the coming years at LHC.

This idea of low string scale physics is deeply related to the proposal of Arkani-Hamed, Dimopoulos and Dvali (ADD) [69,70]. They argue that in the presence of large extra dimensions the electroweak scale can be the only fundamental short distance scale in nature. In this framework gravity and the three fundamental forces of the SM unify at the TeV scale. At low energies fermions and the gauge bosons of the strong and electroweak interactions are confined to a four-dimensional Mikowskian sub-space, while gravity can also propagate into the bulk. The large size of the extra dimensions pushes the four-dimensional Planck scale to its high value and makes gravity look so weak from a four-dimensional point of view. In this sense the theory nullifies the hierarchy problem. Already the case of two extra dimensions is in concordance with all experimental bounds [71] as Newtonian gravity is tested at the moment only down to distances of ≈ 1 mm.

Due to the presence of extra dimensions string theory is the perfect ground to study such models. As mentioned above SM-like gauge groups can be realized in type II compactifications via intersecting D-brane models. A very common attempt in type IIA is to use four stacks of intersecting D6-branes with an intersection pattern as shown in Figure 1.1. This setup gives rise to the gauge group $U(3)_a \times U(2)_b \times U(1)_c \times U(1)_d$. The $U(N)$'s can be further decomposed as $U(N) = SU(N) \times U(1)$. This results in the SM gauge group, where the SM hypercharge is a linear combination of the different $U(1)$'s. New gauge bosons like a heavy Z' can be included by different combinations of the $U(1)$'s[2]. This D-brane setup is embedded in a compactification

---

[2]At the time of writing the origin of the excess in $W \to 2$ jets events measured by CDF [72] is not clear. A physical interpretation could be the existence of a leptophobic Z' which can be realized in such a D-brane model [73].

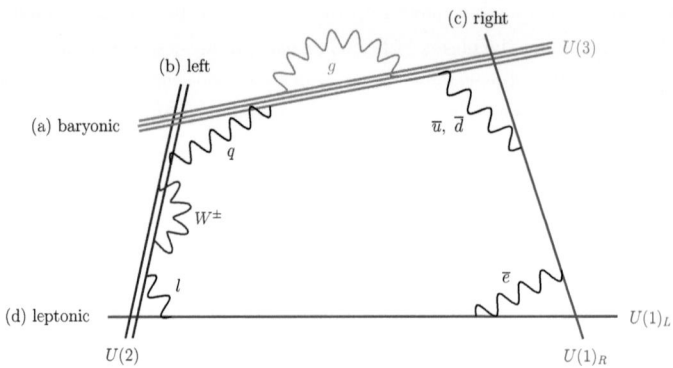

Figure 1.1: Four stacks of intersecting D6-branes resulting in the gauge group of the MSSM, taken from [66].

manifold. The cycles around which the branes wrap have to be small as their size determines the gauge couplings. Other cycles of the manifold, however, must be large such that the total volume of the compactification manifold in total becomes large. This can yield a string scale of the order $M_s = \mathcal{O}(\text{TeV})$. Popular examples of such models are compactifications on "Swiss cheese" CY manifolds. In addition, the dilaton must be stabilized at large values. The string coupling will then be small and we can rely on perturbation theory.

The phenomenology of such SM D-brane constructions is very rich. If the string scale is really that low and the string coupling small the discovery of new, massive gauge bosons at LHC can be expected. These new force-carriers stemming from additional $U(1)$ gauge symmetries could mix with the photons and yield interesting effects. In the same fashion the formation of black holes and other gravity effects could be observed at energies above the string scale. The most promising and so-far best studied phenomena are corrections to hadronic SM processes at energies around the string scale. At these energies the scattering partners can exchange apart from the SM particles also stringy states in the form of Regge recurrences, Kaluza-Klein states and winding modes. The latter two have a model-dependent spectrum and will give a handle on determining the details of the compactification geometry if strings are discovered. A promising discovery channel is the detection of photon production in gluon fusion [74,75] as these processes do not exist at all in the SM. Most interesting are certainly stringy corrections to the hadronic production of dijets [75–77]. As shown in Figure 1.2 in such a case a clear excess over the SM signal in the range of the string scale will be discovered. The CMS experiment has been able to exclude such string resonances up to 1.67 TeV already [78]. Nevertheless, it will remain an interesting channel to look at in the future when the LHC exploits its full potential.

1.3 Scattering Amplitudes

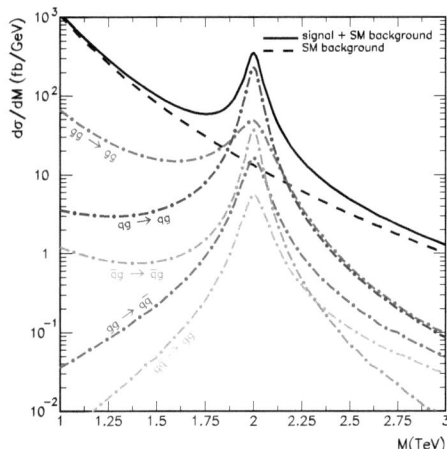

Figure 1.2: Differential cross section vs. invariant mass of the dijet in four parton scattering as calculated in weakly coupled string theory with a string scale of $M_s = 2$ TeV, taken from [76]

Apart from these phenomenological considerations string scattering amplitudes play an important role in many other fields. From the conceptual point of view they turn out to be of great use in string compactifications. Instead of working with the complete theory one usually takes the field theory limit of string theory in order to derive four-dimensional actions. Massive string levels and the extended nature of the strings are hereby neglected. Corrections to the effective action like gauge couplings and metrics for moduli and matter fields can then be derived via string amplitudes [79–81].

The fact that string theory also includes quantum gravity provides a perfect setup to juxtapose gauge and gravity amplitudes. Nowadays recipes are forthcoming how to relate amplitudes involving only open strings with open & closed string amplitudes [82, 83] and learn in this way about the relations between brane-bulk and pure brane couplings. An issues related to this topic are the recently found Bern–Carrasco–Johansson (BCJ) relations, a duality between color and kinematics in field theory amplitudes [84]. These relations have profound consequences as they lead to many new non-trivial relations between distinct amplitudes. Furthermore gravity amplitudes can be obtained in a rather simple way by simply squaring the corresponding gauge theory amplitudes [85]. The same procedure carries over to loop amplitudes via the unitarity method [86]. So far the BCJ relations are just a conjecture, but string theory amplitudes can shed more light onto their structure and existence [83, 87, 88].

A further open issue in theoretical physics deals with the UV finiteness of $\mathcal{N} = 8$ SUGRA [89]. Although of no phenomenological relevance such an investigation might yield unexpected new insight into the structure of quantum gravity and whether enough symmetries can render a point-particle theory of gravity finite. Explicit calculations have proven that the theory is finite up to four-loop order but various arguments suggest that the first divergences occur not until the seventh loop order. In construction of the loop amplitudes and the possible counterterms string theory has proven to a be successful tool [90, 91].

1.4 Motivation & Outline

These topics show that scattering amplitudes in string theory are of high interest. Especially the fact that exact results can be obtained which capture the complete α' behavior of a physical quantity makes them beautiful objects to study.

In the manifestly covariant Ramond–Neveu–Schwarz (RNS) formalism of the superstring the underlying superconformal field theory (SCFT) consists of the Neveu–Schwarz (NS) fermion ψ^m, a Ramond (R) spin field S_A, the string coordinate X^m and further ghost and superghost fields. These fields enter the calculation of a string scattering amplitude through vertex operators creating bosonic and fermionic states. The Lorentz structure of the amplitude is solely determined by the NS fermions and the R spin fields. Unfortunately, these are interacting fields and their correlation functions are therefore difficult to determine. The evaluation of RNS correlators involving ψ^m and S_A is the main topic of this thesis. We consider these quantities both for tree and loop-level scattering in four, six, eight and ten space-time dimensions.

In the case of string compactifications to four dimensions the ten-dimensional spin field splits into an internal and an external part. The interaction of the internal vectors and spin fields is captured by a six-dimensional RNS correlation function, while the interactions of the external spin fields with the NS fermions is contained in the four-dimensional counterpart. In ten dimensions RNS correlators are needed for studying aspects of non-compactified string theory like duality symmetries. In contrast, we study eight-dimensional RNS correlators in this work mainly because of their mathematical beauty.

The outline of this thesis is as follows. In Chapter 2 we lay the groundwork for the calculation of scattering amplitudes in the RNS formalism of the superstring. We introduce the RNS formalism, review the corresponding SCFT and show how to calculate scattering amplitudes. Chapter 3 deals with calculational methods for the evaluation of RNS correlators at tree-level. We first present how RNS correlation functions can in general be calculated and discuss the individual components of the latter. Furthermore we state calculational tools special to each even number of dimensions. Most important we demonstrate how to evaluate any RNS correlator in four dimensions. In Chapter 4 we state results for all RNS tree-level correlators in four-dimensions up to eight-point level. Additionally correlation functions involving arbitrary many NS fermions but only two R spin fields are solved. These are of great use for the calculations of

1.4 Motivation & Outline

hadronic string interactions, which are independent of the compactification scheme. Chapter 5 is devoted to the evaluation of RNS correlators at loop-level. We show how to express the RNS fields in terms of $SO(2)$ spin system operators and calculate loop correlators of these fields. This is used then to derive various loop correlators at least up to six-point level in six, eight and ten dimensions. In addition, we present results for certain general classes of correlation functions, especially such with at most two spin fields. Equipped with these techniques it is then possible to calculate the correlation function of two gauge fields and four gauginos in Chapter 6. We check gauge invariance, write the results in terms of spinor variables and give the field theory limit. This particular purely open string amplitude is of interest as it can be mapped onto an open-closed amplitude involving two gauge fields and four RR bulk moduli fields. In this way it is possible to study the mapping procedure in more detail and determine brane-bulk couplings in terms of pure brane couplings. In Chapter 7 follow our conclusions. Five Appendices finalize this thesis, in which we comment on gamma matrices in higher dimensions, state all necessary index term relations for the previously calculated correlators and give some details on generalized Θ functions. Furthermore we present the details of the calculations in Chapter 6 and review the spinor helicity formalism.

CHAPTER 2

Scattering in String Theory

In this Chapter we lay the groundwork for the succeeding calculations of RNS correlation functions and tree-level scattering amplitudes. We are using the well-known RNS formalism of string theory [92–94], which introduces supersymmetric partners ψ^m to the bosonic string coordinates X^m on the world-sheet. The focus in the following lies on the description of supersymmetric string theory in terms of a SCFT. We discuss the separate actions of matter and ghost fields and introduce vertex operators that create bosonic and fermionic string states in ten and four dimensions. In the last part of this Chapter we present some details on the calculation of scattering amplitudes at tree- and loop-level. We follow the standard textbooks and reviews on string theory [33–36, 95] and conformal field theory [96–98], where additional information can be found.

A drawback of the RNS formalism is the lack of space-time spinors which nevertheless can be incorporated by the inclusion of the R spin field [99]. Alternative approaches are the Green–Schwarz formalism, which contains manifest space-time SUSY, but is difficult to quantize in a covariant way, and the pure spinor formalism [100]. The latter retains Lorentz invariance and space-time SUSY and is a promising approach to the calculation of string scattering amplitudes [87, 88, 101–103].

2.1 Conformal Field Theory on the World-Sheet

In the following we present the actions of bosonic and supersymmetric string theory in the RNS formalism, as well as the ghost and superghost action. We focus on the description in terms of the underlying SCFT.

2.1.1 Matter Fields

Strings are one-dimensional objects sweeping out a two-dimensional surface, the string world-sheet, which is embedded in a D-dimensional space-time. The world-sheet is parame-trized by a time-like coordinate τ and a spatial coordinate σ running from 0 to l, where $l = 2\pi$ for closed strings and $l = \pi$ for open strings. For the moment we focus our attention on closed strings. The embedding of the string world-sheet into the target space is accomplished by the bosonic string coordinate fields $X^m(\tau, \sigma)$, which are subject to the identification $X^m(\tau, \sigma + 2\pi) = X^m(\tau, \sigma)$ under shifting the spatial coordinate by 2π. This physical system is described by the Polyakov action

$$S_P = -\frac{1}{4\pi\alpha'} \int d^2\sigma \sqrt{-\det h}\, h^{ab}\, \partial_a X^m\, \partial_b X^n\, \eta_{mn}\,, \qquad (2.1)$$

where h^{ab} is the metric on the world-sheet and η^{mn} the metric of the flat target space. Apart from D-dimensional Lorentz invariance the symmetries of this action are

- diffeomorphism invariance:

$$\delta h_{ab} = \nabla_a \xi_b + \nabla_b \xi_a\,, \qquad \delta X^m = \xi^a\, \partial_a X^m\,, \qquad (2.2a)$$

- Weyl invariance:

$$\delta h_{ab} = \Lambda\, h_{ab}\,, \qquad \delta X^m = 0 \qquad (2.2b)$$

with arbitrary infinitesimal functions ξ^a, Λ that depend on τ and σ. Making use of three local symmetries (two reparametrizations of the world-sheet and one Weyl scaling) h_{ab} can be set to the flat, two-dimensional Minkowski metric $\eta_{ab} = \mathrm{diag}(-1, 1)$. Note that this does not exhaust the full gauge freedom because one can still perform conformal transformations satisfying

$$\partial^a \xi^b + \partial^a \xi^b = \Lambda\, \eta^{ab}\,, \qquad (2.3)$$

where the change of the metric due to a reparametrization is absorbed by a Weyl scaling. We perform a Wick rotation by introducing Euclidean coordinates $(\sigma^0, \sigma^1) \equiv (i\tau, \sigma)$ and define

$$w \equiv \sigma^0 + i\sigma^1\,, \qquad \bar{w} \equiv \sigma^0 - i\sigma^1\,. \qquad (2.4)$$

Via $z = e^w$ these coordinates on the cylinder are mapped onto to complex plane. Introducing the derivatives $\partial \equiv \partial_z$ and $\bar\partial \equiv \partial_{\bar z}$ the Polyakov action in conformal gauge becomes

$$S = \frac{1}{2\pi\alpha'} \int d^2z\, \partial X^m\, \bar\partial X_m\,. \qquad (2.5)$$

2.1 Conformal Field Theory on the World-Sheet

In the following the coordinates z and $\bar{z}$ are assumed to be independent, although they are related by complex conjugation for real σ^i. On account of (2.3) this action is still invariant under conformal transformations $z \mapsto f(z)$, where f is a holomorphic function.

Upon quantization the Polyakow action unfolds some shortcomings. The spectrum contains a tachyonic ground state and no fermions. To cure these problems one adds world-sheet supersymmetry to the action[1]. The new action involves the superpartners of X^m and h_{ab}, a Majorana spinor $\Psi^m = (\psi^m, \bar{\psi}^m)$ in two dimensions and a gravitino χ_a. Auxiliary scalar fields can be eliminated via their equations of motions. The resulting action is invariant, apart from Lorentz transformations and local SUSY, under diffeomorphisms, Weyl and super-Weyl transformations. As in the bosonic case these symmetries can be used to bring the action in so-called super-conformal gauge, where the degrees of freedom of the world-sheet metric and the gravitino drop out. In complex coordinates the action then reads

$$S = \frac{1}{4\pi} \int dz^2 \left(\frac{2}{\alpha'} \partial X^m \bar{\partial} X_m + \psi^m \bar{\partial} \psi_m + \bar{\psi}^m \partial \bar{\psi}_m \right). \qquad (2.6)$$

Similar to the bosonic case, this action is still invariant under certain diffeomorphisms and local SUSY transformations where the contribution to the world-sheet metric and gravitino can be absorbed by a Weyl and super-Weyl transformation. Indeed it can be shown that the action remains invariant under $z \mapsto f(z)$ if $i(\alpha'/2)^{1/2} \partial X^m$ and ψ^m are primary fields with conformal weight $h = 1$ and $h = 1/2$, respectively.

The equations of motion,

$$\partial(\bar{\partial} X^m) = \bar{\partial}(\partial X^m) = 0, \qquad \bar{\partial}\psi^m = \partial\bar{\psi}^m = 0, \qquad (2.7)$$

imply that $\partial X^m(z)$ and $\psi^m(z)$ are chiral fields, while $\bar{\partial} X^m(\bar{z})$ and $\bar{\psi}^m(\bar{z})$ are anti-chiral. Apart from the equations of motion the variation of the action yields a surface term that also must vanish. From this constraint the periodicity condition of the closed string coordinate under rotating z by $2\pi i$ is recovered. The fermions, however, can satisfy either symmetric or antisymmetric boundary conditions in z. This gives rise to two different fermion sectors:

$$\psi^m(e^{2\pi i} z) = \begin{cases} +\psi^m(z) & : \text{Neveu–Schwarz (NS) sector}, \\ -\psi^m(z) & : \text{Ramond (R) sector}. \end{cases} \qquad (2.8)$$

As the action is invariant under conformal and SUSY transformations Noether's theorem predicts two conserved currents. The energy momentum tensor and the world-sheet supercurrent

---
[1] We just sketch the constructions of such a supersymmetric action. The explicit procedure is nicely explained in [95].

can be derived by varying the non-gauged supersymmetric action with respect to the worldsheet metric and the gravitino. The non-vanishing components of these tensors are holomorphic fields which in super-conformal gauge take the form

$$T(z) = -\frac{1}{\alpha'}\partial X^m(z)\,\partial X_m(z) - \frac{1}{2}\psi^m(z)\,\partial\psi_m(z)\,,$$
$$G(z) = i\sqrt{\frac{2}{\alpha'}}\,\psi^m(z)\,\partial X_m(z) \tag{2.9}$$

and appropriate expressions for their antiholomorphic counter-parts. In the quantum theory these operators must be normal-ordered, which we do not denote explicitly. This conserved currents generate the superconformal algebra. Upon radial quantization we find for their operator product expansions (OPEs)[2]:

$$T(z)\,T(w) \sim \frac{3D}{4(z-w)^4} + \frac{2}{(z-w)^2}T(w) + \frac{1}{z-w}\partial T(w)\,,$$
$$G(z)\,G(w) \sim \frac{D}{(z-w)^3} + \frac{2}{(z-w)}T(w)\,,$$
$$T(z)\,G(w) \sim \frac{3}{2(z-w)^2}G(w) + \frac{1}{z-w}\partial G(w)\,. \tag{2.10}$$

These equations show that $T(z)$ and $G(z)$ are tensors of weight $(2,0)$ and $(3/2,0)$. The central charge of this SCFT can be read off from the first equation. Each boson contributes 1, every fermion $1/2$ and in total we find

$$c = \left(1 + \frac{1}{2}\right)D = \frac{3}{2}D\,. \tag{2.11}$$

Later we will add (super-)ghost fields to the action (2.6) that will also give a contribution to the total central charge. If the total central charge does not vanish, this leads to a superconformal anomaly because the superconformal symmetry of the action is broken at the quantum level. Requiring that the total central charge vanishes will determine the number of dimensions D of the target-space.

In order to derive the OPEs of the primary fields ∂X^m and ψ^m we calculate the corresponding two-point functions on the sphere. These are determined by the conformal properties of the fields and the Dyson-Schwinger equation:

$$\langle\partial X^m(z)\,\partial X^n(w)\rangle_{S^2} = -\frac{\alpha'}{2}\frac{\eta^{mn}}{(z-w)^2}\,, \qquad \langle\psi^m(z)\,\psi^n(w)\rangle_{S^2} = \frac{\eta^{mn}}{z-w}\,,$$
$$\langle X^m(z,\bar z)\,X^n(w,\bar w)\rangle_{S^2} = -\frac{\alpha'}{2}\eta^{mn}\ln|z-w|^2\,, \qquad \langle\psi^m(z)\,\bar\psi^n(w)\rangle_{S^2} = 0\,. \tag{2.12}$$

---
[2]In all following OPEs the sign $\sim$ should be understood as 'equal up to non-singular terms'.

2.1 Conformal Field Theory on the World-Sheet

From these two-point functions the OPEs can easily be read of. One finds:

$$\partial X^m(z)\,\partial X^n(w) \sim -\frac{\alpha'}{2}\frac{\eta^{mn}}{(z-w)^2}\,,\qquad \psi^m(z)\,\psi^n(w) \sim \frac{\eta^{mn}}{z-w}\,. \tag{2.13}$$

The CFT of the fermions in $D=2m$ dimensions has a simple representation in terms of m chiral bosons. This equivalence of the CFTs is known as bosonization [98, 99, 104]. For this purpose we introduce $H(z) = \bigl(H_1(z),\ldots,H_m(z)\bigr)$ containing m chiral boson with the singular behavior $H_i(z)\,H_j(w) \sim -\delta_{ij}\ln|z-w|$. Exponentials of these fields hence suffice the OPE

$$e^{ipH(z)}\,e^{iqH(w)} \sim (z-w)^{pq}\,e^{i(p+q)H(z)}\,, \tag{2.14}$$

where p and q are m-dimensional lattice vectors with entries ± 1. This yields indeed the fermion OPEs if the exponentials are identified with the Cartan-Weyl elements of the ψ's in the following way:

$$e^{\pm i H_j(z)} \equiv \psi^{\pm j}(z) \equiv \frac{1}{\sqrt{2}}\bigl(\psi^{2j}(z)\pm i\psi^{2j+1}(z)\bigr)\,. \tag{2.15}$$

Correlation functions of these operators on the sphere are easily calculated with the formula

$$\left\langle \prod_i e^{ip_i H(z_i)}\right\rangle_{S^2} = \delta\Bigl(\sum_i p\Bigr)\prod_{i<j} z_{ij}^{p_i p_j} \tag{2.16}$$

with $z_{ij} \equiv z_i - z_j$, which also holds for more general vectors p_i.

2.1.2 The Spin Field

Let us have a closer look on the boundary conditions of the fermions. These fields live in the double cover of the complex plane because they are only defined up to a sign as shown in (2.8). Their Laurent expansion is therefore

$$\psi^m(z) = \sum_r \psi_r^m\, z^{-r-1/2}\,,\qquad r \in \begin{cases} \mathbb{Z}+\tfrac{1}{2} & : \text{NS sector}\,,\\ \mathbb{Z} & : \text{R sector}\,.\end{cases} \tag{2.17}$$

We see that in the R sector ψ^m introduces a branch cut due to the presence of $z^{-1/2}$ in its expansion. The OPE (2.13) implies that the Laurent modes satisfy the anti-commutation relations

$$\{\psi_m^m,\psi_n^n\} = \eta^{mn}\,\delta_{m+n}\,. \tag{2.18}$$

The spectra of the R and NS sector are entirely different due to the integer and half-integer mode numbering. In the R sector there is no zero mode and the ground state is defined to be annihilated by all positive modes:

$$\psi_r^m |0\rangle_{\text{NS}} = 0 \qquad \forall\, r > 0\,. \tag{2.19}$$

The modes with $r < 0$ act as creation operators. Every mode can be excited at most once because they square to zero as can be seen from (2.18).

The ground state of the R sector $|0\rangle_{\text{R}}$ has more structure due to the presence of the zero mode ψ_0^m. Again, it is annihilated by all modes ψ_r^m with $r > 0$, but not by the zero mode. In fact $\psi_0^m |0\rangle_{\text{R}}$ is another ground state because it is annihilated by the positive modes on account of $\{\psi_r^m, \psi_0^n\} = 0$. Thus, the ground state is degenerate. In fact it is a representation of the Clifford algebra with dimension $2^{D/2}$ because the ψ_0's satisfy the Clifford algebra for $\Gamma^m = i\sqrt{2}\,\psi_0^m$. States created from the R vacuum are then space-time fermions as the creation operators ψ_r^m, $r < 0$, change the spin by integers. States in NS sector in contrast have bosonic character.

Operators creating R ground states out of the NS vacuum are called spin fields S_A [99,105]:

$$|A\rangle_{\text{R}} = \lim_{z \to 0} S_A(z) |0\rangle_{\text{NS}}\,. \tag{2.20}$$

A is a spinor index in the target space and spin fields thus transform as space-time spinors. As these fields intertwine the R and NS sector and thereby change the boundary conditions, their action leads to an opening and closing of a branch cut on the string world-sheet. Their conformal weight can be understood by calculating the Laurent modes of the energy momentum tensor. These are given by

$$L_m = \frac{1}{2} \sum_n \alpha_{m-n}^m \alpha_{mn} + \frac{1}{4} \sum_r (2r - m)\, \psi_{m-r}^m \psi_{mr} + a\, \delta_{m,0}\,, \tag{2.21}$$

where α_n^m, $n \in \mathbb{Z}$, are the expansion modes of the bosonic string coordinate. The constant a arrises from normal ordering. It vanishes in the NS sector, but takes the value

$$a = \frac{D}{16} \tag{2.22}$$

in the R sector. Therefore, S_A has conformal weight $D/16$ because $L_0|A\rangle_{\text{R}} = D/16\,|A\rangle_{\text{R}}$. The two-point function of two spin fields on the sphere hence becomes

$$\langle S_A(z)\, S_B(w) \rangle_{S^2} = \frac{\mathcal{C}_{AB}}{(z-w)^{D/8}}\,. \tag{2.23}$$

Here we have introduced the charge conjugation matrix $\mathcal{C}_{AB}$ in order to obtain Poincaré invariant

2.1 Conformal Field Theory on the World-Sheet

results[3]. The OPEs of spin fields and fermions can be determined by considering the three-point function $\langle S_A(z_1)\,\psi^m(z_2)\,S_B(z_3)\rangle$ in the limits $z_1 \to \infty$, $z_3 \to 0$. The findings can be summarized as

$$\psi^m(z)\,S_A(w) \sim \frac{(\Gamma^m)_A{}^B\,S_B(w)}{\sqrt{2}\,(z-w)^{1/2}}\,,$$
$$S_A(z)\,S_B(w) \sim \frac{\mathcal{C}_{AB}}{(z-w)^{D/8}} + \frac{(\Gamma^m \mathcal{C})_{AB}\,\psi_m(w)}{\sqrt{2}\,(z-w)^{D/8-1/2}}\,. \qquad (2.24)$$

Spin fields in $D=2m$ can also be presented by m chiral bosons. The leading singularity in the OPE of one fermion and one spin field in (2.24) suggests in comparison with (2.14) that the lattice vector of a spin field must contain half integer values $p = (\pm\frac{1}{2},\ldots,\pm\frac{1}{2})$. In even dimensions it is always possible to decompose a Dirac spinor into left- and right-handed Weyl spinors. The spin field S_A thus decomposes into left- and right-handed spin fields S_a and $S^{\dot b}$. We follow the convention that in bosonized form the lattice vector of a left-handed spin fields contains an even number of $-1/2$ entries, while the number of negative entries is odd for a right-handed spin field.

2.1.3 Ghost Fields

We have mentioned above that the non-gauged action of the supersymmetric string is invariant under diffeomorphisms and (super-)Weyl transformations. Therefore, it is not possible to naively plug this action into the path integral. One must only integrate over configurations of the world-sheet metric and the gravitino which are not related by these symmetry transformations. Otherwise this leads to a massive over-counting. Usually, this is settled by introducing a Fadeev–Popov determinant to the partition function. As can be shown this is equivalent to adding a ghost action to (2.6) which becomes in superconformal gauge

$$S_{\text{gh}} = \frac{1}{4\pi}\int \mathrm{d}^2 z \big(b\,\bar{\partial}c + \bar{b}\,\partial\bar{c} + \beta\,\bar{\partial}\gamma + \bar{\beta}\,\partial\bar{\gamma}\big)\,. \qquad (2.25)$$

The fields b, c are anticommuting ghost fields, which are necessary for the quantization of the bosonic action (2.5), while the commuting super-ghost fields β, γ are required in addition for the supersymmetric action (2.6). The equations of motion, derived from the action above, identify these fields as chiral or anti-chiral, respectively:

$$\bar{\partial}b = \bar{\partial}c = 0\,, \qquad \bar{\partial}\beta = \bar{\partial}\gamma = 0\,. \qquad (2.26)$$

---

[3] A detailed discussion of spinors in higher dimensions and the charge conjugation matrix follows in Chapter 3 and Appendix A.

We restrict our discussion to the chiral fields in the following and quickly summarize their conformal properties. The energy momentum tensors

$$T^{b,c}(z) = -2\,b(z)\,\partial c(z) - \partial b(z)\,c(z)\,,$$
$$T^{\beta,\gamma}(z) = -\frac{3}{2}\,\beta(z)\,\partial\gamma(z) - \frac{1}{2}\partial\beta(z)\,\gamma(z)\,, \tag{2.27}$$

imply that the ghost fields have the conformal weights

$$h(c) = -1\,, \qquad h(b) = 2\,, \qquad h(\gamma) = -\frac{1}{2}\,, \qquad h(\beta) = \frac{3}{2}\,. \tag{2.28}$$

The central charges of the ghost and superghost CFTs can be obtained from the OPEs of the energy momentum tensors. One finds that $c^{b,c} = -26$ and $c^{\beta,\gamma} = +11$. If the central charge of the matter system (2.11) is also taken into account, the total central charge vanishes for $D = 10$:

$$c^{X,\psi} + c^{b,c} + c^{\beta,\gamma} = \frac{3}{2}D - 26 + 11 = 0\,. \tag{2.29}$$

As previously described the quantum theory does not suffer in this case from a superconformal anomaly. The cb- and $\gamma\beta$-propagators, derived from the action (2.25), demand that the ghost fields satisfy the OPEs

$$c(z)\,b(w) \sim \frac{1}{z-w}\,, \qquad \gamma(z)\,\beta(w) \sim \frac{1}{z-w}\,. \tag{2.30}$$

As in the case of the matter fields we can perform a Laurent expansion of the ghost and superghost fields. We obtain for the former

$$c(z) = \sum_n c_n\,z^{-n+1}\,, \qquad b(z) = \sum_n b_n\,z^{-n-2}\,, \tag{2.31}$$

where the modes have to satisfy the anticommutation relations

$$\{b_m, c_n\} = \delta_{m+n}\,, \qquad \{b_m, b_n\} = \{c_m, c_n\} = 0 \tag{2.32}$$

due to the OPE of $c(z)$ and $b(w)$. The operator-state correspondence implies that vacuum state of the ghost system is annihilated by all b_n, $n > -2$ and c_n, $n > 1$, but not by the mode c_1:

$$\lim_{z \to 0} c(z)\,|0\rangle_{b,c} = c_1|0\rangle_{b,c} \equiv |1\rangle_{b,c} \neq 0\,. \tag{2.33}$$

Due to $[L_0, c_1] = -1$, where L_0 is the zero mode of the energy momentum tensor $T^{b,c}(z)$, the

2.1 Conformal Field Theory on the World-Sheet

state $|1\rangle_{b,c}$ is the state with lowest energy and therefore the proper ground state of the ghost system. It is also annihilated by c_1 because of $\{c_1, c_1\} = 0$.

We now discuss the superghost fields. These are associated to the fermions ψ^m and hence satisfy the same periodicity conditions. Therefore, the mode expansion also yields an NS and an R sector,

$$\gamma(z) = \sum_r \gamma_r\, z^{-r+1/2}, \quad \beta(z) = \sum_r \beta_r\, z^{-r-3/2}, \quad r \in \begin{cases} \mathbb{Z} + \tfrac{1}{2} & : \text{NS sector}, \\ \mathbb{Z} & : \text{R sector}. \end{cases} \quad (2.34)$$

The modes must satisfy the commutation relations

$$[\gamma_r, \beta_s] = \delta_{r+s}, \quad [\beta_r, \beta_s] = [\gamma_r, \gamma_s] = 0 \quad (2.35)$$

because the superghost fields are of bosonic type. As in the case above, the shift in the mode expansion implies that the vacuum in the NS sector is not a highest weight state. In fact, due to $[L_0, \gamma_{1/2}] = -1/2\, \gamma_{1/2}$, the vacuum can be lowered to arbitrary negative energies because $\gamma_{1/2}$ does not square to zero. In the R sector an operator analogous to the spin field is needed creating a branch cut and interpolating between the different boundary conditions. As shown in [36] by bosonizing the superghost system, the proper ground states for the two sectors are

$$e^{-\phi(0)} |0\rangle_{\beta,\gamma} \equiv |q = -\tfrac{1}{2}\rangle_{\beta,\gamma} \quad : \text{NS sector},$$

$$e^{-\phi(0)/2} |0\rangle_{\beta,\gamma} \equiv |q = -1\rangle_{\beta,\gamma} \quad : \text{R sector}. \quad (2.36)$$

These states are annihilated by $\gamma_{1/2}$ and γ_1 as required.

The action (2.25) is invariant under two chiral $U(1)$ symmetries generated by the currents

$$j^{b,c} = -b(z)\, c(z), \quad j^{\beta,\gamma} = -\beta(z)\, \gamma(z). \quad (2.37)$$

The OPE of these currents with the respective energy momentum tensor $T^{b,c}$ or $T^{b,j}$

$$T(z)\, j(w) \sim \frac{Q}{(z-w)^3} + \frac{j(w)}{(z-w)^2} + \frac{\partial j(w)}{z-w} \quad (2.38)$$

exhibits an anomaly. The charge Q takes the value -3 for the ghost and $+2$ for the superghost system. The anomalous conservation law of the currents reads

$$\bar\partial j(z) = \frac{1}{4} Q \sqrt{h} R, \quad (2.39)$$

where h is the determinant of the world-sheet metric and R the corresponding curvature scalar.

It can be shown that the anomaly arises from the presence of (super-)ghost zero modes. It is possible to calculate their number from (2.39) using the Riemann-Roch theorem:

$$N_c - N_b = 3 - 3g, \qquad N_\gamma - N_\beta = 2 - 2g. \tag{2.40}$$

This has profound consequences. For string scattering at g loops the string world-sheet is a Riemann surface of genus g. The vertex operators creating string states have to be inserted with the right superghost factors in order to cancel the superghost background charge of $2 - 2g$. Furthermore, at tree-level, i.e. $g = 0$, the presence of three ghost zero modes follows from the three globally defined diffeomorphisms on the sphere. In order to cancel this residual gauge freedom three vertex operators positions can be fixed in the calculation of the amplitude.

2.1.4 Open strings

So far our discussion has been centered on closed strings. The closed string nature entered the considerations through the boundary conditions of the fields X^m and ψ^m at $\sigma^1 = 0, 2\pi$. These were chosen in such a way to make the surface term vanish which arises from the variation of the action (2.6) apart from the equations of motion. Let us have a closer look on this surface term for the open string case:

$$\int d^2\sigma \left(-\frac{2}{\alpha'} \delta X^m \, \partial_1 X_m + \delta\psi^m \, \psi_m - \delta\bar{\psi}^m \, \bar{\psi}_m \right) \Bigg|_{\sigma^1=0}^{\sigma^1=\pi} \equiv 0. \tag{2.41}$$

It is obvious that the respective terms for the fermions vanish if $\psi^m = \pm \bar{\psi}^m$ at either end of the string. The overall relative sign between ψ and $\bar{\psi}$ is a matter of convention and we set

$$\psi^m(\sigma^0, 0) = \bar{\psi}^m(\sigma^0, 0). \tag{2.42}$$

The sign at the other end becomes meaningful and we again obtain two sectors, similar to the closed string case:

$$\begin{aligned} \psi^m(\sigma^0, \pi) &= -\bar{\psi}^m(\sigma^0, \pi) &&: \text{NS sector}, \\ \psi^m(\sigma^0, \pi) &= +\bar{\psi}^m(\sigma^0, \pi) &&: \text{R sector}. \end{aligned} \tag{2.43}$$

Via the doubling trick the fields ψ^m and $\bar{\psi}^m$ can be combined into a single field Ψ^m with range $0 \leq \sigma^1 \leq 2\pi$. We define

$$\Psi^m(\sigma^0, \sigma^1) \equiv \begin{cases} \psi^m(\sigma^0, \sigma^1) & : 0 \leq \sigma^1 \leq \pi, \\ \bar{\psi}^m(\sigma^0, 2\pi - \sigma^1) & : \pi \leq \sigma^1 \leq 2\pi. \end{cases} \tag{2.44}$$

2.1 Conformal Field Theory on the World-Sheet   23

The field Ψ^m is on the same footing as ψ^m in the closed string case. First, they are both holomorphic in z and they satisfy the same boundary conditions (2.8). The CFT results for the closed string sector therefore also apply to the open string sector. In the following discussion of the open string we simply denote Ψ^m by ψ^m, which has then the same properties as the world-sheet fermion in the closed string action.

The boundary term for the open string coordinates vanishes if X^m satisfies at the endpoints $\sigma^1 = 0, \pi$ one of the following conditions:

$$\begin{aligned} \delta X^m &= 0 \quad : \text{Dirichlet conditions}, \\ \partial_1 X^m &= 0 \quad : \text{Neumann conditions}. \end{aligned} \quad (2.45)$$

Dirichlet boundary conditions are solved by $X^m(\sigma^0, \sigma^1 = 0, \pi) = \text{const}$ and thereby state that the endpoints of the string in these directions are fixed to hypersurfaces, so-called D-branes [106]. Neumann boundary conditions imply that strings are always perpendicular to these surfaces. Historically, only Neumann conditions were considered, but T duality forces the inclusion of Dirichlet conditions as well [107]. At first glimpse the concept of D-branes seems to break Lorentz invariance because they single out certain regions of space-time. However, D-branes are dynamically objects. Under the influence of gravity, i.e. the interaction of closed strings with the open strings ending on their world-volume, they can fluctuate in form and position.

At the endpoints of an open string one can introduce new degrees of freedom i, j which run from 1 to n and label the states of the two endpoints. A general string wavefunction can thus be decomposed as

$$|\phi, a\rangle = \sum_a T^a_{ij} |\phi, ij\rangle, \quad (2.46)$$

where the Chan–Paton factors T^a_{ij} form a complete set of $n \times n$ hermitian matrices. These Chan–Paton degrees of freedom have trivial world-sheet dynamics and therefore do not change superconformal nor Poincaré invariance. However, they have profound impact on space-time physics as they add gauge degrees of freedom to scattering amplitudes. For the scattering of oriented strings the underlying gauge group is found to be $U(n)$, while for unoriented strings one obtains the groups $SO(n)$ or $USp(n)$. An operator creating gauge bosons from the vacuum must therefore contain T^a_{ij}.

In the D-brane picture the Chan–Paton degrees of freedom have a nice geometric interpretation. If n branes coincide there exist open strings stretching between them. These are massless because they can have vanishing length. Quantization of an oriented theory yields n^2 massless vectors that form the adjoint of a gauge group $U(n)$. Hence, the Chan–Paton degrees simply count on which branes the string starts and ends.

2.1.5 Vertex Operators

With all the previous results it is now possible to write down vertex operators which create string states from the vacuum in ten dimensions. These are local operators on the world-sheet. As a particular position z on the world-sheet has no physical meaning one must integrate over all vertex operator positions. The requirement that the integrated vertex operators are independent of coordinate transformations on the world-sheet requires that the vertex operators should have conformal weight $h = 1$.

In the supersymmetric string a vertex operator $V(z)$ can create bosons in the NS sector and fermions in the R sector from the vacuum. The operator splits into a plane wave e^{ikX} and a remaining conformal field $v(z)$. The plane wave generates eigenstates of the momentum generator

$$P^m = \frac{1}{2\alpha'} \oint \frac{\mathrm{d}z}{2\pi i} i\,\partial X^m \qquad (2.47)$$

with eigenvalues proportional to the momentum k^m. With respect to the energy momentum tensor (2.9) the plane wave has conformal weight $h(e^{ikX}) = \alpha' k^2$. In the following we only consider massless string states for which k^2 vanishes. From our previous considerations $v(z)$ must then have conformal weight 1. The exact form of v depends on whether a fermionic or bosonic string state is created by the vertex operator. For bosons, i.e. in the NS sector, the fermion ψ^m is combined with the corresponding superghost contribution $e^{-\phi/2}$ from (2.36). In the R sector fermions are created by the spin field as shown in (2.20). Combining this field with the superghost in bosonized form e^ϕ from (2.36) yields the correct expression for v. The most general open string vertex operators for bosonic and fermionic states are therefore

$$V_{A^a}^{(-1)}(z,\xi,k) = g_A\, T^a\, \xi_m\, \psi^m(z)\, e^{-\phi(z)}\, e^{ikX(z)} \qquad : \text{NS sector}\,,$$

$$V_{\lambda^a}^{(-1/2)}(z,u,k) = g_\lambda\, T^a\, u_A\, S^A(z)\, e^{-\phi(z)/2}\, e^{ikX(z)} \qquad : \text{R sector}\,, \qquad (2.48)$$

where we have included coupling constants g_A, g_λ, Chan–Paton factors T^a and a polarization vector and spinor ξ_m, u_A. These vertex operators create gauge bosons and gauginos, respectively. Although ψ^m is a world-sheet fermion it appears in the vertex operator creating space-time bosons, while the spin field excites space-time fermions. In the following we call ψ^m an NS fermion and S_A the R spin field. Due to the presence of the exponential $e^{p\phi(z)}$ stemming from the superghost fields the vertex operators (2.48) are said to be in the (-1) or $(-1/2)$ ghost picture. Keeping in mind that $e^{p\phi(z)}$ has conformal weight $h = -p^2/2 - p$ and $h(\psi) = 1/2$, $h(S) = 5/8$, this shows that the vertex operators have indeed conformal weight 1 in total.

A more rigorous way to derive the vertex operators (2.48) makes use of the BRST operator Q_{BRST} as shown in [98]. After gauge fixing the RNS world-sheet action still has a symmetry that mixes ghost and matter degrees of freedom. The integral over the associated current gives

2.1 Conformal Field Theory on the World-Sheet

the BRST operator. The cancellation of anomalies in $D = 10$ translates into $Q_{\text{BRST}}^2 = 0$, but the operator has another task. Unphysical states must be decoupled from the spectrum. The BRST operator therefore annihilates all physical states. This implies for the vertices of the NS and R sector that

$$[Q_{\text{BRST}}, V] = 0 \tag{2.49}$$

up to total derivatives in z which vanish upon integration over the world-sheet. This BRST condition constrains the operators (2.48) to be on-shell, i.e. the polarization vector must be transversal to the momentum, $\xi k = 0$, and the polarization spinor has to satisfy the massless Dirac equation, $u^\alpha k_{\alpha\dot\beta} = 0$. Yet in another way the BRST operator proves to be very useful. Given a vertex operator in ghost picture q, one can show that

$$V^{(q+1)}(z) \equiv -2\left[Q_{\text{BRST}}, \xi(z) V^{(q)}(z)\right], \tag{2.50}$$

where ξ is from the bosonization of the superghosts, is not BRST exact but is a vertex operator with ghost charge $q+1$. Higher ghost pictures are particularly necessary for calculating higher-point amplitudes at tree-level because the vertex operators have to be inserted with such ghost contributions in order to cancel the background ghost charge of $2 - 2g$. For the purpose of this thesis only the boson vertex operators in the 0 ghost picture is needed. One finds from (2.50):

$$V_{A^a}^{(0)}(z,\xi,k) = \frac{g_A}{(2\alpha')^{1/2}} T^a \xi_m \left[i\,\partial X^m(z) + 2\alpha'\left(k\,\psi(z)\right)\psi^m(z)\right] e^{ikX(z)}. \tag{2.51}$$

Space-time SUSY requires that the vertex operators (2.48) are related. Acting with the supercharge

$$Q_A^{(-1/2)} \equiv \alpha'^{-1/4} \oint S_A(z)\, e^{-\phi(z)/2} \tag{2.52}$$

and its counter-part in the $+1/2$ ghost picture on the vertex operators shows that the coupling are related, $g_\lambda = g_A\, \alpha'^{-1/4}$, but more important:

$$[Q^{(-1/2)}(\eta), V_{\lambda^a}^{(-1/2)}(u)] = V_{A^a}^{(-1)}(\xi), \qquad \xi^m = \frac{1}{\sqrt{2}} \eta^a (\gamma^m C)_{ab}\, u^b,$$

$$[Q^{(1/2)}(\eta), V_{A^a}^{(-1)}(\xi)] = V_{\lambda^a}^{(-1/2)}(u), \qquad u^\beta = \frac{1}{\sqrt{2}} \eta^a (\gamma^{mn})_a{}^b\, k_m\, \xi_n. \tag{2.53}$$

These relations are major ingredients for SUSY Ward identities [108] which allow to relate different string scattering amplitudes, where the scattering partners sit in the same multiplet.

Our considerations so far have been devoted to string theory in ten space-time dimensions. We focus now on string compactifications to four dimensions, which are highly interesting from a phenomenological point of view. Under the decomposition of the ten-dimensional Lorentz group $SO(1,9) \to SO(1,3) \times SO(6)$ vectors split into direct sums, $X^m = X^\mu \oplus X^i$, while spinors decompose like $\chi_A = (\chi_\alpha \otimes \chi^I) \oplus (\bar\chi^{\dot\alpha} \otimes \bar\chi_J)$. Here χ and $\bar\chi$ are left- and right-handed spinors, whereas μ and α are external, i and I internal Lorentz indices. The conformal fields ∂X^m, ψ^m and the left- and right-handed parts of S_A decompose as

$$\partial X^m = (\partial X^\mu, \partial Z^i), \qquad S_a = S_\alpha \Sigma^I \oplus S^{\dot\alpha} \bar\Sigma_I,$$
$$\psi^m = (\psi^\mu, \Psi^i), \qquad S^{\dot a} = S_\alpha \bar\Sigma_I \oplus S^{\dot\alpha} \Sigma^I. \qquad (2.54)$$

The external fields $\partial^\mu X$, ψ^μ as well as the left- and right-handed spin fields S_α and $S^{\dot\beta}$ transforming under the Lorentz group $SO(1,3)$ form a SCFT with central charge 6 which decouples completely from the SCFT of their internal counter-parts ∂Z^i, Ψ^i, Σ^I and $\bar\Sigma_I$ with charge 9. In addition, the decomposition yields the following conformal weights:

$$h(\partial X^\mu) = h(\partial Z^i) = 1, \qquad h(\psi^\mu) = h(\Psi^i) = \frac{1}{2}, \qquad h(S_\alpha) = \frac{1}{4}, \qquad h(\Sigma^I) = \frac{3}{8}. \qquad (2.55)$$

From these it is easy to determine the OPEs of the external and internal CFT. In fact, they coincide with the chiral version of (2.24) for $D=4$ and $D=6$[4]:

$$\psi^\mu(z)\,S_\alpha \sim \frac{1}{\sqrt{2}\,(z-w)^{1/2}}\,\sigma^\mu_{\alpha\dot\beta}\,S^{\dot\beta}(w),$$
$$S_\alpha(z)\,S_\beta \sim \frac{\varepsilon_{\alpha\beta}}{(z-w)^{1/2}},$$
$$S_\alpha(z)\,S^{\dot\beta} \sim \frac{1}{\sqrt{2}}\,(\sigma^\mu \varepsilon)_\alpha{}^{\dot\beta}\,\psi^\mu(w), \qquad (2.56\text{a})$$

$$\Psi^i(z)\,\Sigma_\alpha \sim \frac{1}{\sqrt{2}\,(z-w)^{1/2}}\,\gamma^{IJ}_i\,\bar\Sigma_J(w),$$
$$\Sigma^I(z)\,\bar\Sigma_J \sim \frac{C^I{}_J}{(z-w)^{3/4}},$$
$$\Sigma^I(z)\,\Sigma^J \sim \frac{1}{\sqrt{2}(z-w)^{1/4}}\,(\gamma_k C)^{IJ}\,\Psi_k(w). \qquad (2.56\text{b})$$

Maximally supersymmetric toroidal compactification of type I or type II string theory yield a SYM theory in four dimensions with $\mathcal{N}=4$ supercharges. The gauge vector multiplet consists of three complex scalars ϕ^I, four gauginos λ^I and one gauge field A^μ. The corresponding vertex operators in the canonical ghost pictures have the form

$$V^{(-1)}_{\phi^{a,i}}(z,k) = g_\phi\,T^a\,\Phi_i\,\Psi^i(z)\,e^{-\phi(z)}\,e^{ikX(z)},$$
$$V^{(-1/2)}_{\lambda^{a,I}}(z,u,k) = g_\lambda\,T^a\,u^\alpha_I\,S_\alpha(z)\,\Sigma^I(z)\,e^{-\phi(z)/2}\,e^{ikX(z)},$$

---
[4]These OPEs look very distinct because the chirality structure of the charge conjugation matrix differs in four and six dimensions, which will be explained in Chapter 3.1. Please also note that the internal space has Euclidean signature.

2.2 String Scattering Amplitudes

$$V_{A^a}^{(-1)}(z,\xi,k) = g_A T^a \xi_\mu \psi^\mu(z) e^{-\phi(z)} e^{ikX(z)} \quad (2.57)$$

with the polarizations Φ_i, u_α and ξ_μ. A comparison of field theory and string theory scattering amplitudes in the limit $\alpha' \to 0$ shows that the couplings are related by

$$g_A = (2\alpha')^{1/2} g_{YM}, \qquad g_\lambda = (2\alpha')^{1/2} \alpha'^{1/4} g_{YM}, \qquad g_\phi = (2\alpha')^{1/2} g_{YM}. \quad (2.58)$$

The $D = 4$ gauge coupling g_{YM} can be expressed in terms of the ten-dimensional coupling g_{10} and the dilaton field ϕ_{10} through the relation $g_{YM} = g_{10} e^{\phi_{10}/2}$ [36].

The breaking of SUSY can be incorporated into the SCFT by orbifold projections acting on the internal fields Z^i and ψ^i. Only the fields which are invariant under the projection remain in the spectrum. For $\mathcal{N} = 1$ SUSY all scalars and three of the internal spin fields are projected out. Therefore one gaugino and one gauge field remain as required. The vertex operators take the simple form

$$V_{\lambda^a}^{(-1/2)}(z,u,k) = g_\lambda T^a u^\alpha S_\alpha(z) \Sigma(z) e^{-\phi(z)/2} e^{ikX(z)},$$

$$V_{A^a}^{(-1)}(z,\xi,k) = g_A T^a \xi_\mu \psi^\mu(z) e^{-\phi(z)} e^{ikX(z)}. \quad (2.59)$$

Apart from the vector multiplet there can be additional massless fields like chiral scalars and fermions stemming from D-brane intersections. Their vertex operators are very similar to (2.57), however, the internal fields are replaced by bosonic and fermionic twist fields [66]. These have an angular dependence on the D-brane intersection angle and are more difficult to handle [109–112].

An important fact to notice is that no internal fields ∂Z^i, Ψ^i or Σ^I enter the four-dimensional gluon vertex operator in (2.57) or (2.59). Tree-level scattering amplitudes in string theory involving only gluons are thus completely independent of the compactification details and even hold if SUSY is completely broken [113–115]. A further class of correlation functions which do not depend on the compactification are correlators involving arbitrary many gluons and at most two scalars, gauginos or chiral fermions. Under these circumstances the arising two-point function of the internal fields is completely determined by their conformal weights, but no further compactification details enter at tree-level. The analysis of such amplitudes is therefore a promising attempt to bypass the landscape problem [116].

2.2 String Scattering Amplitudes

So far our CFT description was centered on the free string. We come now to string interactions which result from strings splitting and joining. On account of the diffeomorphism invariance

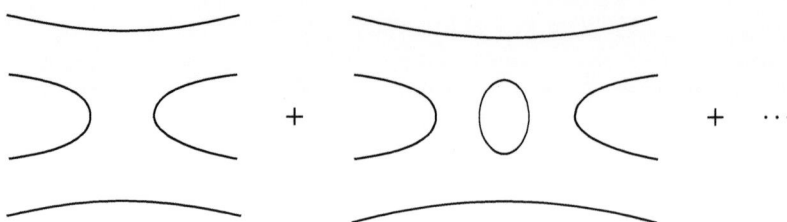

Figure 2.1: Summing over different world-sheet topologies.

of gravity one cannot calculate off-shell amplitudes in string theory. We can only evaluate entries of the S-matrix, i.e. scattering processes with asymptotic (on-shell) states. Conformal invariance allows to bring the initial and final string states to finite distances. The quantum numbers of the external states are then created by local operators on this compact world-sheet. These operators are exactly the vertex operators introduced above which can be understood as conformal projections of the initial and final string states.

In the Polyakov approach the amplitude is given by the functional integration over the world-sheet metric, gravitino and the matter fields weighted by the exponential of the action. The world-sheet of open string scattering at g loops is a two-dimensional surface with g holes and one boundary. For the total amplitude we must now also sum over different topologies of the world-sheet as depicted in Figure 2.1. A general n-point amplitude can then be computed by the following path integral:

$$\begin{aligned} \mathcal{M}_n &= \sum_g \mathcal{M}_n^{(g)} \\ &= \sum_g \int \mathcal{D}X \, \mathcal{D}\psi \, \mathcal{D}h \, \mathcal{D}\xi \int dz_1 \dots dz_n \, \mathcal{V}^{-1} \, V_1(z_1) \dots V_n(z_n) \, e^{-S[X,\psi,h,\xi]} \\ &= \sum_g \int dz_1 \dots dz_n \, \mathcal{V}^{-1} \left\langle V_1(z_1) \dots V_n(z_n) \right\rangle. \end{aligned} \qquad (2.60)$$

We mentioned before that the action of the superstring is invariant under diffeomorphisms and (super-)Weyl transformations. The integrals in (2.60) are therefore divergent. The measure of the integral must be changed in such a way that only configurations are taken into account which are not related to another by a symmetry transformation. This is achieved by the inclusion of the volume factor $\mathcal{V}^{-1}$ of the symmetry group. Its derivation is, e.g., greatly discussed in [95], we just sketch the results. At this point the ghost fields b and c enter the stage again. Their zero modes determine the number of moduli of the world-sheet and the number of conformal

2.2 String Scattering Amplitudes

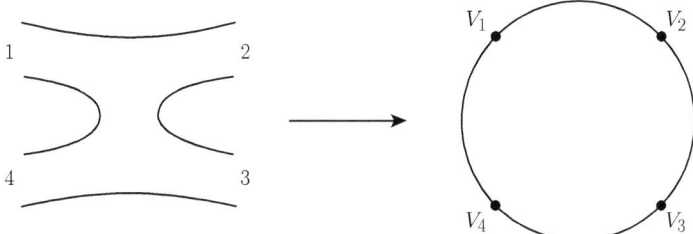

Figure 2.2: Mapping the string world-sheet onto the unit disk $\mathbb{D}$.

Killing vectors. For different genus they take the values:

$$N_b = \begin{cases} 0 & : g = 0, \\ 1 & : g = 1, \\ 3g-3 & : g \geq 2, \end{cases} \qquad N_c = \begin{cases} 3 & : g = 0, \\ 1 & : g = 1, \\ 0 & : g \geq 2. \end{cases} \qquad (2.61)$$

A detailed discussion shows that the volume factor demands to additionally integrate over N_b moduli in (2.60), while N_c integrated vertex operators $\int \mathrm{d}z_i\, V(z_i)$ can be replaced by $c(w_i)\,V(w_i)$ at a fixed position w_i. At this point we furthermore would like to mention that the vertex operators in (2.60) must be inserted in appropriate ghost pictures in order to cancel the background ghost charge of $2+2g$ given in (2.40).

2.2.1 Tree-level Amplitudes

Let us now discuss the easiest case $g = 0$ in more detail, i.e. tree-level scattering. The following discussion will find application in Chapter 6, when we calculate the tree-level amplitude of two gauge fields and four gauginos. The Riemann mapping theorem states that there exists a biholomorphic mapping of the tree-level world-sheet onto the unit disk $\mathbb{D} = \{z \in \mathbb{C} : |z| \leq 1\}$. The string states are then inserted by the vertex operators on the boundary of the disk as shown in Figure 2.2. Along the boundary the vertex operators have a definite ordering. For the calculation of the complete amplitude it is therefore necessary to sum over all cyclic non-equivalent permutations of the vertex operators:

$$\mathcal{M} = \sum_{\rho_i \in S_{n-1}} \int \mathrm{d}w_1 \ldots \mathrm{d}w_n\, \mathcal{V}^{-1} \left\langle V_1(w_1)\, V_{\rho(2)}(w_{\rho(2)}) \ldots V_{\rho(n)}(w_{\rho(n)}) \right\rangle. \qquad (2.62)$$

For the integration over the vertex positions it is more convenient to map the unit disk via the

Möbius transformation
$$w \mapsto z = i\frac{1-w}{1+w} \tag{2.63}$$
onto the upper half plane $\mathbb{H} = \{z \in \mathbb{C} : \Im(z) \geq 0\}$ as shown in Figure 2.3. The boundary is mapped onto the real axis, where the vertex operators then sit. The integration over each position z_i runs over $\mathbb{R}$, where the cyclic ordering of the positions must be kept. Depending on the purpose either the formulation of tree-level scattering on the disk or on the upper half plane can be more convenient.

The volume factor $\mathcal{V}$ rendering the integration finite has an easy expression for tree-level scattering. At genus 0 the world-sheet has no moduli which have to be integrated over. However, three c ghost zero modes exist. These are due to the fact that the disk has three globally defined conformal transformations that respect the boundary. The symmetry group is $SL(2,\mathbb{R})/\mathbb{Z}_2$ and acts on the coordinates $z \in \mathbb{H}$ in the following way:

$$z \mapsto \frac{az+b}{cz+d} \quad \text{with} \quad \begin{pmatrix} a & b \\ c & d \end{pmatrix} = -\begin{pmatrix} a & b \\ c & d \end{pmatrix}, \quad \det\begin{pmatrix} a & b \\ c & d \end{pmatrix} = 1. \tag{2.64}$$

The volume factor is then the volume of the conformal Killing group $\mathcal{V}_{\text{CKG}}$ which is canceled by fixing three vertex operator positions and including three c ghosts. The tree-level amplitude for a certain ordering takes the form

$$\mathcal{M}_\rho = \int_\mathbb{R} dz_4 \ldots dz_n \left\langle cV_1(z_1)\, cV_2(z_2)\, cV_3(z_3)\, V_4(z_4) \ldots V_n(z_n) \right\rangle. \tag{2.65}$$

The ghost correlator is easy to calculate. These fields decouple from the rest and their conformal weight $h(c) = -1$ determines

$$\left\langle c(z_1)\, c(z_2)\, c(z_3) \right\rangle = z_{12}\, z_{13}\, z_{23}. \tag{2.66}$$

The first step in the calculation of the amplitude is to evaluate the correlation function of the vertex operators. From the vertex operators (2.48), (2.57) and (2.59) it is clear that the matter fields, among them in particular the RNS fields ψ^μ and S_A, as well as the scalars bosonizing the superghost fields enter the amplitude. Only the RNS fields form an interacting CFT as shown in (2.56), all others simply decouple. The correlation function of the vertex operators therefore factorizes, written schematically as

$$\left\langle V_1 \ldots V_n \right\rangle \sim \mathcal{C}_X\, \mathcal{C}_\phi\, \mathcal{C}_{\psi,S}\, \mathcal{C}_{\Phi,\Sigma}, \tag{2.67}$$

where the last correlator only appears in the case of string compactifications. The correlation

2.2 String Scattering Amplitudes

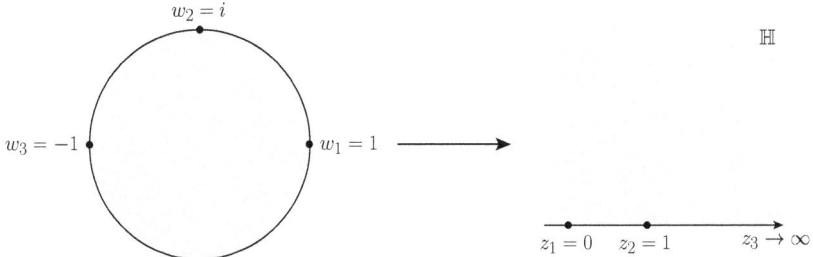

Figure 2.3: Mapping the unit disk $\mathbb{D}$ onto the upper half plane $\mathbb{H}$ via the Möbius transformation $w \mapsto z = i\frac{1-w}{1+w}$.

functions involving the string coordinates X^μ or superghost scalars ϕ are rather trivial to calculate as they are free fields. The RNS correlators are much more difficult to evaluate due to their interacting nature. The calculation of such n-point functions for massless interactions is the topic of the following Chapters 3-5. RNS correlators for higher spin scattering have been considered in [68, 117].

For gauge theory amplitudes also the Chan–Paton matrices T^a_{ij} occur in the amplitude. As previously explained these arise from a stack of D-branes, where the indices i, j label the particular D-branes to which the endpoints are attached. An interaction, i.e. the joining of two string ends, can only occur if these are attached to the same D-brane. This in turn implies that the indices of the Chan–Paton factors of the two strings must coincide: $T^{a_1}_{ij}$ and $T^{a_1}_{jk}$. Repeating this argument for all other string ends implies that the amplitude contains the trace over all Chan–Paton factors as illustrated in Figure 2.4. The full amplitude then reads

$$\mathcal{M} = \sum_{\rho \in S_{n-1}} \mathrm{Tr}(T^{a_1} T^{a_{\rho(2)}} \ldots T^{a_{\rho(n)}}) \, \mathcal{A}\big(1, \rho(2), \ldots, \rho(n)\big). \tag{2.68}$$

The sub-amplitude $\mathcal{A}$ is calculated by stripping of the Chan–Paton factors off the vertex operators and integration over the positions of the latter. It is called the color-ordered or color-stripped amplitude and satisfies among others the cyclic property

$$\mathcal{A}(1, 2, \ldots, n) = \mathcal{A}(n, 1, 2, \ldots, n-1). \tag{2.69}$$

The next step in the calculation is the integration over the world-sheet positions z, which is very challenging for higher-point amplitudes. The six-gluon amplitude for example contains hypergeometric functions and Euler-Zagier sums [118]. Let us discuss as an easier example the

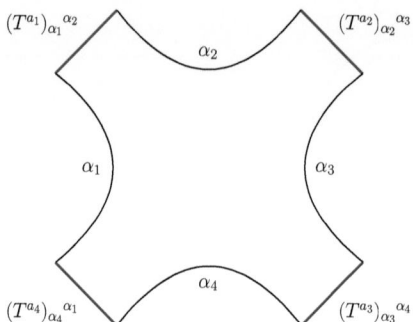

Figure 2.4: Chan–Paton factors in a four-string interactions. String can only join or split when the respective endpoints have the same Chan–Paton label.

partial four gluon amplitude [113, 114]:

$$\mathcal{M}(g_1^-, g_2^-, g_3^+, g_4^+) = 4\,g^2\,\mathrm{Tr}(T^{a_1}\,T^{a_2}\,T^{a_3}\,T^{a_4})\,\frac{\langle 12\rangle^4}{\langle 12\rangle\,\langle 23\rangle\,\langle 34\rangle\,\langle 41\rangle}\,V(k_1, k_2, k_3, k_4)\,. \quad (2.70)$$

The integration over the vertex operator positions leads to the formfactor

$$V(k_1, k_2, k_3, k_4) = \frac{su}{s+u}\,B(s, u)\,, \quad (2.71)$$

where B is the Euler beta function and s, u, t are the stringy Mandelstam variables

$$s \equiv 2\alpha' k_1\,k_2\,, \qquad t \equiv 2\alpha' k_1\,k_3\,, \qquad u \equiv 2\alpha' k_1\,k_4\,. \quad (2.72)$$

An expansion of the formfactor in α',

$$V(k_1, k_2, k_3, k_4) \sim 1 - \zeta(2)\,su + \zeta(3)\,stu + \ldots\,, \quad (2.73)$$

reveals that the kth order in α' is accompanied by the zeta value $\zeta(k)$, where ambiguities occur at higher order due to multi-zeta values [90, 118]. At level five for example one can have $\zeta(3, 1, 1) = 2\,\zeta(5) - \zeta(2)\,\zeta(3)$. In the limit $\alpha' \to 0$ the formfactor reduces to 1 and the amplitude (2.70) simply becomes the Park-Taylor amplitude of four gluons [119]. In the field theory limit only massless strings participate in the interaction, while interactions of massive string states sum up to finite terms of order α'^2 and higher. In an effective field theory description these can be interpreted as new contact interactions. Another way of looking at (2.70) is to expand V as

$$V(k_1, k_2, k_3, k_4) \sim \sum_{n=0}^{\infty} \frac{\gamma(n)}{s - \frac{n}{\alpha'}}\,, \qquad \gamma(n) \equiv \frac{1}{n!}\,\frac{\Gamma(-u+n)}{\Gamma(-u)} \quad (2.74)$$

2.2 String Scattering Amplitudes

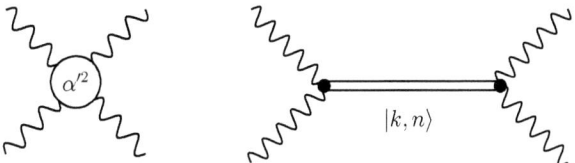

Figure 2.5: String contact interaction and exchange of Regge resonances, taken from [66].

with an infinite number of s-channel poles. The physical interpretation of this expression is that the scattering partners can exchange string Regge excitations of mass $M^2 = n/\alpha'$, which is depicted together with the contact interaction in Figure 2.5.

2.2.2 Loop-Level Amplitudes

The schematic form of an n-point g-loop amplitude of open strings has the form [120–122]:

$$\mathcal{M}_b = \int \frac{\mathrm{d}^{N_b}\Omega}{\det \Omega} \int \frac{\prod_{i=1}^{N} \mathrm{d}z^i}{\mathcal{V}_{\mathrm{CKG}}^g} \sum_{(\vec{a},\vec{b})} \mathcal{Z}^{(\vec{a},\vec{b})} \langle V_1(z_1,\Omega) \ldots V_n(z_n,\Omega) \rangle_{\vec{b}}^{\vec{a}}. \qquad (2.75)$$

In this expression $\mathcal{Z}$ is the genus g partition function, i.e. the vacuum to vacuum amplitude. The first integral in (2.75) runs over the moduli space of dimension N_b as stated in (2.61), the second one over the vertex operator positions on the boundary of the disk with g holes. At $g = 1$ the volume factor of the conformal Killing group is taken care of by fixing one position and including the corresponding c ghost. For higher genus no position can be fixed as dictated by (2.61).

The correlation function of the vertex operators factorizes again into correlators involving the fields X, the external RNS fields, internal fields and superghosts. Correlators of the latter are well understood [123], whereas the internal correlation functions can be described by some character valued partition function or elliptic genus [124–129]. The ghost charges of the vertex operators must sum up to $2g - 2$ in order to cancel the ghost background charge (2.40).

In addition, g-dimensional vectors $(\vec{a},\vec{b})$ with entries 0 or 1 enter the calculation of loop amplitudes. They are called spin structures. Let us explain their meaning with the help of the RNS fields. The doubling trick (2.44) extends ψ^m and S_A to their counterparts of the closed string. The chiral field $\psi^m(z)$ is only defined on the upper half-plane. Combining it with the anti-chiral field $\bar{\psi}^m(\bar{z})$ yields a new field that is defined on the full complex plane. Therefore, the resulting RNS fields do not have support on the disk with g holes but in fact on the sphere with g handles. Such a Riemann surface of genus g has $2g$ homology cycles α_I and β_I, $I = 1, \ldots, g$, as illustrated in Figure 2.6. The fermions must either be periodic

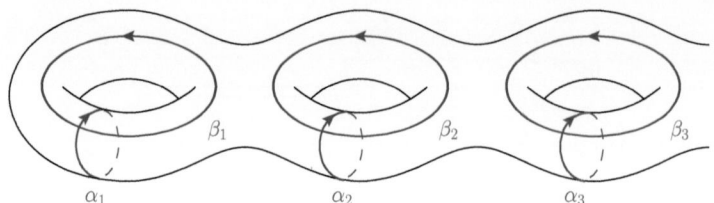

Figure 2.6: One-cycles α_I and β_I of a Riemann surface.

or antiperiodic if they are shifted around the one-cycles. These periodicity properties define the entries of the spin structure vectors $(\vec{a}, \vec{b})$. As we have learned before the spin fields S_A create branch cuts on the world-sheet. The brunch cut extends all along the cycle if these fields are shifted around. Hence, shifting the spin fields changes the complete spin structure of the RNS correlator. Further details can be found in [130] and Appendix C. The summation over different spin structures is a delicate business. In the RNS correlators the spin vectors $(\vec{a}, \vec{b})$ enter through generalized Θ functions. So the spin sum requires various relations between Θ functions like generalized Riemann identities [121, 131–135].

In this work we do not attempt to calculate a full loop amplitude. We concentrate on calculating the analogs of the tree-level RNS correlation functions which are needed for loop scattering. These are presented in Chapter 5.

CHAPTER 3

Ramond–Neveu–Schwarz Correlators at Tree-Level

We consider in this Chapter correlation functions involving the RNS fields ψ^μ and S_α, $S^{\dot\beta}$ at tree-level in arbitrary even space–time dimensions $D = 2m$, $m \in \mathbb{N}_0$. As the Ramond spin fields are of fermionic type we review first the Clifford algebra and spinors in higher dimensions. Then we discuss the CFT and the OPEs which govern the short-distance behavior of the RNS fields and show how correlation functions can be calculated therefrom. Before we present special techniques how to calculate correlators in the dimensions $D = 4, 6, 8$ or 10 and give some results for general D, we comment on the index terms, which are important ingredients in the RNS correlators. The work presented here has been published in [1, 2].

3.1 Prerequisites

Before we discuss RNS correlators at tree-level we have to lay the groundwork for their calculation. The conformal field theory of the RNS fields plays an important role here, as it provides sufficient input to calculate correlation functions. Additionally, we comment on spinors in higher dimensions as the R spin fields are of this type. Further information is found in Appendix A and in [36, 136–138].

3.1.1 Spinors in Higher Dimensions

Dirac spinors in D space-time dimensions form a complex vector space of dimension $2^{D/2}$. They are representations of the Clifford algebra

$$\{\Gamma^\mu, \Gamma^\nu\} = -2\,\eta^{\mu\nu}\,. \tag{3.1}$$

The $2^{D/2} \times 2^{D/2}$ matrices Γ^μ are known as gamma matrices. In even dimensions - as we consider here - Dirac spinors can be decomposed into two irreducible Weyl representations. This implies for the R spin fields

$$S_A \equiv S_\alpha \oplus S^{\dot\alpha}\,. \tag{3.2}$$

We call the Weyl spinors S_α and $S^{\dot\alpha}$ in the following left- and right-handed spin field. In component notation the gamma matrices read

$$(\Gamma^\mu)_A{}^B = \begin{pmatrix} 0 & \gamma^\mu_{\alpha\dot\beta} \\ \bar\gamma^{\mu\,\dot\alpha\beta} & 0 \end{pmatrix}\,, \tag{3.3}$$

with the generalized Pauli matrices γ^μ and $\bar\gamma^\mu$ as off-diagonal blocks. In terms of these the Clifford algebra becomes

$$\gamma^\mu_{\alpha\dot\beta}\bar\gamma^{\nu\,\dot\beta\gamma} + \gamma^\nu_{\alpha\dot\beta}\bar\gamma^{\mu\,\dot\beta\gamma} = -2\,\delta^\gamma_\alpha\,\eta^{\mu\nu}\,, \qquad \bar\gamma^{\mu\,\dot\alpha\beta}\gamma^\nu_{\beta\dot\gamma} + \bar\gamma^{\nu\,\dot\alpha\beta}\gamma^\mu_{\beta\dot\gamma} = -2\,\delta^{\dot\alpha}_{\dot\gamma}\,\eta^{\mu\nu}\,. \tag{3.4}$$

The charge conjugation matrix $\mathcal{C}$ acts like a metric on the space of spinors. It is defined in even dimensions by the equations

$$\mathcal{C}^t = (-1)^{\frac{m}{2}(m+1)}\,\mathcal{C}\,, \tag{3.5a}$$

$$(\Gamma^\mu)^t = -\mathcal{C}^{-1}\,\Gamma^\mu\,\mathcal{C}\,. \tag{3.5b}$$

The chirality structure of the charge conjugation matrix depends on D because of the representation theory of the associated Lorentz group $SO(1, D-1)$. One finds in component notation:

$$D = 0 \bmod 4: \quad \mathcal{C}_{AB} = \begin{pmatrix} C_{\alpha\beta} & 0 \\ 0 & C^{\dot\alpha\dot\beta} \end{pmatrix}\,,$$

$$D = 2 \bmod 4: \quad \mathcal{C}_{AB} = \begin{pmatrix} 0 & C_\alpha{}^{\dot\beta} \\ C^{\dot\alpha}{}_\beta & 0 \end{pmatrix}\,. \tag{3.6}$$

The inverse matrix $\mathcal{C}^{-1}$ reads then

$$D = 0 \bmod 4: \quad (\mathcal{C}^{-1})^{AB} = \begin{pmatrix} (C^{-1})^{\alpha\beta} & 0 \\ 0 & (C^{-1})_{\dot\alpha\dot\beta} \end{pmatrix}\,,$$

$$D = 2 \bmod 4: \quad (\mathcal{C}^{-1})^{AB} = \begin{pmatrix} 0 & (C^{-1})^\alpha{}_{\dot\beta} \\ (C^{-1})_{\dot\alpha}{}^\beta & 0 \end{pmatrix}\,. \tag{3.7}$$

3.1 Prerequisites

In the following we refrain from writing out C^{-1} explicitly because the position and order of the spinor indices distinguish C and C^{-1} from each other.

Equation (3.5a) states that C is antisymmetric in four dimensions. For appropriate representations of the Clifford algebra, in particular where γ^μ are the standard Pauli matrices σ^μ, it coincides with the ε tensor. We adopt the same conventions and denote in the following

$$\sigma^\mu_{\alpha\dot\alpha} \equiv \gamma^\mu_{\alpha\dot\alpha}\big|_{D=4}\,, \qquad \varepsilon_{\alpha\beta} \equiv C_{\alpha\beta}\big|_{D=4} \tag{3.8}$$

for the calculation of RNS correlation functions in four space-time dimensions.

3.1.2 The Underlying Conformal Field Theory

The short distance behavior of the RNS fields is given by their OPEs [98, 99, 104]. As we see below this is sufficient input to calculate RNS correlation function at tree-level. The OPEs of the NS fermions and R spin fields have been stated in (2.13) and (2.24), we present now their chiral versions involving left- and right-handed spin fields S_α and $S^{\dot\beta}$.

The singular behavior of the fermions as well as the interaction between one fermion and one spin field does not depend on the number of dimensions:

$$\psi^\mu(z)\,\psi^\nu(w) \sim \frac{\eta^{\mu\nu}}{z-w}\,, \tag{3.9a}$$

$$\psi^\mu(z)\,S_\alpha(w) \sim \frac{1}{\sqrt{2}}\,(z-w)^{-1/2}\,\gamma^\mu_{\alpha\dot\beta}\,S^{\dot\beta}(w)\,. \tag{3.9b}$$

In contrast, conformal weights and therefore the OPEs of two spin fields depend on the dimensionality. Furthermore, as the chiral structure of the charge conjugation matrix C depends on D, one encounters two distinct scenarios:

- $D = 0 \bmod 4$:

$$S_\alpha(z)\,S_\beta(w) \sim (z-w)^{-D/8}\,C_{\alpha\beta}\,, \tag{3.10a}$$

$$S_\alpha(z)\,S^{\dot\beta}(w) \sim \frac{1}{\sqrt{2}}\,(z-w)^{-D/8+1/2}\,(\gamma^\mu C)_\alpha{}^{\dot\beta}\,\psi_\mu(w)\,, \tag{3.10b}$$

- $D = 2 \bmod 4$:

$$S_\alpha(z)\,S^{\dot\beta}(w) \sim (z-w)^{-D/8}\,C_\alpha{}^{\dot\beta}\,, \tag{3.11a}$$

$$S_\alpha(z)\,S_\beta(w) \sim \frac{1}{\sqrt{2}}\,(z-w)^{-D/8+1/2}\,(\gamma^\mu C)_{\alpha\beta}\,\psi_\mu(w) + \ldots\,. \tag{3.11b}$$

Using these OPEs it is in principle possible to construct every correlator involving RNS fields as we describe in Chapter 3.2.

The reader should note that in contrast to [1] the spin fields in (3.10) have been redefined by a factor of i to avoid proliferation of minus signs. In order to be consistent with the original literature, we use (3.10) and (3.11) for the calculation of correlators in six, eight and ten dimensions following [2], while in four dimensions we use the conventions from [1],

$$S_\alpha(z)\, S_\beta(w) \sim -(z-w)^{-1/2}\, \varepsilon_{\alpha\beta}\,, \tag{3.12a}$$

$$S_{\dot\alpha}(z)\, S_{\dot\beta}(w) \sim +(z-w)^{-1/2}\, \varepsilon_{\dot\alpha\dot\beta}\,, \tag{3.12b}$$

$$S_\alpha(z)\, S_{\dot\beta}(w) \sim \frac{1}{\sqrt{2}}\, \sigma^\mu_{\alpha\dot\beta}\, \psi_\mu(w)\,, \tag{3.12c}$$

together with the identifications (3.8). It is easy to check that these OPEs are consistent with (3.9).

3.2 The Evaluation of Correlators

We now present various methods to calculate RNS correlators at tree-level. The starting point for the evaluation are the OPEs of the RNS fields.

3.2.1 The Iterative Procedure

One possible way to calculate correlation functions involving NS fermions and R spin fields

$$\left\langle \psi^{\mu_1}(z_1)\ldots\psi^{\mu_n}(z_n)\, S_{\alpha_1}(x_1)\ldots S_{\alpha_r}(x_r)\, S^{\dot\beta_1}(y_1)\ldots S^{\dot\beta_s}(y_s) \right\rangle \tag{3.13}$$

with $n,r,s \in \mathbb{N}_0$ is by considering the correlator in all possible limits $z_i \to z_j$, $z_i \to x_j$, $z_i \to y_j$, ... and applying the respective OPEs from (3.9), (3.10) and (3.11), where we only keep the most singular part. In this way (3.13) reduces to a lower-point correlation function. If the expression for this function is known one can match the findings from the different limits to construct the final result. Using this iterative procedure we can determine higher-point correlation functions from already known correlators. As an example let us discuss the correlation function $\langle \psi^\mu(z_1)\, S_\alpha(z_2)\, S_{\dot\beta}(z_3) \rangle$ in four space-time dimensions. By examining this

3.2 The Evaluation of Correlators

correlator in all possible limits $z_i \to z_j$ we find with $z_{ij} \equiv z_i - z_j$,

$$\langle \psi^\mu(z_1) \, S_\alpha(z_2) \, S_{\dot\beta}(z_3) \rangle \sim \begin{cases} \frac{1}{\sqrt{2}} \sigma^\mu_{\alpha\dot\beta} \, (z_{12} \, z_{13})^{-1/2} & : z_1 \to z_2 \,, \\ \frac{1}{\sqrt{2}} \sigma^\mu_{\alpha\dot\beta} \, (z_{12} \, z_{13})^{-1/2} & : z_1 \to z_3 \,, \\ \frac{1}{\sqrt{2}} \sigma^\mu_{\alpha\dot\beta} \, (z_{12})^{-1} & : z_2 \to z_3 \,, \end{cases} \quad (3.14)$$

where the OPEs (3.9), (3.12) and the two-point functions of the fermions and spin fields have been used. The result of the correlation function should reduce to these terms in the respective limits. We therefore find

$$\langle \psi^\mu(z_1) \, S_\alpha(z_2) \, S_{\dot\beta}(z_3) \rangle = \frac{1}{\sqrt{2}} \sigma^\mu_{\alpha\dot\beta} \, (z_{12} \, z_{13})^{-1/2} \,, \quad (3.15)$$

where we have replaced z_{12}^{-1} by $(z_{12} \, z_{13})^{-1/2}$ in the limit $z_2 \to z_3$ in (3.14). From this result we can deduce the general structure of RNS correlation functions. They will always consist of terms that carry all Lorentz indices of the involved RNS fields, in the case above $\sigma^\mu_{\alpha\dot\beta}$. We denote such expressions in the following as *index terms*. These are accompanied by a coefficient which depends on the positions of the fields z_i. The index terms play an important role in the evaluation of correlation functions and are discussed in great detail in Chapter 3.3.

As a more complicated example let us have a look at the correlation function of four left-handed spin fields in four dimensions, $\langle S_\alpha(z_1) \, S_\beta(z_2) \, S_\gamma(z_3) \, S_\delta(z_4) \rangle$, which is known from [108]. With the OPE (3.12a) we find

$$\langle S_\alpha(z_1) \, S_\beta(z_2) \, S_\gamma(z_3) \, S_\delta(z_4) \rangle \sim \begin{cases} \varepsilon_{\alpha\beta} \, \varepsilon_{\gamma\delta} \, (z_{12} \, z_{34})^{-1/2} & : z_1 \to z_2 \text{ or } z_3 \to z_4 \,, \\ \varepsilon_{\alpha\gamma} \, \varepsilon_{\beta\delta} \, (z_{13} \, z_{24})^{-1/2} & : z_1 \to z_3 \text{ or } z_2 \to z_4 \,, \\ \varepsilon_{\alpha\delta} \, \varepsilon_{\beta\gamma} \, (z_{14} \, z_{23})^{-1/2} & : z_1 \to z_4 \text{ or } z_2 \to z_3 \,. \end{cases} \quad (3.16)$$

Three different index terms arise in the limits and at this point it is simply impossible to match the different contribution to obtain the final result. However, not all index terms are independent, one can be eliminated:

$$\varepsilon_{\alpha\gamma} \, \varepsilon_{\beta\delta} = \varepsilon_{\alpha\beta} \, \varepsilon_{\gamma\delta} + \varepsilon_{\alpha\delta} \, \varepsilon_{\beta\gamma} \,. \quad (3.17)$$

Only with this identity it is possible to match the z coefficients in the different limits and derive the following result:

$$\langle S_\alpha(z_1) \, S_\beta(z_2) \, S_\gamma(z_3) \, S_\delta(z_4) \rangle = \left(\frac{z_{12} \, z_{14} \, z_{23} \, z_{34}}{z_{13} \, z_{24}} \right)^{1/2} \left(\frac{\varepsilon_{\alpha\beta} \, \varepsilon_{\gamma\delta}}{z_{12} z_{34}} - \frac{\varepsilon_{\alpha\delta} \, \varepsilon_{\gamma\beta}}{z_{14} \, z_{32}} \right) \,. \quad (3.18)$$

To check the consistency of this expression with the separate limits in (3.16) the z-crossing identity

$$z_{ij}\, z_{kl} = z_{ik}\, z_{jl} + z_{il}\, z_{kj} \tag{3.19}$$

proves to be useful. This four-point function is our first example of an RNS correlator, where the possible index terms are not independent from each other. Determining the number of independent index terms is a crucial task in order to calculate the correlation function and we address this problem in Chapter 3.3.1.

Some care is required to incorporate complex phases which arise upon performing the OPEs in the limits like (3.14) and (3.16). Since OPEs are defined by the action of the involved fields on the vacuum state $|0\rangle$, it is necessary to "shift" the respective fields first to the right end of the correlation function before applying the OPE. Commuting the RNS fields with each other results in factors of i or -1 due to the different powers in $(z-w)$ in (3.9), (3.10) and (3.11).

Particular easy results can be obtained for the $2n$-point function involving only NS fermions in even dimensions and also for the correlator consisting of only left-handed spin fields in 0 mod 4 dimensions. This is due to the OPEs (3.9a) and (3.10a) which state that in these cases ψ^μ and S_α are free fields and do not interact with the other RNS operators. Such correlation functions can hence easily be determined by Wick's theorem [139].

We come now to the proof that the matching procedure of the different limits indeed yields the correct result for the correlation function. It relies on Liouville's theorem stating that every holomorphic bounded function has to be constant. Let R denote the result of (3.13) which has been obtained by forming all limits, applying the OPEs and matching the different results. In the following we keep all positions of the RNS fields fixed but arbitrary apart from z_1 and consider the function

$$\langle \psi^{\mu_1}(z_1)\ldots\psi^{\mu_n} S_{\alpha_1}\ldots S_{\alpha_r} S^{\dot\beta_1}\ldots S^{\dot\beta_s}\rangle - R(z_1)\,. \tag{3.20}$$

Poles in the correlator originate from the OPEs as z_1 approaches the position of another RNS field. By construction $R(z_1)$ has the same poles and therefore (3.20) has no singularities and is a holomorphic function. The correlator and R vanish both for $z_1 \to \infty$. In fact they vanish with $z_1^{-2h_1}$ as required by CFT, where h_1 is the conformal weight of ψ^{μ_1}. Therefore the function (3.20) is bounded. Altogether, Liouville's theorem then implies that the function vanishes. Repeating this argument for all other field positions instead of z_1 completes the proof. Thus the iterative method indeed gives the correct result for a tree-level correlation function.

3.2.2 From Fermions to Spin Fields

Another way of calculating RNS correlators is to reduce them to correlation functions involving only spin fields. Indeed, by using the relation $\text{Tr}\{\gamma^\mu \bar{\gamma}^\nu\} = -2^{(D-2)/2}\eta^{\mu\nu}$ the equations (3.10b) and (3.11b) can be inverted:

$$\psi^\mu(w) = -2^{(3-D)/2} \lim_{z \to w} (z-w)^{D/8-1/2} \times \begin{cases} (C^{-1}\bar{\gamma}^\mu)_{\dot{\beta}}{}^\alpha S_\alpha(z) S^{\dot{\beta}}(w) & : D = 0 \bmod 4, \\ (C^{-1}\bar{\gamma}^\mu)^{\dot{\beta}\alpha} S_\alpha(z) S_{\dot{\beta}}(w) & : D = 2 \bmod 4. \end{cases} \quad (3.21)$$

Hence, every fermion ψ^μ appearing in a correlation function can be replaced by two spin fields $S_\alpha, S^{\dot{\beta}}$ or $S_\alpha, S_{\dot{\beta}}$ depending on the number of dimensions. So the following correlators can be calculated from the respective spin field correlators via (3.21):

- $D = 0 \bmod 4$:

$$\left\langle \psi^{\mu_1} \ldots \psi^{\mu_n} S_{\alpha_1} \ldots S_{\alpha_r} S^{\dot{\beta}_1} \ldots S^{\dot{\beta}_s} \right\rangle \longrightarrow \left\langle S_{\alpha_1} \ldots S_{\alpha_{r+n}} S^{\dot{\beta}_1} \ldots S^{\dot{\beta}_{s+n}} \right\rangle, \quad (3.22a)$$

- $D = 2 \bmod 4$:

$$\left\langle \psi^{\mu_1} \ldots \psi^{\mu_{k+l}} S_{\alpha_1} \ldots S_{\alpha_r} S^{\dot{\beta}_1} \ldots S^{\dot{\beta}_s} \right\rangle \longrightarrow \left\langle S_{\alpha_1} \ldots S_{\alpha_{r+2k}} S^{\dot{\beta}_1} \ldots S^{\dot{\beta}_{s+2l}} \right\rangle, \quad (3.22b)$$

where also $k, l \in \mathbb{N}_0$. This method of reducing an arbitrary correlator to a pure spin field correlator leads to higher-point expressions that seem at first to be more difficult to calculate. Yet for $D = 4$ the correlators appearing on the r.h.s. of (3.22) can be calculated for arbitrary many spin fields, while for $D = 6$ one can at least evaluate the correlation function with the same number of left- and right-handed spin fields. The reason for this is that such correlators can be expressed by index terms which are only products of charge conjugation matrices but no γ matrices enter. In this sense replacing the fermions with spin fields turns out to be a useful method. Furthermore, (3.22) provides nice consistency checks if the correlation functions on both sides are known.

Let us discuss this in more details. In Chapter 3.4 it is shown that in the case of $D = 4$ dimensions every RNS correlator can be reduced to the correlation function consisting of $2M$ left-handed spin fields:

$$\left\langle S_{\alpha_1}(z_1) \ldots S_{\alpha_{2M}}(z_{2M}) \right\rangle. \quad (3.23)$$

A solution for this correlation function can be derived by induction and hence we are able to calculate every RNS correlator in four dimensions at tree-level. The only index terms that enter

3. Ramond–Neveu–Schwarz Correlators at Tree-Level

M	$D=4$	$D=8$	$(2M-1)!!$
1	1	1	1
2	2	3	3
3	5	15	15
4	14	106	105
5	42	981	945

Table 3.1: Number of independent index terms for the correlator $\langle S_{\alpha_1} \ldots S_{\alpha_{2M}}\rangle$ in four and eight dimensions, as well as the number of possible index terms of the type (3.24).

the result of (3.23) are products of the charge conjugation matrix:

$$C_{\alpha_1 \alpha_2} \ldots C_{\alpha_{2M-1}\alpha_{2M}} \,. \tag{3.24}$$

The number of index terms of this type is calculated by simple combinatorics. Distributing the $2M$ spinor indices to the tensors C yields

$$\binom{2M}{2}\binom{2M-2}{2}\ldots\binom{2}{2} = \prod_{i=0}^{M}\binom{2M-2i}{2} \tag{3.25}$$

possibilities. This must be divided by $M!$ to account for permutations of the C's. In total one finds $(2M-1)!!$ terms of the type (3.24). These are sufficient because only $2M!/M!\,(M+1)!$ are independent in four dimensions due to relations like (3.17) as we show in Chapter 3.4.1. Meanwhile the situation in $D=8$ dimensions is different. In Chapter 3.3 it is discussed that more index terms then the ones from (3.24) have to be taken into account for $M \geq 4$. The respective numbers are juxtaposed in Table 3.1. In addition to the tensors (3.24) terms involving γ-matrices like

$$(\gamma^\mu \bar{\gamma}^\nu C)_{\alpha\beta}\,(\gamma_\mu \bar{\gamma}_\nu C)_{\gamma\delta} \tag{3.26}$$

are now required, where the vector indices are contracted as pure spin field correlator do not carry indices of this type. Therefore we are not able to directly construct an easy expression for the correlation function of $2M$ left-handed spin fields in eight dimensions. Even more complicated correlators, like the ones appearing on the l.h.s. in (3.22a), have to be calculated by hand. Still, $SO(8)$ triality, as we describe in Chapter 3.6, is a powerful tool to relate different correlators in eight dimensions at tree-level.

In the case of $D=6$ and $D=10$ dimensions we encounter a similar scenario. Due to the different chirality structure of the charge conjugation matrix $C_\alpha{}^{\dot\beta}$ we consider here the correlators

$$\langle S_{\alpha_1}(z_1)\, S^{\dot\beta_1}(z_2)\ldots S_{\alpha_M}(z_{2M-1})\, S^{\dot\beta_M}(z_{2M})\rangle \tag{3.27}$$

3.2 The Evaluation of Correlators

M	$D=6$	$D=10$	$M!$
1	1	1	1
2	2	3	2
3	6	19	6
4	24	210	24
5	119	3514	120

Table 3.2: Number of independent index terms for the correlator $\langle S_{\alpha_1} S^{\dot\beta_1} \ldots S_{\alpha_M} S^{\dot\beta_M} \rangle$ in six and ten dimensions, as well as the number of possible index terms of the type (3.28).

from (3.22b) with $r + 2k = s + 2l = M$. In $D = 6$ space-time dimensions one is able to derive a general expression for these $2M$-point functions, where only index terms of the form

$$C_{\alpha_1}{}^{\dot\beta_1} \ldots C_{\alpha_M}{}^{\dot\beta_M} \tag{3.28}$$

enter. In total there are $M!$ terms of this type, arising from permuting the $\dot\beta$'s while keeping the α's fixed. These index terms are sufficient in six space-time dimensions and thus we are able to calculate (3.27). In contrast, for $D = 10$ the number of independent index terms is greater then $M!$ for $M \geq 2$ as discussed in Chapter 3.3. Apart from the expressions (3.28) additional terms are needed which involve Lorentz contractions of γ-matrices, like

$$(\gamma^\mu C)_{\alpha\beta} (\bar\gamma_\mu C)^{\dot\gamma\dot\delta} . \tag{3.29}$$

The respective numbers of index terms are summarized in Table 3.2. Thus, all RNS correlation functions in ten dimensions have to be calculated by hand.

However, the situation in eight and ten dimensions is not as unpromising as it first seems. We derive in Chapter 5.3 results for the correlation functions

$$\langle \psi^{\mu_1} \ldots \psi^{\mu_n} S_\alpha S_\beta \rangle , \qquad \langle \psi^{\mu_1} \ldots \psi^{\mu_n} S_\alpha S^{\dot\beta} \rangle \tag{3.30}$$

with arbitrary many NS fermions and two spin fields in four, six, eight and ten dimensions. These results even hold at loop-level.

3.2.3 Alternative Methods

In the literature RNS correlators have also been calculated using different methods. One technique we like to mention is bosonization as described in Chapter 2.1. For this purpose the fermion and spin fields in $D = 2m$ dimensions are expressed through exponentials

$$\psi^\mu(z) = e^{ipH(z)} c , \qquad S_A(z) = e^{ipH(z)/2} c' \tag{3.31}$$

of a vector $H = (H_1, \ldots, H_m)$ with m free bosons and the m-dimensional lattice vector p with entries ± 1. The free bosons fulfill the normalization convention:

$$\langle H_i(z_i) H_j(z_j) \rangle = \delta_{ij} \ln(z-w) \,. \tag{3.32}$$

The factors c in (3.31) are so-called cocycle operators which ensure that the correct (anti-) commutation properties of the RNS fields arise. Bosonization of the RNS fields does not yield results in manifestly Lorentz covariant form. For this purpose the cocyle factors have to be related to the index terms via the OPEs (3.9), (3.10) and (3.11). This has been achieved up to six-point level e.g. in [104, 140, 141]. In the following however, we pursuit our method which directly yields results in covariant form.

Another possibility for the evaluation of tree-level correlation functions with several NS fermions inserted at the same position on the string world-sheet is via the $o(2m)$ current algebra. The operators

$$J^{\mu\nu}(z) \equiv \psi^{[\mu}(z) \psi^{\nu]}(z) \tag{3.33}$$

realize the $SO(1, D-1)$ current algebra at level $k=1$ [142], see also [98, 99, 104]. The action of the current on the other fields is determined by the OPEs

$$\begin{aligned} J^{\mu\nu}(z) \psi^\lambda(w) &\sim -\frac{2}{z-w} \eta^{\lambda[\mu} \psi^{\nu]}(w) \,, \\ J^{\mu\nu}(z) S_\alpha(w) &\sim -\frac{1}{2(z-w)} \gamma^{\mu\nu}{}_\alpha{}^\beta S_\beta(w) \end{aligned} \tag{3.34}$$

and the central term of the current-current OPE

$$\begin{aligned} J^{\mu\nu}(z) J^{\lambda\rho}(w) &\sim \frac{1}{(z-w)^2} \left(\eta^{\mu\rho} \eta^{\nu\lambda} - \eta^{\mu\lambda} \eta^{\nu\rho} \right) \\ &+ \frac{1}{z-w} \left[\eta^{\mu\lambda} J^{\nu\rho}(w) - \eta^{\mu\rho} J^{\nu\lambda}(w) - \eta^{\nu\lambda} J^{\mu\rho}(w) + \eta^{\nu\rho} J^{\mu\lambda}(w) \right] \,. \end{aligned} \tag{3.35}$$

Hence, any correlator including $J^{\mu\nu}$ can be reduced to a correlation function with one current insertion less:

$$\begin{aligned} &\left\langle J^{\mu\nu}(z) \psi_{\lambda_1}(z_1) \ldots \psi_{\lambda_n}(z_n) S_{\alpha_1}(x_1) \ldots S_{\alpha_r}(x_r) S_{\dot\beta_1}(y_1) \ldots S_{\dot\beta_s}(y_s) \right\rangle \\ &= -\sum_{j=1}^n \frac{2}{z-z_j} \\ &\quad \times \delta^{[\mu}_{\lambda_j} \left\langle \psi_{\lambda_1}(z_1) \ldots \psi^{\nu]}(z_j) \ldots \psi_{\lambda_n}(z_n) S_{\alpha_1}(x_1) \ldots S_{\alpha_r}(x_r) S_{\dot\beta_1}(y_1) \ldots S_{\dot\beta_s}(y_s) \right\rangle \\ &\quad - \sum_{j=1}^r \frac{1}{2(z-z_j)} \gamma^{\mu\nu}{}_{\alpha_j}{}^\kappa \end{aligned}$$

$$\times \left\langle \psi_{\lambda_1}(z_1) \ldots \psi_{\lambda_n}(z_n) S_{\alpha_1}(x_1) \ldots S_\kappa(x_j) \ldots S_{\alpha_r}(x_r) S_{\dot{\beta}_1}(y_1) \ldots S_{\dot{\beta}_s}(y_s) \right\rangle$$

$$+ \sum_{j=1}^{s} \frac{1}{2(z-z_j)} \bar{\gamma}^{\mu\nu\dot{\kappa}}{}_{\dot{\beta}_j}$$

$$\times \left\langle \psi_{\lambda_1}(z_1) \ldots \psi_{\lambda_n}(z_n) S_{\alpha_1}(x_1) \ldots S_{\alpha_r}(x_r) S_{\dot{\beta}_1}(y_1) \ldots S_{\dot{\kappa}}(y_j) \ldots S_{\dot{\beta}_s}(y_s) \right\rangle. \qquad (3.36)$$

However, the goal of our work goes far beyond the application of (3.36). All the correlation functions in the following will be derived in full generality without any coinciding arguments. Of course, by a posteriori moving fermion positions together, one can obtain nice consistency checks for the results in the following Chapters.

3.3 The Index Terms

We have seen above that the index terms and their relations play an important role in the calculation of RNS correlation functions. Therefore we discuss them in the following in more detail.

3.3.1 The Group Theory Behind the Correlators

Let us have a look at the ψ–S correlators from a group theoretical point of view. The NS fields ψ^μ transform as vectors under the Lorentz group $SO(1, D-1)$, while the R spin fields $S_\alpha, S^{\dot\beta}$ as left- and right-handed spinors. Therefore an arbitrary correlation function made up of these fields lies in the corresponding tensor product

$$\left\langle \psi^{\mu_1} \ldots \psi^{\mu_n} S_{\alpha_1} \ldots S_{\alpha_r} S_{\dot{\beta}_1} \ldots S_{\dot{\beta}_s} \right\rangle \in (V)^{\otimes n} \otimes (S)^{\otimes r} \otimes (\dot{S})^{\otimes s}, \qquad (3.37)$$

where (V) is the vector representation, (S) and $(\dot{S})$ are the left- and right-handed spinor representations of $SO(1, D-1)$. From the results of the correlators (3.15) and (3.18) we conclude that the coefficients depending on the vertex operator positions z_i transform as scalars with respect to the D-dimensional Lorentz group. The index terms must then be *Clebsch-Gordan coefficients* associated with the particular scalar representation.

The decomposition of the tensor product (3.37) does not help in finding the precise expressions for the Clebsch-Gordan coefficients. Yet it is possible to determine the number of independent index terms from the appropriate tensor product. This is simply given by the number of scalar representations. If a correlator has no scalar representations, i.e. there exist no Clebsch–Gordan coefficients, then the whole expression has to vanish[1]. This has non-trivial con-

---
[1] In bosonization or in the loop methods displayed in Chapter 5 conservation of Ramond charge ensures that these correlators vanish as well

sequences for full string amplitudes in which these correlators enter. Certain string amplitudes involving bosons and fermions then simply vanish.

In Table 3.3 we state the number of independent index terms for certain correlators in four, six, eight and ten space-time dimensions. For $D = 4$ these number can be calculated analytically as we show in Chapter 3.4.1. For other dimensions these are difficult to evaluate [143], so we have used the computer program LiE [144]. The cases $D = 4, 8$ and $D = 6, 10$ are treated separately because of the different chirality structure of the spinor representations in these dimensions. One observes that the number of independent Clebsch–Gordan coefficients for a given correlation function increases with the number of dimensions. We comment on this circumstance in Chapter 3.3. This makes the calculation of seven- and higher-point correlators troublesome as more and more index terms have to be taken into account. We refrain from calculating any RNS correlation function beyond eight-point level in $D = 4$ and six-point level in $D = 6, 8, 10$, although this is in principle possible with the techniques presented in this work.

3.3.2 Relations between Index Terms

It was just explained how the number of independent index terms can be obtained from the tensor product (3.37). In general it is possible to write down more Clebsch–Gordan coefficients for a given correlator then there are independent index terms. The easiest example of such a case is the already familiar correlation function of four left-handed spin fields in four dimensions,

$$\langle S_\alpha(z_1)\, S_\beta(z_2)\, S_\gamma(z_3)\, S_\delta(z_4) \rangle. \tag{3.38}$$

The tensor structure of this correlator suggests that the result can be expressed in terms of $\varepsilon_{\alpha\beta}\,\varepsilon_{\gamma\delta}$, $\varepsilon_{\alpha\gamma}\,\varepsilon_{\beta\delta}$ and $\varepsilon_{\alpha\delta}\,\varepsilon_{\beta\gamma}$. However, Table (3.63) for $M = 2$ tells that only two independent index terms exist. Indeed, the identity

$$\varepsilon_{\alpha\gamma}\,\varepsilon_{\beta\delta} = \varepsilon_{\alpha\beta}\,\varepsilon_{\gamma\delta} + \varepsilon_{\alpha\delta}\,\varepsilon_{\beta\gamma} \tag{3.39}$$

allows to reduce one index term. This relation is crucial for the calculation of this correlator. Otherwise, deriving a result for (3.38) by analyzing its singularity structure and applying the OPEs (3.12a) leads to inconsistencies as the contributions to the different index terms cannot properly be separated. Therefore it is very important to start with a minimal set of index terms and know how they can be related to the Clebsch–Gordan coefficients which are not included in this set. This turns out to be also most important for the calculation of RNS correlators at loop-level.

The identity (3.39) is easily derived by noticing that a Weyl spinor in $D = 4$ dimensions has only $2^{D/2-1} = 2$ independent (complex) components. Therefore, all expressions which are

3.3 The Index Terms

	$D=4$	$D=8$		$D=6$	$D=10$
$\langle S_\alpha S_\beta \rangle$	1	1	$\langle S_\alpha S^{\dot\beta} \rangle$	1	1
$\langle \psi^\mu S_\alpha S^{\dot\beta} \rangle$	1	1	$\langle \psi^\mu S_\alpha S_\beta \rangle$	1	1
$\langle \psi^\mu \psi^\nu S_\alpha S_\beta \rangle$	2	2	$\langle \psi^\mu \psi^\nu S_\alpha S^{\dot\beta} \rangle$	2	2
$\langle \psi^\mu \psi^\nu \psi^\lambda S_\alpha S^{\dot\beta} \rangle$	4	4	$\langle \psi^\mu \psi^\nu \psi^\lambda S_\alpha S_\beta \rangle$	4	4
$\langle \psi^\mu \psi^\nu \psi^\lambda \psi^\rho S_\alpha S_\beta \rangle$	10	10	$\langle \psi^\mu \psi^\nu \psi^\lambda \psi^\rho S_\alpha S^{\dot\beta} \rangle$	10	10
$\langle \psi^\mu \psi^\nu \psi^\lambda \psi^\rho \psi^\tau S_\alpha S^{\dot\beta} \rangle$	25	26	$\langle \psi^\mu \psi^\nu \psi^\lambda \psi^\rho \psi^\tau S_\alpha S_\beta \rangle$	26	26
$\langle \psi^\mu \psi^\nu \psi^\lambda \psi^\rho \psi^\tau \psi^\xi S_\alpha S_\beta \rangle$	70	76	$\langle \psi^\mu \psi^\nu \psi^\lambda \psi^\rho \psi^\tau \psi^\xi S_\alpha S^{\dot\beta} \rangle$	76	76
$\langle S_\alpha S_\beta S^{\dot\gamma} S^{\dot\delta} \rangle$	1	2	$\langle S_\alpha S_\beta S_\gamma S_\delta \rangle$	1	2
$\langle S_\alpha S_\beta S_\gamma S_\delta \rangle$	2	3	$\langle S_\alpha S_\beta S^{\dot\gamma} S^{\dot\delta} \rangle$	2	3
$\langle \psi^\mu S_\alpha S_\beta S_\gamma S^{\dot\delta} \rangle$	2	4	$\langle \psi^\mu S_\alpha S_\beta S_\gamma S^{\dot\delta} \rangle$	3	5
$\langle \psi^\mu \psi^\nu S_\alpha S_\beta S^{\dot\gamma} S^{\dot\delta} \rangle$	4	9	$\langle \psi^\mu S_\alpha S_\beta S_\gamma S_\delta \rangle$	6	11
$\langle \psi^\mu \psi^\nu S_\alpha S_\beta S_\gamma S_\delta \rangle$	5	10	$\langle \psi^\mu \psi^\nu S_\alpha S_\beta S^{\dot\gamma} S^{\dot\delta} \rangle$	7	12
$\langle \psi^\mu \psi^\nu \psi^\lambda S_\alpha S_\beta S_\gamma S^{\dot\delta} \rangle$	10	24	$\langle \psi^\mu \psi^\nu \psi^\lambda S_\alpha S_\beta S_\gamma S^{\dot\delta} \rangle$	17	31
$\langle \psi^\mu \psi^\nu \psi^\lambda \psi^\rho S_\alpha S_\beta S^{\dot\gamma} S^{\dot\delta} \rangle$	25	68	$\langle \psi^\mu \psi^\nu \psi^\lambda \psi^\rho S_\alpha S_\beta S_\gamma S_\delta \rangle$	45	88
$\langle \psi^\mu \psi^\nu \psi^\lambda \psi^\rho S_\alpha S_\beta S_\gamma S_\delta \rangle$	28	71	$\langle \psi^\mu \psi^\nu \psi^\lambda \psi^\rho S_\alpha S_\beta S^{\dot\gamma} S^{\dot\delta} \rangle$	48	91
$\langle S_\alpha S_\beta S_\gamma S_\delta S^{\dot\epsilon} S^{\dot\zeta} \rangle$	2	10	$\langle S_\alpha S_\beta S_\gamma S_\delta S_\epsilon S^{\dot\zeta} \rangle$	4	16
$\langle S_\alpha S_\beta S_\gamma S_\delta S_\epsilon S_\zeta \rangle$	5	15	$\langle S_\alpha S_\beta S_\gamma S^{\dot\delta} S^{\dot\epsilon} S^{\dot\zeta} \rangle$	6	19
$\langle \psi^\mu S_\alpha S_\beta S_\gamma S^{\dot\delta} S^{\dot\epsilon} S^{\dot\zeta} \rangle$	4	24	$\langle \psi^\mu S_\alpha S_\beta S_\gamma S_\delta S_\epsilon S_\zeta \rangle$	9	40
$\langle \psi^\mu S_\alpha S_\beta S_\gamma S_\delta S_\epsilon S^{\dot\zeta} \rangle$	5	26	$\langle \psi^\mu S_\alpha S_\beta S_\gamma S_\delta S^{\dot\epsilon} S^{\dot\zeta} \rangle$	12	45
$\langle \psi^\mu \psi^\nu S_\alpha S_\beta S_\gamma S_\delta S^{\dot\epsilon} S^{\dot\zeta} \rangle$	10	68	$\langle \psi^\mu \psi^\nu S_\alpha S_\beta S_\gamma S_\delta S_\epsilon S^{\dot\zeta} \rangle$	29	125
$\langle \psi^\mu \psi^\nu S_\alpha S_\beta S_\gamma S_\delta S_\epsilon S_\zeta \rangle$	14	76	$\langle \psi^\mu \psi^\nu S_\alpha S_\beta S_\gamma S^{\dot\delta} S^{\dot\epsilon} S^{\dot\zeta} \rangle$	32	130
$\langle S_\alpha S_\beta S_\gamma S_\delta S^{\dot\epsilon} S^{\dot\zeta} S^{\dot\theta} S^{\dot\iota} \rangle$	4	71	$\langle S_\alpha S_\beta S_\gamma S_\delta S_\epsilon S_\zeta S_\theta S_\iota \rangle$	14	175
$\langle S_\alpha S_\beta S_\gamma S_\delta S_\epsilon S_\zeta S^{\dot\theta} S^{\dot\iota} \rangle$	5	76	$\langle S_\alpha S_\beta S_\gamma S_\delta S_\epsilon S_\zeta S^{\dot\theta} S^{\dot\iota} \rangle$	19	196
$\langle S_\alpha S_\beta S_\gamma S_\delta S_\epsilon S_\zeta S_\theta S_\iota \rangle$	14	106	$\langle S_\alpha S_\beta S_\gamma S_\delta S^{\dot\epsilon} S^{\dot\zeta} S^{\dot\theta} S^{\dot\iota} \rangle$	24	210

Table 3.3: Number of independent Clebsch–Gordan coefficients for various correlation functions in $D = 4, 8$ and $D = 6, 10$ space-time dimensions.

antisymmetric in three or more Weyl indices, have to vanish. This is indeed the case for (3.39) as it can be written as

$$\varepsilon_{\alpha[\beta}\,\varepsilon_{\gamma\delta]} = 0\,. \tag{3.40}$$

We want to stress that the same relations does not hold in $D = 8$ dimensions,

$$C_{\alpha[\beta}\,C_{\gamma\delta]} \neq 0\,, \tag{3.41}$$

because a Weyl spinor then has eight independent components. Therefore all three index terms $C_{\alpha\beta}\,C_{\gamma\delta}$, $C_{\alpha\gamma}\,C_{\beta\delta}$ and $C_{\alpha\delta}\,C_{\beta\gamma}$ will appear in the correlator of four left-handed spin fields, which is consistent with the corresponding entry in Table 3.3. Hence, the number of independent index terms for a given correlator increases in general with the number of dimensions.

Additional relations for correlation functions involving fermions can be obtained in the same fashion. The anti-symmetrization argument proves e.g. that in four dimensions

$$(\sigma^{\mu_1}\,\bar{\sigma}^{\mu_2}\ldots\sigma^{\mu_{2n-1}}\,\bar{\sigma}^{\mu_{2n}}\,\varepsilon)_{\alpha[\beta}\,\varepsilon_{\sigma\delta]} = 0\,, \tag{3.42}$$

which is needed for the correlator involving $2n$ fermions and four left-handed spin fields. This procedure can easily be extended to higher dimensions D. Then however, one has to anti-symmetrize over at least $(2^{D/2-1} + 1)$ Weyl indices as these run from 1 to $2^{D/2-1}$.

Multiplying known equations with further γ matrices gives also rise to new relations. If (3.39) for example is multiplied with $\bar{\sigma}^{\mu\dot{\delta}\delta}\,\varepsilon_{\dot{\epsilon}\dot{\delta}}$ we obtain

$$\sigma^\mu_{\beta\dot{\epsilon}}\,\varepsilon_{\alpha\gamma} = \sigma^\mu_{\gamma\dot{\epsilon}}\,\varepsilon_{\alpha\beta} + \sigma^\mu_{\alpha\dot{\epsilon}}\,\varepsilon_{\beta\gamma}\,, \tag{3.43}$$

which is needed for the calculation of the correlator consisting of one fermion, three left-handed and one right-handed spin fields in four dimensions.

Further relations between different index terms can be derived from Fierz identities [136, 145, 146]. The antisymmetric γ-products $\Gamma^{\mu_1\ldots\mu_n}$ form a complete set of $2^{D/2} \times 2^{D/2}$ matrices. Therefore any bi-spinor $\psi_\alpha\,\chi_\beta$ or $\psi_\alpha\,\bar{\chi}^{\dot{\beta}}$ can be expanded in terms of these forms. The expansion prescriptions are referred to as Fierz identities. One arrives again at two scenarios due to the different chirality structure of the charge conjugation matrix in different dimensions:

- $D = 0$ mod 4:

$$\begin{aligned}\psi_\alpha\,\chi_\beta &= 2^{-D/2} \sum_{\substack{n=0 \\ n\text{ even}}}^{D/2-2} \frac{1}{n!}\,(\gamma^{\mu_1\ldots\mu_n}\,C)_{\beta\alpha}\,(\psi\,C^{-1}\,\gamma_{\mu_n\ldots\mu_1}\,\chi) \\ &\quad + \frac{2^{-D/2}}{2\,(D/2)!}\,(\gamma^{\mu_1\ldots\mu_{D/2}}\,C)_{\beta\alpha}\,(\psi\,C^{-1}\,\gamma_{\mu_{D/2}\ldots\mu_1}\,\chi)\,,\end{aligned} \tag{3.44a}$$

3.3 The Index Terms

$$\psi_\alpha \bar{\chi}^{\dot\beta} = -2^{-D/2} \sum_{\substack{n=0 \\ n \text{ odd}}}^{D/2-1} \frac{1}{n!} \left(\bar\gamma^{\mu_1\ldots\mu_n} C\right)^{\dot\beta}{}_\alpha \left(\psi\, C^{-1} \gamma_{\mu_n\ldots\mu_1} \bar\chi\right), \tag{3.44b}$$

- $D = 2 \bmod 4$:

$$\psi_\alpha \chi_\beta = -2^{-D/2} \sum_{\substack{n=0 \\ n \text{ odd}}}^{D/2-2} \frac{1}{n!} \left(\gamma^{\mu_1\ldots\mu_n} C\right)_{\beta\alpha} \left(\psi\, C^{-1} \bar\gamma_{\mu_n\ldots\mu_1} \chi\right)$$
$$-\frac{2^{-D/2}}{2\,(D/2)!} \left(\gamma^{\mu_1\ldots\mu_{D/2}} C\right)_{\beta\alpha} \left(\psi\, C^{-1} \bar\gamma_{\mu_{D/2}\ldots\mu_1} \chi\right), \tag{3.44c}$$

$$\psi_\alpha \bar{\chi}^{\dot\beta} = 2^{-D/2} \sum_{\substack{n=0 \\ n \text{ even}}}^{D/2-1} \frac{1}{n!} \left(\bar\gamma^{\mu_1\ldots\mu_n} C\right)^{\dot\beta}{}_\alpha \left(\psi\, C^{-1} \bar\gamma_{\mu_n\ldots\mu_1} \bar\chi\right). \tag{3.44d}$$

Note that only forms up to degree $D/2$ appear as every n-fold product $\gamma^{\mu_1\ldots\mu_n}$ is related to $(D-n)$-fold products via Hodge duality. By making clever choices for the spinors ψ and χ in (3.44) one obtains relations between various index terms. Choosing $\chi_\alpha = \varepsilon_{\alpha\gamma}$ and $\chi_\beta = \varepsilon_{\beta\delta}$ for $D = 4$ (3.44a) implies that

$$(\sigma^{\mu\nu})_\alpha{}^\beta (\sigma_{\mu\nu})_\gamma{}^\delta = 4\,\delta_\alpha^\beta \delta_\gamma^\delta - 8\,\delta_\alpha^\delta \delta_\gamma^\beta, \tag{3.45}$$

while (3.44b) directly yields

$$(\sigma^\mu)_{\alpha\dot\beta} (\sigma_\mu)_{\gamma\dot\delta} = -2\,\varepsilon_{\alpha\gamma}\,\varepsilon_{\dot\beta\dot\delta}. \tag{3.46}$$

Equations (3.44) are discussed separately for $D = 4, 6, 8$ and $D = 10$ in Appendix A.

In eight dimensions there exists another way of deriving relations between different index terms, namely $SO(8)$ triality. Details on this can be found in Chapter 3.6.

3.3.3 Manipulation of Index Terms

In the calculation of correlators it is sometimes necessary to manipulate Clebsch–Gordan coefficients, especially interchanging vector and spinor indices. The Clifford algebra (3.4) for the Weyl blocks γ^μ, $\bar\gamma^\mu$ plays an important role here. It can be used to interchange γ matrices in longer γ-chains, e.g.

$$\gamma^\nu \bar\gamma^\mu \gamma^\lambda = -\gamma^\mu \bar\gamma^\nu \gamma^\lambda - 2\,\eta^{\mu\nu} \gamma^\lambda. \tag{3.47}$$

Relations of this type provide the basis to use only index terms with ordered γ-products, where the Lorentz indices appear in ascending order. In the same way antisymmetric γ-products like $\gamma^{\mu\nu}$ can be reduced to ordinary γ-products and η terms.

Often it is also important to know how chains of γ matrices like $\gamma^{\mu_1\ldots\mu_p} C$ and $\bar\gamma^{\mu_1\ldots\mu_p} C$ behave under interchanging their spinor indices. These $2^{m-1} \times 2^{m-1}$ matrices appear as blocks in the $2^m \times 2^m$ matrix $\Gamma^{\mu_1\mu_2\cdots\mu_p}\mathcal{C}$,

- $D = 0 \mod 4$:

$$(\Gamma^{\mu_1\cdots\mu_p}\mathcal{C})_{AB} = \begin{cases} \begin{pmatrix} (\gamma^{\mu_1\cdots\mu_p}C)_{\alpha\beta} & 0 \\ 0 & (\bar{\gamma}^{\mu_1\cdots\mu_p}C)^{\dot{\alpha}\dot{\beta}} \end{pmatrix} & : p \text{ even}, \\ \begin{pmatrix} 0 & (\gamma^{\mu_1\cdots\mu_p}C)_{\alpha}{}^{\dot{\beta}} \\ (\bar{\gamma}^{\mu_1\cdots\mu_p}C)^{\dot{\alpha}}{}_{\beta} & 0 \end{pmatrix} & : p \text{ odd}, \end{cases} \quad (3.48a)$$

- $D = 2 \mod 4$:

$$(\Gamma^{\mu_1\cdots\mu_p}\mathcal{C})_{AB} = \begin{cases} \begin{pmatrix} 0 & (\gamma^{\mu_1\cdots\mu_p}C)_{\alpha}{}^{\dot{\beta}} \\ (\bar{\gamma}^{\mu_1\cdots\mu_p}C)^{\dot{\alpha}}{}_{\beta} & 0 \end{pmatrix} & : p \text{ even}, \\ \begin{pmatrix} (\gamma^{\mu_1\cdots\mu_p}C)_{\alpha\beta} & 0 \\ 0 & (\bar{\gamma}^{\mu_1\cdots\mu_p}C)^{\dot{\alpha}\dot{\beta}} \end{pmatrix} & : p \text{ odd}. \end{cases} \quad (3.48b)$$

By analyzing how these matrices behave under transposition one finds relations for the (off-)diagonal blocks. From Chapter 3.1.1 we know that

$$(\Gamma^\mu)^t = -\mathcal{C}^{-1}\Gamma^\mu\mathcal{C}, \qquad (3.49a)$$
$$\mathcal{C}^t = (-1)^{\frac{1}{2}m(m+1)}\mathcal{C}. \qquad (3.49b)$$

Taking these together the anti-symmetric Γ-chains behave like

$$(\Gamma^{\mu_1\cdots\mu_p}\mathcal{C})^t = (-1)^{\frac{1}{2}[m(m+1)+p(p+1)]}(\Gamma^{\mu_1\cdots\mu_p}\mathcal{C}). \qquad (3.50)$$

We can translate this relation back to the level of Weyl blocks and find e.g. in four dimensions for $p = 1$ and $p = 2$:

$$(\sigma^\mu\varepsilon)_\alpha{}^{\dot{\beta}} = (\bar{\sigma}^\mu\varepsilon)^{\dot{\beta}}{}_\alpha, \qquad (3.51a)$$
$$(\sigma^{\mu\nu}\varepsilon)_{\alpha\beta} = (\sigma^{\mu\nu}\varepsilon)_{\beta\alpha}. \qquad (3.51b)$$

Together with the Clifford algebra (3.4) this implies

$$(\sigma^\mu\bar{\sigma}^\nu\varepsilon)_{\beta\alpha} = (\sigma^\mu\bar{\sigma}^\nu\varepsilon)_{\alpha\beta} + 2\eta^{\mu\nu}\varepsilon_{\alpha\beta}. \qquad (3.52)$$

Relations between index terms similar to (3.51) can easily be derived from (3.50) for higher

3.4 Techniques in Four Dimensions

$D=4$	$D=6$
$C_{\alpha\beta} = -C_{\beta\alpha}$	$C_\alpha{}^{\dot\beta} = +C^{\dot\beta}{}_\alpha$
$(\gamma^\mu C)_\alpha{}^{\dot\beta} = +(\bar\gamma^\mu C)^{\dot\beta}{}_\alpha$	$(\gamma^\mu C)_{\alpha\beta} = -(\gamma^\mu C)_{\beta\alpha}$
$(\gamma^{\mu\nu} C)_{\alpha\beta} = +(\gamma^{\mu\nu} C)_{\beta\alpha}$	$(\gamma^{\mu\nu} C)_\alpha{}^{\dot\beta} = -(\bar\gamma^{\mu\nu} C)^{\dot\beta}{}_\alpha$
$(\gamma^{\mu\nu\lambda} C)_\alpha{}^{\dot\beta} = -(\bar\gamma^{\mu\nu\lambda} C)^{\dot\beta}{}_\alpha$	$(\gamma^{\mu\nu\lambda} C)_{\alpha\beta} = +(\gamma^{\mu\nu\lambda} C)_{\beta\alpha}$
$(\gamma^{\mu\nu\lambda\rho} C)_{\alpha\beta} = -(\gamma^{\mu\nu\lambda\rho} C)_{\beta\alpha}$	$(\gamma^{\mu\nu\lambda\rho} C)_\alpha{}^{\dot\beta} = +(\bar\gamma^{\mu\nu\lambda\rho} C)^{\dot\beta}{}_\alpha$
$D=8$	$D=10$
$C_{\alpha\beta} = +C_{\beta\alpha}$	$C_\alpha{}^{\dot\beta} = -C^{\dot\beta}{}_\alpha$
$(\gamma^\mu C)_\alpha{}^{\dot\beta} = -(\bar\gamma^\mu C)^{\dot\beta}{}_\alpha$	$(\gamma^\mu C)_{\alpha\beta} = +(\gamma^\mu C)_{\beta\alpha}$
$(\gamma^{\mu\nu} C)_{\alpha\beta} = -(\gamma^{\mu\nu} C)_{\beta\alpha}$	$(\gamma^{\mu\nu} C)_\alpha{}^{\dot\beta} = +(\bar\gamma^{\mu\nu} C)^{\dot\beta}{}_\alpha$
$(\gamma^{\mu\nu\lambda} C)_\alpha{}^{\dot\beta} = +(\bar\gamma^{\mu\nu\lambda} C)^{\dot\beta}{}_\alpha$	$(\gamma^{\mu\nu\lambda} C)_{\alpha\beta} = -(\gamma^{\mu\nu\lambda} C)_{\beta\alpha}$
$(\gamma^{\mu\nu\lambda\rho} C)_{\alpha\beta} = +(\gamma^{\mu\nu\lambda\rho} C)_{\beta\alpha}$	$(\gamma^{\mu\nu\lambda\rho} C)_\alpha{}^{\dot\beta} = -(\bar\gamma^{\mu\nu\lambda\rho} C)^{\dot\beta}{}_\alpha$

Table 3.4: Symmetry properties of certain tensors built from γ^μ and C in different dimensions.

dimensions and higher p. The relevant cases for the following calculations are summed up in Table 3.4.

We summarize that knowing relations between the relevant Clebsch–Gordan coefficients for a given RNS correlation function is crucial for its calculation. These can either be obtained by anti-symmetrization techniques, Fierz identities or multiplying known identities with further γ matrices. In Appendix B we give all necessary index term relations for the correlators that are calculated in Chapters 4 and 5. Obtaining these relations can be a very tedious task. We have checked these identities in all conscience and furthermore tested them with Mathematica [147] for a particular representation of the Clifford algebra, which is given in Appendix A.4.

In the remainder of this Chapter we present special methods how to calculate RNS correlation functions in four, six, eight and ten space-time dimensions. Concrete results follow in Chapters 4 and 5.

3.4 Techniques in Four Dimensions

Let us start the discussion with the case $D = 4$. Four-dimensional RNS correlators play an important role in the calculation of string amplitudes, where ten-dimensional space-time is compactified to four dimensions. In such a scenario it is possible to connect string theory with four-dimensional particle physics and string scattering amplitudes might describe corrections to SM processes [66–68]. Before the calculations we shed some light on the number of independent index terms.

3.4.1 The Number of Index Terms

In Chapter 3.3.1 we have seen that the number of independent index terms for a certain correlator is given by the number of scalar representations in the respective tensor product (3.37). This tensor product can be evaluated analytically in four dimensions. It is well known that the Lorentz algebra $so(1,3)$ decomposes into a direct sum of two $su(2)$ subalgebras, a left- and a right-handed one. General representations of $SO(1,3)$ with spins j_1, j_2 with respect to the left- and right-handed $SU(2)$ are denoted by $(\mathbf{j_1}, \mathbf{j_2})$. The fermions ψ^μ then transform as $\left(\frac{1}{2}, \frac{1}{2}\right)$ under $SO(1,3)$, whereas the spin fields S_α and $S_{\dot\beta}$ transform as $\left(\frac{1}{2}, 0\right)$ and $\left(0, \frac{1}{2}\right)$ respectively. The tensor product then becomes

$$\left\langle \psi^{\mu_1} \ldots \psi^{\mu_n} S_{\alpha_1} \ldots S_{\alpha_r} S_{\dot\beta_1} \ldots S_{\dot\beta_s} \right\rangle \in \left(\tfrac{1}{2}, \tfrac{1}{2}\right)^{\otimes n} \otimes \left(\tfrac{1}{2}, 0\right)^{\otimes r} \otimes \left(0, \tfrac{1}{2}\right)^{\otimes s}, \quad (3.53)$$

where as usual

$$\mathbf{0} \otimes \tfrac{1}{2} = \tfrac{1}{2} \quad \text{and} \quad \tfrac{i}{2} \otimes \tfrac{1}{2} = \tfrac{i+1}{2} \oplus \tfrac{i-1}{2}, \quad i > 1. \quad (3.54)$$

The product in (3.53) then reads

$$\left\langle \psi^{\mu_1} \ldots \psi^{\mu_n} S_{\alpha_1} \ldots S_{\alpha_r} S_{\dot\beta_1} \ldots S_{\dot\beta_s} \right\rangle \in \left(\tfrac{1}{2}, \mathbf{0}\right)^{\otimes(n+r)} \otimes \left(\mathbf{0}, \tfrac{1}{2}\right)^{\otimes(n+s)}. \quad (3.55)$$

This is a first hint on the fact that RNS correlators in four dimensions factorize into correlation functions that only contain left- or right handed spin fields. This is achieved by replacing each fermions with a left- and a right-handed spin field as described in (3.21) and then splitting the correlator up. We discuss this in more details below. For the moment it is thus satisfactory to consider

$$\left\langle S_{\alpha_1} \ldots S_{\alpha_N} \right\rangle \in \left(\tfrac{1}{2}, \mathbf{0}\right)^{\otimes N} = \left(\tfrac{1}{2}^{\otimes N}, \mathbf{0}\right). \quad (3.56)$$

Applying (3.54) this can be expanded as

$$\left(\tfrac{1}{2}, \mathbf{0}\right)^{\otimes N} = \bigoplus_{i=0}^{N} q(i, N) \left(\tfrac{i}{2}, \mathbf{0}\right). \quad (3.57)$$

Finding the integer coefficient $q(i, N)$ is a common counting problem in combinatorics which is e.g. equivalent to a random walk with step size $1/2$ on the positive real axis [148, 149]. It is obvious that $q(0, 0)$ has to fulfill

$$q(0, 0) = 1. \quad (3.58)$$

Following (3.54) the representation $\left(\tfrac{i}{2}, \mathbf{0}\right)$ in (3.57) stems either from the representations $\left(\tfrac{i-1}{2}, \mathbf{0}\right)$ or $\left(\tfrac{i+1}{2}, \mathbf{0}\right)$ in the lower tensor product $\left(\tfrac{1}{2}, \mathbf{0}\right)^{\otimes(N-1)}$. Therefore, the second defining equation

3.4 Techniques in Four Dimensions

for $q(i, N)$ is
$$q(i, N) = q(i - 1, N - 1) + q(i + 1, N - 1). \tag{3.59}$$

By induction one can show that
$$q(i, N) \equiv \frac{i+1}{N+1} \binom{N+1}{\frac{N-i}{2}} \tag{3.60}$$

fulfills the defining equations (3.58) and (3.59). For $(N - i) \notin 2\,\mathbb{N}_0$ the binomial coefficient is not defined and in this case we set $q(i, N)$ to zero. In Table 3.5 we list some values of $q(i, N)$. By replacing i with $n - k$ and N with $n + k$ (3.60) yields the standard form of the numbers appearing in the Catalan triangle:
$$c_{n,k} \equiv q(n - k, n + k) = \frac{(n+k)!\,(n-k+1)}{k!\,(n+1)!}. \tag{3.61}$$

From (3.60) we can read off the number of scalar representations in the tensor product (3.56). It is given by
$$q(0, N) = \frac{1}{N+1} \binom{N+1}{N/2}, \tag{3.62}$$

which is only non-zero if N is an even number. Then $q(0, N = 2M)$ takes the well known form of the Catalan numbers:
$$q(0, 2M) = \frac{1}{2M+1} \binom{2M+1}{M} = \frac{2M!}{M!\,(M+1)!}. \tag{3.63}$$

We conclude that $q(0, 2M)$ yields the number of index terms of the RNS correlator in four dimensions consisting of $2M$ left-handed spin fields. For an odd number of spin fields there exists no scalar representation and hence this correlator has to vanish:
$$\langle S_{\alpha_1}(x_1) \ldots S_{\alpha_{2r-1}}(x_{2r-1}) \rangle = 0. \tag{3.64}$$

Together with (3.55) this yields that the following correlators vanish as well $(n, r, s \in \mathbb{N}_0)$:
$$\langle \psi^{\mu_1}(z_1) \ldots \psi^{\mu_{2n-1}}(z_{2n-1})\, S_{\alpha_1}(x_1) \ldots S_{\alpha_{2r}}(x_{2r})\, S_{\dot{\alpha}_1}(y_1) \ldots S_{\dot{\alpha}_s}(y_s) \rangle = 0,$$
$$\langle \psi^{\mu_1}(z_1) \ldots \psi^{\mu_{2n-1}}(z_{2n-1})\, S_{\dot{\alpha}_1}(y_1) \ldots S_{\dot{\alpha}_{2s}}(y_{2s})\, S_{\alpha_1}(x_1) \ldots S_{\alpha_r}(x_r) \rangle = 0,$$
$$\langle \psi^{\mu_1}(z_1) \ldots \psi^{\mu_{2n}}(z_{2n})\, S_{\alpha_1}(x_1) \ldots S_{\alpha_{2r-1}}(x_{2r-1})\, S_{\dot{\alpha}_1}(y_1) \ldots S_{\dot{\alpha}_s}(y_s) \rangle = 0,$$
$$\langle \psi^{\mu_1}(z_1) \ldots \psi^{\mu_{2n}}(z_{2n})\, S_{\dot{\alpha}_1}(y_1) \ldots S_{\dot{\alpha}_{2s-1}}(y_{2s-1})\, S_{\alpha_1}(x_1) \ldots S_{\alpha_r}(x_r) \rangle = 0. \tag{3.65}$$

N,i	0	1	2	3	4	5	6
0	1						
1		1					
2	1		1				
3		2		1			
4	2		3		1		
5		5		4		1	
6	5		9		5		1

Table 3.5: Values of the coefficients $q(i, N)$ in the tensor product (3.57).

Let us give two examples of tensor product decompositions and their help in determining the linear independent set of Clebsch–Gordan coefficients. Let us start with the familiar correlation function in four dimensions consisting of four left-handed spin fields. The tensor product

$$\langle S_\alpha S_\beta S_\gamma S_\delta \rangle \in (\tfrac{1}{2}, 0)^{\otimes 4} = 1\,(\mathbf{2, 0}) \oplus 3\,(\mathbf{1, 0}) \oplus 2\,\underline{(\mathbf{0, 0})} \qquad (3.66)$$

contains two scalar representations and therefore the correlator can be written in terms of two Clebsch–Gordan coefficients. This coincides with our previous result (3.18). The correlator formed by three fermions and one left- and right-handed spin field each

$$\begin{aligned}\langle \psi^\mu \psi^\nu \psi^\lambda S_\alpha S_{\dot\beta} \rangle &\in (\tfrac{1}{2}, \tfrac{1}{2})^{\otimes 3} \otimes (\tfrac{1}{2}, 0) \otimes (0, \tfrac{1}{2}) \\ &= 1\,(\mathbf{2, 2}) \oplus 3\,(\mathbf{2, 1}) \oplus 3\,(\mathbf{1, 2}) \oplus 2\,(\mathbf{2, 0}) \oplus 2\,(\mathbf{0, 2}) \\ &\quad \oplus 9\,(\mathbf{1, 1}) \oplus 6\,(\mathbf{1, 0}) \oplus 6\,(\mathbf{0, 1}) \oplus 4\,\underline{(\mathbf{0, 0})}\end{aligned} \qquad (3.67)$$

can be written in terms of four index terms.

3.4.2 Replacing Fermions with Spin Fields

We turn now to the problem of finding a result for the correlation function

$$\langle \psi^{\mu_1}(z_1) \ldots \psi^{\mu_n}(z_n) \, S_{\alpha_1}(x_1) \ldots S_{\alpha_r}(x_r) \, S_{\dot\beta_1}(y_1) \ldots S_{\dot\beta_s}(y_s) \rangle \qquad (3.68)$$

with arbitrary many vector and spin fields. Luckily, this can be solved in full generality at tree-level in four space-time dimensions. Applying (3.21), which reads in four dimensions

$$\psi^\mu(z) = -\frac{1}{\sqrt{2}} \bar\sigma^{\mu\dot\kappa\kappa} S_\kappa(z)\, S_{\dot\kappa}(z)\,, \qquad (3.69)$$

3.4 Techniques in Four Dimensions

it is possible to replace all NS fermions in (3.68):

$$\langle \psi^{\mu_1}(z_1) \ldots \psi^{\mu_n}(z_n) S_{\alpha_1}(x_1) \ldots S_{\alpha_r}(x_r) S_{\dot{\beta}_1}(y_1) \ldots S_{\dot{\beta}_s}(y_s) \rangle = \prod_{i=1}^{n} \left(-\frac{\bar{\sigma}^{\mu_i \dot{\kappa}_i \kappa_i}}{\sqrt{2}} \right)$$
$$\times \langle S_{\kappa_1}(z_1) \ldots S_{\kappa_n}(z_n) S_{\alpha_1}(x_1) \ldots S_{\alpha_r}(x_r) S_{\dot{\kappa}_1}(z_1) \ldots S_{\dot{\kappa}_n}(z_n) S_{\dot{\beta}_1}(y_1) \ldots S_{\dot{\beta}_s}(y_s) \rangle . \quad (3.70)$$

We see that an arbitrary correlation function can be written as a pure spin field correlator contracted with some σ matrices. The next step is to systematically determine these correlators.

The correlator consisting of two left- and two-right handed spin fields has been calculated in [108] to determine the scattering of four gauginos. It is given by the expression

$$\langle S_\alpha(z_1) S_{\dot{\beta}}(z_2) S_\gamma(z_3) S_{\dot{\delta}}(z_4) \rangle = -\frac{\varepsilon_{\alpha\gamma} \varepsilon_{\dot{\beta}\dot{\delta}}}{(z_{13} z_{24})^{1/2}} . \quad (3.71)$$

One can identify this result as the product of the two-point functions

$$\langle S_\alpha(z_1) S_\gamma(z_3) \rangle = -\frac{\varepsilon_{\alpha\gamma}}{z_{13}^{1/2}} , \qquad \langle S_{\dot{\beta}}(z_2) S_{\dot{\delta}}(z_4) \rangle = \frac{\varepsilon_{\dot{\beta}\dot{\delta}}}{z_{24}^{1/2}} . \quad (3.72)$$

We prove now that this factorization property holds for an arbitrary number of spin fields. In order to do this it is most convenient to treat them in bosonized form. The left- and right-handed spin fields in four dimensions can be represented by two boson $H_{i=1,2}(z)$

$$S_{\alpha=1,2}(z) \sim e^{\pm\frac{i}{2}[H_1(z)+H_2(z)]} \equiv e^{ipH(z)} ,$$
$$S_{\dot{\beta}=1,2}(z) \sim e^{\pm\frac{i}{2}[H_1(z)-H_2(z)]} \equiv e^{iqH(z)} \quad (3.73)$$

with the notation $H(z) = (H_1(z), H_2(z))$ for the bosons and the weight vectors $p = (\pm 1/2, \pm 1/2)$, $q = (\pm 1/2, \mp 1/2)$. Note that the weight vectors of distinct chiralities are orthogonal, $p\,q = 0$. The two bosons fulfill the normalization convention:

$$\langle H_i(z) H_j(w) \rangle = \delta_{ij} \ln(z-w) . \quad (3.74)$$

Cocycle factors which yield complex phases upon moving spin fields across each other are irrelevant for the following discussion and are therefore neglected. The necessary OPEs and n-point functions are

$$e^{ipH(z)} e^{iqH(w)} \sim (z-w)^{pq} e^{i(p+q)H(w)} , \quad (3.75a)$$

$$\left\langle \prod_{k=1}^{n} e^{ip_k H(z_k)} \right\rangle \sim \delta\left(\sum_{k=1}^{n} p_k \right) \prod_{\substack{i,j=1 \\ i<j}}^{n} z_{ij}^{p_i p_j} . \quad (3.75b)$$

Hence the correlation function of r left-handed and s right-handed spin fields becomes:

$$\langle S_{\alpha_1}(z_1)\ldots S_{\alpha_r}(z_r) S_{\dot\beta_1}(w_1)\ldots S_{\dot\beta_s}(w_s)\rangle = \left\langle \prod_{k=1}^{r} e^{ip_k H(z_k)} \prod_{l=1}^{s} e^{iq_l H(w_l)} \right\rangle$$

$$= \delta\left(\sum_{k=1}^{r} p_k + \sum_{l=1}^{s} q_l\right) \prod_{\substack{i,j=1 \\ i<j}}^{r} z_{ij}^{p_i p_j} \prod_{\substack{i,j=1 \\ i<j}}^{s} w_{ij}^{q_i q_j} \underbrace{\prod_{m=1}^{r}\prod_{n=1}^{s}(z_m - w_n)^{p_m q_n}}_{=1}$$

$$= \delta\left(\sum_{k=1}^{r} p_k\right) \prod_{\substack{i,j=1 \\ i<j}}^{r} z_{ij}^{p_i p_j} \, \delta\left(\sum_{l=1}^{s} q_l\right) \prod_{\substack{i,j=1 \\ i<j}}^{s} w_{ij}^{q_i q_j}$$

$$= \left\langle \prod_{k=1}^{r} e^{ip_k H(z_k)} \right\rangle \left\langle \prod_{l=1}^{s} e^{iq_l H(w_l)} \right\rangle$$

$$= \langle S_{\alpha_1}(z_1)\ldots S_{\alpha_r}(z_r)\rangle \langle S_{\dot\beta_1}(w_1)\ldots S_{\dot\beta_s}(w_s)\rangle. \quad (3.76)$$

From the second to the third line we have used that $p_m\, q_n = 0$ and the δ-function has been split into the linearly independent p and q contributions. So we see that a general spin field correlation function in four dimensions splits indeed into two correlators involving only left- and right-handed spin fields.

We would like to stress that this factorization property does not hold for tree-level correlators in other dimensions. The reason for this lies in the fact that the weight vectors p, q of left- and right-handed spin fields then do not satisfy anymore $p\,q = 0$ and thus the whole argument (3.76) breaks down.

Using the factorization property (3.76) our previous result (3.70) becomes:

$$\langle \psi^{\mu_1}(z_1)\ldots\psi^{\mu_n}(z_n) S_{\alpha_1}(x_1)\ldots S_{\alpha_r}(x_r) S_{\dot\beta_{n+1}}(y_{n+1})\ldots S_{\dot\beta_s}(y_s)\rangle = \prod_{i=1}^{n}\left(-\frac{\bar\sigma^{\mu_i\,\dot\kappa_i\kappa_i}}{\sqrt{2}}\right)$$

$$\times \langle S_{\kappa_1}(z_1)\ldots S_{\kappa_n}(z_n) S_{\alpha_1}(x_1)\ldots S_{\alpha_r}(x_r)\rangle \langle S_{\dot\kappa_1}(z_1)\ldots S_{\dot\kappa_n}(z_n) S_{\dot\beta_1}(y_1)\ldots S_{\dot\beta_s}(y_s)\rangle. \quad (3.77)$$

This formula shows how correlators involving NS fermions factorize into a product of correlators involving only left- or right-handed spin fields. Hence, if the latter correlators are known for an arbitrary number of spin fields it is possible to calculate in principle any correlator. We address the calculation of these correlators in the following.

3.4.3 Pure Spin Field Correlators

It has been shown in Chapter 3.2.2 that pure spin field correlator in four dimensions can be stated in terms of Clebsch–Gordan coefficients that are products of ε tensors. For $2M$ spin fields there are $(2M-1)!!$ possible index configurations of this type, whereas only $q(0, 2M) =$

3.4 Techniques in Four Dimensions 57

$(2M)!/M!(M+1)!$ are independent due to (3.63). The necessary relations arise from generalization of the Fierz identity (3.40),

$$\varepsilon_{[\alpha_1\underline{\alpha_2}}\varepsilon_{\alpha_3\underline{\alpha_4}}\ldots\varepsilon_{\alpha_{2M-1}\underline{\alpha_{2M}}]} = 0, \tag{3.78}$$

where we antisymmetrize over the underlined indices. Yet we will show that the results assume a nicer form if we use a special non-minimal basis of $M!$ index terms.

The RNS correlator consisting of four left-handed spin fields has previously been calculated:

$$\langle S_\alpha(z_1) S_\beta(z_2) S_\gamma(z_3) S_\delta(z_4) \rangle = \left(\frac{z_{12} z_{14} z_{23} z_{34}}{z_{13} z_{24}} \right)^{1/2} \left(\frac{\varepsilon_{\alpha\beta}\varepsilon_{\gamma\delta}}{z_{12}z_{34}} - \frac{\varepsilon_{\alpha\delta}\varepsilon_{\gamma\beta}}{z_{14}z_{32}} \right). \tag{3.79}$$

From the $3!! = 3$ possible index terms we have eliminated the term $\varepsilon_{\alpha\gamma}\varepsilon_{\beta\delta}$ using the Fierz identity (3.39). The remaining two terms are independent which coincides with (3.63) for $M = 2$. For the six point correlator $M = 3$ there exist $5!! = 15$ possible index terms, however, only five are independent. Taking into account all possible OPEs one finds:

$$\langle S_\alpha(z_1) S_\beta(z_2) S_\gamma(z_3) S_\delta(z_4) S_\epsilon(z_5) S_\zeta(z_6) \rangle = -\prod_{i<j}^{6} z_{ij}^{-1/2} \Big[\varepsilon_{\alpha\beta}\varepsilon_{\gamma\delta}\varepsilon_{\epsilon\zeta}\, z_{14}\,z_{15}\,z_{23}\,z_{26}\,z_{36}\,z_{45}$$

$$+ \varepsilon_{\alpha\beta}\varepsilon_{\gamma\zeta}\varepsilon_{\epsilon\delta}\, z_{14}\,z_{23}\,z_{56}\,(z_{15}\,z_{26}\,z_{34} - z_{12}\,z_{35}\,z_{46}) + \varepsilon_{\alpha\delta}\varepsilon_{\gamma\zeta}\varepsilon_{\epsilon\beta}\, z_{12}\,z_{13}\,z_{23}\,z_{45}\,z_{46}\,z_{56}$$

$$+ \varepsilon_{\alpha\delta}\varepsilon_{\gamma\beta}\varepsilon_{\epsilon\zeta}\, z_{12}\,z_{36}\,z_{45}\,(z_{15}\,z_{26}\,z_{34} - z_{13}\,z_{24}\,z_{56}) + \varepsilon_{\alpha\zeta}\varepsilon_{\gamma\beta}\varepsilon_{\epsilon\delta}\, z_{12}\,z_{14}\,z_{24}\,z_{35}\,z_{36}\,z_{56} \Big]. \tag{3.80}$$

However, the result assumes a more symmetric form and has a less complicated z dependence if we introduce a sixth index term $\varepsilon_{\alpha\zeta}\varepsilon_{\gamma\delta}\varepsilon_{\epsilon\beta}$:

$$\langle S_\alpha(z_1) S_\beta(z_2) S_\gamma(z_3) S_\delta(z_4) S_\epsilon(z_5) S_\zeta(z_6) \rangle = -\left(\frac{z_{12} z_{14} z_{16} z_{23} z_{25} z_{34} z_{36} z_{45} z_{56}}{z_{13} z_{15} z_{24} z_{26} z_{35} z_{46}} \right)^{1/2}$$

$$\times \left(\frac{\varepsilon_{\alpha\beta}\varepsilon_{\gamma\delta}\varepsilon_{\epsilon\zeta}}{z_{12}z_{34}z_{56}} - \frac{\varepsilon_{\alpha\beta}\varepsilon_{\gamma\zeta}\varepsilon_{\epsilon\delta}}{z_{12}z_{36}z_{54}} + \frac{\varepsilon_{\alpha\delta}\varepsilon_{\gamma\zeta}\varepsilon_{\epsilon\beta}}{z_{14}z_{36}z_{52}} - \frac{\varepsilon_{\alpha\delta}\varepsilon_{\gamma\beta}\varepsilon_{\epsilon\zeta}}{z_{14}z_{32}z_{56}} + \frac{\varepsilon_{\alpha\zeta}\varepsilon_{\gamma\beta}\varepsilon_{\epsilon\delta}}{z_{16}z_{32}z_{54}} - \frac{\varepsilon_{\alpha\zeta}\varepsilon_{\gamma\delta}\varepsilon_{\epsilon\beta}}{z_{16}z_{34}z_{52}} \right). \tag{3.81}$$

Comparing (3.79) and (3.81) a number of similarities become visible. In both cases the prefactor consists of all possible terms of the schematic form $(z_{\text{odd even}} z_{\text{even odd}})^{1/2}$ in the numerator and $(z_{\text{odd odd}} z_{\text{even even}})^{1/2}$ in the denominator. Furthermore, the first index at every ε tensor belongs to a spin field of position z_{odd} whereas the second index stems from a spin field located at z_{even}. Finally every ε tensor comes with the corresponding factor $(z_{\text{odd}} - z_{\text{even}})^{-1}$. The overall sign can be traced back to $(-1)^M$ coming from the OPE (3.12a), whereas the relative signs between the index terms can be understood as the sign of the respective permutation of the spinor indices.

The results (3.79) and (3.81) suggest the following expression for the $2M$ point function of left-handed spin fields:

$$\langle S_{\alpha_1}(z_1) S_{\alpha_2}(z_2) \ldots S_{\alpha_{2M-1}}(z_{2M-1}) S_{\alpha_{2M}}(z_{2M}) \rangle$$

$$= (-1)^M \left(\frac{\prod_{i \leq j}^M z_{2i-1,2j} \prod_{\bar{i} < \bar{j}}^M z_{2\bar{i},2\bar{j}-1}}{\prod_{k<l}^M z_{2k-1,2l-1} z_{2k,2l}} \right)^{1/2} \sum_{\rho \in S_M} \text{sgn}(\rho) \prod_{m=1}^M \frac{\varepsilon_{\alpha_{2m-1}\alpha_{\rho(2m)}}}{z_{2m-1,\rho(2m)}}. \qquad (3.82)$$

We prove this expression by induction. For the base case $M = 1$ this gives correctly the two-point function of left-handed spin fields. The inductive step makes use of the fact that the $2M - 2$ correlator should appear from the $2M$ correlator if we replace two spin fields by the OPE in the corresponding limit $z_i \to z_j$. As every spin field can be permuted to the very right in the correlator we study without loss of generality the case $z_{2M-1} \to z_{2M}$:

$$\langle S_{\alpha_1}(z_1) \ldots S_{\alpha_{2M-2}}(z_{2M-2}) S_{\alpha_{2M-1}}(z_{2M-1}) S_{\alpha_{2M}}(z_{2M}) \rangle \Big|_{z_{2M-1} \to z_{2M}}$$

$$= -\frac{\varepsilon_{\alpha_{2M-1}\alpha_{2M}}}{z_{2M-1,2M}^{1/2}} \langle S_{\alpha_1}(z_1) \ldots S_{\alpha_{2M-2}}(z_{2M-2}) \rangle + \mathcal{O}(z_{2M-1,2M})$$

$$= -\frac{\varepsilon_{\alpha_{2M-1}\alpha_{2M}}}{z_{2M-1,2M}} z_{2M-1,2M}^{1/2} (-1)^{M-1} \left(\frac{\prod_{i \leq j}^{M-1} z_{2i-1,2j} \prod_{\bar{i}<\bar{j}}^{M-1} z_{2\bar{i},2\bar{j}-1}}{\prod_{k<l}^{M-1} z_{2k-1,2l-1} z_{2k,2l}} \right)^{1/2}$$

$$\times \underbrace{\left(\frac{\prod_{p=1}^{M-1} z_{2p-1,2M} z_{2p,2M-1}}{\prod_{q=1}^{M-1} z_{2q-1,2M-1} z_{2q,2M}} \right)^{1/2}}_{=1+\mathcal{O}(z_{2M-1,2M})} \sum_{\rho \in S_{M-1}} \text{sgn}(\rho) \prod_{m=1}^{M-1} \frac{\varepsilon_{\alpha_{2m-1}\alpha_{\rho(2m)}}}{z_{2m-1,\rho(2m)}} + \mathcal{O}(z_{2M-1,2M})$$

$$= (-1)^M \left(\frac{\prod_{i \leq j}^M z_{2i-1,2j} \prod_{\bar{i}<\bar{j}}^M z_{2\bar{i},2\bar{j}-1}}{\prod_{k<l}^M z_{2k-1,2l-1} z_{2k,2l}} \right)^{1/2} \sum_{\rho \in S_M} \text{sgn}(\rho) \prod_{m=1}^M \delta_{\rho(2M),2M} \frac{\varepsilon_{\alpha_{2m-1}\alpha_{\rho(2m)}}}{z_{2m-1,\rho(2m)}}$$

$$+ \mathcal{O}(z_{2M-1,2M}). \qquad (3.83)$$

The most singular piece of (3.82) in $z_{2M-1,2M}$ is the subset of permutations $\rho \in S_m$ with $\rho(2M) = 2M$. This is precisely what we obtain by applying the OPE for $S_{\alpha_{2M-1}}(z_{2M-1})$ and $S_{\alpha_{2M}}(z_{2M})$ and then assuming the claimed expression for $\langle S_{\alpha_1}(z_1) \ldots S_{\alpha_{2M-2}}(z_{2M-2}) \rangle$. This completes the proof of (3.82).

The correlator of $2M$ right-handed spin fields can easily be read off from (3.82). The factor $(-1)^M$ drops out due to the different sign in the OPE (3.12b) and all ε tensors carry dotted indices instead:

$$\langle S_{\dot{\alpha}_1}(z_1) S_{\dot{\alpha}_2}(z_2) \ldots S_{\dot{\alpha}_{2M-1}}(z_{2M-1}) S_{\dot{\alpha}_{2M}}(z_{2M}) \rangle$$

$$= \left(\frac{\prod_{i\leq j}^M z_{2i-1,2j} \prod_{\bar{i}<\bar{j}}^M z_{2\bar{i},2\bar{j}-1}}{\prod_{k<l}^M z_{2k-1,2l-1} z_{2k,2l}} \right)^{1/2} \sum_{\rho \in S_M} \text{sgn}(\rho) \prod_{m=1}^M \frac{\varepsilon_{\dot{\alpha}_{2m-1}\dot{\alpha}_{\rho(2m)}}}{z_{2m-1,\rho(2m)}} . \qquad (3.84)$$

By plugging (3.82) and (3.84) into (3.77) it is now possible to calculate any RNS correlation function involving arbitrary many fermions ψ^μ and spin fields S_α, $S_{\dot{\alpha}}$ in four dimensions.

3.5 Techniques in Six Dimensions

Next we focus our attention on RNS correlators in $D = 6$ space-time dimensions. They enter the calculation of string amplitudes, when ten-dimensional space-time is compactified to six dimensions. Furthermore they describe the interaction of the internal fields appearing in the vertex operators of scalars and gauginos (2.57) and (2.59) in the phenomenological interesting case of compatifications to four space-time dimensions. For this purpose however we have to switch from Minkowsian to Euclidean signature. Let us start the discussion with the special class of correlation functions consisting of r left- and right-handed spin fields:

$$\left\langle S_{\alpha_1}(z_1) S^{\dot{\beta}_1}(z_2) \ldots S_{\alpha_r}(z_{2r-1}) S^{\dot{\beta}_r}(z_{2r}) \right\rangle . \qquad (3.85)$$

In Chapter 3.2 it has been discussed how to relate these expressions to correlators involving fermions. Indeed, using equation (3.11b) for $D = 6$

$$\psi^\mu(w) = -\frac{1}{2\sqrt{2}} \lim_{z \to w} (z-w)^{1/4} (C^{-1}\bar{\gamma}^\mu)^{\beta\alpha} S_\alpha(z) S_\beta(w) \qquad (3.86)$$

and its complex conjugate, every fermion ψ^μ can be replaced by two spin fields of the same chirality. Therefore the correlator (3.85) can be used to derive expressions for the large class of correlation functions

$$\left\langle \psi^{\mu_1}(z_1) \ldots \psi^{\mu_{k+l}}(z_{k+l}) S_{\alpha_1}(x_1) \ldots S_{\alpha_{r-2k}}(x_{r-2k}) S^{\dot{\beta}_1}(y_1) \ldots S^{\dot{\beta}_{r-2l}}(y_{r-2l}) \right\rangle . \qquad (3.87)$$

Before giving a general formula for (3.85) we show the results for lower-point expressions. In the case $r = 2$, i.e. two left- and two right-handed spin fields, one finds

$$\left\langle S_\alpha(z_1) S^{\dot{\beta}}(z_2) S_\gamma(z_3) S^{\dot{\delta}}(z_4) \right\rangle = \left(\frac{z_{12} z_{14} z_{23} z_{34}}{z_{13} z_{24}} \right)^{1/4} \left(\frac{C_\alpha{}^{\dot{\beta}} C_\gamma{}^{\dot{\delta}}}{z_{12} z_{34}} - \frac{C_\alpha{}^{\dot{\delta}} C_\gamma{}^{\dot{\beta}}}{z_{14} z_{32}} \right) . \qquad (3.88)$$

The correlator consisting of three spin fields of each chirality is given by

$$\left\langle S_\alpha(z_1) S^{\dot{\beta}}(z_2) S_\gamma(z_3) S^{\dot{\delta}}(z_4) S_\epsilon(z_5) S^{\dot{\zeta}}(z_6) \right\rangle = \left(\frac{z_{12} z_{14} z_{16} z_{23} z_{25} z_{34} z_{36} z_{45} z_{56}}{z_{13} z_{15} z_{24} z_{26} z_{35} z_{46}} \right)^{1/4}$$

$$\times \left(\frac{C_\alpha{}^{\dot\beta} C_\gamma{}^{\dot\delta} C_\epsilon{}^{\dot\zeta}}{z_{12}\, z_{34}\, z_{56}} - \frac{C_\alpha{}^{\dot\beta} C_\gamma{}^{\dot\zeta} C_\epsilon{}^{\dot\delta}}{z_{12}\, z_{36}\, z_{54}} + \frac{C_\alpha{}^{\dot\delta} C_\gamma{}^{\dot\zeta} C_\epsilon{}^{\dot\beta}}{z_{14}\, z_{36}\, z_{52}} \right.$$
$$\left. - \frac{C_\alpha{}^{\dot\delta} C_\gamma{}^{\dot\beta} C_\epsilon{}^{\dot\zeta}}{z_{14}\, z_{32}\, z_{56}} + \frac{C_\alpha{}^{\dot\zeta} C_\gamma{}^{\dot\beta} C_\epsilon{}^{\dot\delta}}{z_{16}\, z_{32}\, z_{54}} - \frac{C_\alpha{}^{\dot\zeta} C_\gamma{}^{\dot\delta} C_\epsilon{}^{\dot\beta}}{z_{16}\, z_{34}\, z_{52}} \right). \tag{3.89}$$

These last two results have a similar structure as the four and six spin field correlators (3.79) and (3.81) in four dimensions. They only differ in the chirality of the charge conjugation matrix and the power of the overall coefficient due to the different conformal weights of the spin fields. Hence, also the general formula for an arbitrary number of left- and right-handed spin fields is very similar to the four-dimensional expression (3.82). It is given by

$$\left\langle \prod_{i=1}^{M} S_{\alpha_i}(z_{2i-1})\, S^{\dot\beta_i}(z_{2i}) \right\rangle = \left(\prod_{i \leq j}^{M} z_{2i-1,2j} \prod_{\bar i < \bar j}^{M} z_{2\bar i, 2\bar j - 1} \right)^{1/4} \left(\prod_{i<j}^{M} z_{2i-1,2j-1}\, z_{2i,2j} \right)^{-1/4}$$
$$\times \sum_{\rho \in S_M} \operatorname{sgn}(\rho) \prod_{m=1}^{M} \frac{C_{\alpha_{2m-1}}{}^{\dot\beta_{\rho(2m)}}}{z_{2m-1,\rho(2m)}}, \tag{3.90}$$

where the proof proceeds in the same way as in Chapter 3.4.3. Note that the factor $(-1)^M$ vanishes as the corresponding OPE in six dimensions (3.11a) comes in our conventions with a plus sign.

In contrast to four dimensions, knowing (3.90) is not sufficient to solve all RNS correlation function with arbitrary many fields. Therefore only the class of correlators (3.87) can be evaluated in six dimensions using (3.90). Every other correlation function up to six points is collected in [2]. As these have been calculated using methods for loop correlators we present them in Chapter 5.3.2.

3.6 Techniques in Eight Dimensions

RNS correlators in eight dimensions are of less phenomenological relevance than their relatives in four, six and ten dimensions. However, they come with a nice mathematical peculiarity which we present in the following.

3.6.1 $SO(8)$ Triality

The S_3 permutation symmetry of the "Mercedes star-shaped" $SO(8)$ Dynkin diagram in Figure 3.1 – also referred to as triality – plays an important role for the RNS CFT. In eight dimensions fermions and spin fields have equal conformal dimension $h = D/16 = 1/2$. Therefore, the OPEs (3.9) and (3.11) become particularly symmetric and we will make use of $SO(8)$ triality to rewrite them in unified fashion.

3.6 Techniques in Eight Dimensions

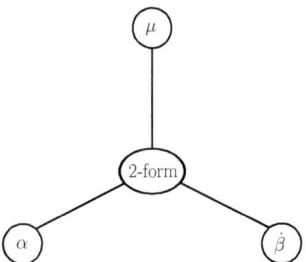

Figure 3.1: Dynkin diagram of the group $SO(8)$.

The short distance behavior of conformal fields is sufficient input to determine their correlations on the sphere. This is why triality covariance of OPEs is inherited by tree-level correlators. However, at higher genus, the different global properties of the ψ^μ and $S_\alpha, S^{\dot\beta}$ fields under transport around the world-sheet's homology cycles will break this covariance. Hence, triality does not hold for correlators at loop-level.

Before we apply $SO(8)$ triality to RNS correlators we have to introduce some notation. Firstly, it is convenient to work with generalized fields P^i, Q^j, R^k of conformal dimension $h = 1/2$ that can be either ψ^μ, S_α or $S^{\dot\beta}$:

$$(P^i, Q^j, R^k) = \bigl(\rho(\psi^\mu), \rho(S_\alpha), \rho(S^{\dot\beta})\bigr) \quad \text{for } \rho \in S_3\,. \tag{3.91}$$

On the level of Clebsch–Gordan coefficients, we introduce a universal metric g

$$g^{ij} \equiv \begin{cases} \eta^{\mu\nu} & : (i,j) = (\mu, \nu)\,, \\ C_{\alpha\beta} & : (i,j) = (\alpha, \beta)\,, \\ C^{\dot\alpha\dot\beta} & : (i,j) = (\dot\alpha, \dot\beta)\,, \\ 0 & : \text{otherwise}\,, \end{cases} \qquad g_{ij} \equiv \begin{cases} \eta_{\mu\nu} & : (i,j) = (\mu, \nu)\,, \\ C^{\alpha\beta} & : (i,j) = (\alpha, \beta)\,, \\ C_{\dot\alpha\dot\beta} & : (i,j) = (\dot\alpha, \dot\beta)\,, \\ 0 & : \text{otherwise}\,, \end{cases} \tag{3.92}$$

and the three-point couplings G

$$G^{ijk} \equiv \begin{cases} \frac{1}{\sqrt{2}} (\gamma^\mu C)_\alpha{}^{\dot\beta} & : (i,j,k) = \bigl(\rho(\mu), \rho(\alpha), \rho(\dot\beta)\bigr) \text{ with } \rho \in S_3\,, \\ 0 & : \text{otherwise}\,, \end{cases} \tag{3.93}$$

where the general indices fulfill $i, j, k \in \{\mu, \alpha, \dot\beta\}$. The above definitions allow us to rewrite the $D = 8$ OPEs (3.9) and (3.10) in unified fashion:

$$P^i(z)\, P^j(w) \sim \frac{g^{ij}}{z - w}\,, \tag{3.94a}$$

$$P^i(z) \, Q^j(w) \sim \frac{G^{ijk}}{(z-w)^{1/2}} \, g_{kl} \, R^l(w) \, . \tag{3.94b}$$

As before the OPEs are the only input that is necessary to calculate the tree-level correlation function $\langle P^{i_1}(x_1) \ldots P^{i_n}(x_n) \, Q^{j_1}(y_1) \ldots Q^{j_r}(y_r) \, R^{k_1}(z_1) \ldots R^{k_s}(z_s) \rangle$. The arising index terms are then built out of the metric (3.92) and the three-point coupling (3.93). Upon specifying the generalized fields as in (3.91) we can derive from the generalized result the six RNS correlators, written schematically as

$$\langle \psi^n \, S^r \, \dot{S}^s \rangle \, , \quad \langle \psi^n \, S^s \, \dot{S}^r \rangle \, , \quad \langle \psi^r \, S^n \, \dot{S}^s \rangle \, , \quad \langle \psi^r \, S^s \, \dot{S}^n \rangle \, , \quad \langle \psi^s \, S^n \, \dot{S}^r \rangle \, , \quad \langle \psi^s \, S^r \, \dot{S}^n \rangle \, . \tag{3.95}$$

Triality can also be used to derive new correlators from known ones. Suppose the correlation function

$$\langle \psi^{\mu_1} \ldots \psi^{\mu_n} \, S_{\alpha_1} \ldots S_{\alpha_r} \, S^{\dot{\beta}_1} \ldots S^{\dot{\beta}_s} \rangle \tag{3.96}$$

is known. The result can then be written in triality covariant form making use of the generalized fields P^i, Q^j, R^k and re-writing the index terms using (3.92) and (3.93). We can then play the same game as above and re-assign $(P^i, Q^j, R^k) = (\rho(\psi^\mu), \rho(S_\alpha), \rho(S^{\dot{\beta}}))$ for some $\rho \in S_3$ and thus obtain the missing correlation functions in (3.95). This method of course fails for triality invariant correlators, i.e. $n = r = s$. Such correlators have to be calculated by first principles.

Triality implies that correlators involving only one type of field, like $\langle S^r \rangle$, are related to $\langle \psi^r \rangle$. Therefore each field $\psi^\mu, S_\alpha, S^{\dot{\beta}}$ by itself behaves like a free world-sheet fermion. The respective correlators in eight dimensions can be calculated using Wick's theorem and are simply given by products of two-point functions.

3.6.2 Examples

Let us apply the triality-based methods to the five-point functions $\langle P \, P \, P \, Q \, R \rangle$. In Chapter 5.3 we show that in eight dimensions

$$\begin{aligned}\langle \psi^\mu(z_1) \, \psi^\nu(z_2) \, \psi^\lambda(z_3) \, S_\alpha(z_4) \, S^{\dot{\beta}}(z_5) \rangle &= \frac{1}{\sqrt{2} \, (z_{14} \, z_{15} \, z_{24} \, z_{25} \, z_{34} \, z_{35} \, z_{45})^{1/2}} \\ &\times \left[\frac{\eta^{\mu\nu}}{z_{12}} (\gamma^\lambda C)_\alpha{}^{\dot{\beta}} \, z_{14} \, z_{25} - \frac{\eta^{\mu\lambda}}{z_{13}} (\gamma^\nu C)_\alpha{}^{\dot{\beta}} \, z_{14} \, z_{35} \right. \\ &\left. + \frac{\eta^{\nu\lambda}}{z_{23}} (\gamma^\mu C)_\alpha{}^{\dot{\beta}} \, z_{24} \, z_{35} + \frac{z_{45}}{2} (\gamma^\mu \, \bar{\gamma}^\nu \, \gamma^\lambda C)_\alpha{}^{\dot{\beta}} \right] \, . \end{aligned} \tag{3.97}$$

3.6 Techniques in Eight Dimensions

By inserting the generalized fields $P = \psi$, $Q = S$ and $R = \dot{S}$ this can be translated into triality covariant notation,

$$\begin{aligned}\left\langle P^{i_1}(z_1)\, P^{i_2}(z_2)\, P^{i_3}(z_3)\, Q^{j_4}(z_4)\, R^{k_5}(z_5)\right\rangle &= \frac{1}{(z_{14}\, z_{15}\, z_{24}\, z_{25}\, z_{34}\, z_{35}\, z_{45})^{1/2}} \\ &\times \left[\frac{g^{i_1 i_2}}{z_{12}}\, G^{i_3 j_4 k_5}\, z_{14}\, z_{25} - \frac{g^{i_1 i_3}}{z_{13}}\, G^{i_2 j_4 k_5}\, z_{14}\, z_{35}\right.\\ &\left. +\frac{g^{i_2 i_3}}{z_{23}}\, G^{i_1 j_4 k_5}\, z_{24}\, z_{35} - G^{i_1 j_4 k}\, g_{kk'} G^{i_2 j k'}\, g_{jj'}\, G^{i_3 j' k_5}\, z_{45}\right]\,, \end{aligned}$$
(3.98)

where we have replaced the index terms in (3.97) by (3.92) and (3.93). Note that

$$(\gamma^\mu\, \bar{\gamma}^\nu\, \gamma^\lambda\, C)_\alpha{}^{\dot{\beta}} = -2\sqrt{2}\, G^{i_1 j_4 k}\, g_{kk'}\, G^{i_2 j k'}\, g_{jj'}\, G^{i_3 j' k_5}\,.$$
(3.99)

The minus sign arises from $(\bar{\gamma}^\nu\, C)^{\dot{\gamma}}{}_\gamma = -(\gamma^\nu\, C)_\gamma{}^{\dot{\gamma}} \equiv -G^{i_2 j k'}$. Choosing now another configuration, e.g. $P = S$, $Q = \dot{S}$ and $R = \psi$ yields the new correlator

$$\begin{aligned}\left\langle \psi^\mu(z_1)\, S_\alpha(z_2)\, S_\beta(z_3)\, S_\gamma(z_4)\, S^{\dot{\delta}}(z_5)\right\rangle &= \frac{-1}{(z_{12}\, z_{13}\, z_{14}\, z_{15}\, z_{25}\, z_{35}\, z_{45})^{1/2}} \\ &\times \left[\frac{C_{\alpha\beta}}{z_{23}}\, (\gamma^\mu\, C)_\gamma{}^{\dot{\delta}}\, z_{13}\, z_{25} - \frac{C_{\alpha\gamma}}{z_{24}}\, (\gamma^\mu\, C)_\beta{}^{\dot{\delta}}\, z_{14}\, z_{25}\right.\\ &\left. +\frac{C_{\beta\gamma}}{z_{34}}\, (\gamma^\mu\, C)_\alpha{}^{\dot{\delta}}\, z_{14}\, z_{35} + \frac{z_{15}}{2}\, (\gamma^\mu\, \bar{\gamma}^\nu\, C)_{\gamma\beta}\, (\gamma_\nu\, C)_\alpha{}^{\dot{\delta}}\right].\end{aligned}$$
(3.100)

In the same way, by specifying $P = \dot{S}$, $Q = S$ and $R = \psi$, we can derive the correlation function

$$\begin{aligned}\left\langle \psi^\mu(z_1)\, S_\alpha(z_2)\, S^{\dot{\beta}}(z_3)\, S^{\dot{\gamma}}(z_4)\, S^{\dot{\delta}}(z_5)\right\rangle &= \frac{-1}{(z_{12}\, z_{13}\, z_{14}\, z_{15}\, z_{23}\, z_{24}\, z_{25})^{1/2}} \\ &\times \left[\frac{C^{\dot{\beta}\dot{\gamma}}}{z_{34}}\, (\gamma^\mu\, C)_\alpha{}^{\dot{\delta}}\, z_{14}\, z_{23} - \frac{C^{\dot{\beta}\dot{\delta}}}{z_{35}}\, (\gamma^\mu\, C)_\alpha{}^{\dot{\gamma}}\, z_{15}\, z_{23}\right.\\ &\left. +\frac{C^{\dot{\gamma}\dot{\delta}}}{z_{45}}\, (\gamma^\mu\, C)_\alpha{}^{\dot{\beta}}\, z_{15}\, z_{24} + \frac{z_{12}}{2}\, (\bar{\gamma}^\mu\, \gamma^\nu\, C)^{\dot{\delta}\dot{\gamma}}\, (\gamma_\nu\, C)_\alpha{}^{\dot{\beta}}\right].\end{aligned}$$
(3.101)

This exhausts all correlators that can be derived from (3.97) using triality.

A further application of the triality techniques lies in the derivation of relations between index terms. Using (3.92) and (3.93) one finds that the Clifford algebra (3.4) and the Fierz identity (B.39a)

$$C_{\alpha\beta}\, C^{\dot{\gamma}\dot{\delta}} = \frac{1}{2}(\gamma^\mu\, C)_\alpha{}^{\dot{\gamma}} (\gamma_\mu\, C)_\beta{}^{\dot{\delta}} + \frac{1}{2}(\gamma^\mu\, C)_\beta{}^{\dot{\gamma}} (\gamma_\mu\, C)_\alpha{}^{\dot{\delta}}$$
(3.102)

are special cases of the triality covariant tensor equation

$$g^{i_1 i_2}\, g^{j_1 j_2} = G^{i_1 j_1 k_1}\, G^{i_2 j_2 k_2}\, g_{k_1 k_2} + G^{i_1 j_2 k_1}\, G^{i_2 j_1 k_2}\, g_{k_1 k_2}\,.$$
(3.103)

3. Ramond–Neveu–Schwarz Correlators at Tree-Level

Further identities are found in a similar way. Generalizing

$$2\eta^{\mu\nu}(\gamma^\lambda C)_\alpha{}^{\dot\beta} = -(\gamma^\mu \bar\gamma^\nu \gamma^\lambda C)_\alpha{}^{\dot\beta} - (\gamma^\nu \bar\gamma^\mu \gamma^\lambda C)_\alpha{}^{\dot\beta}, \tag{3.104}$$

which easily follows from the Clifford algebra, to

$$G^{i_4 j_3 k}\, g_{kk'}\, G^{i j_1 k'}\, g_{ii'}\, G^{i' j_2 k_5} + G^{i_4 j_3 k}\, g_{kk'}\, G^{i j_2 k'}\, g_{ii'}\, G^{i' j_1 k_5} = g^{j_1 j_2}\, G^{i_4 j_3 k_5}, \tag{3.105}$$

and then making the choice $(j_1, j_2, j_3) = (\alpha, \beta, \gamma)$ and $i_4 = \mu$, $k_5 = \dot\delta$ yields

$$C_{\alpha\beta}(\gamma^\mu C)_\gamma{}^{\dot\delta} = -\frac{1}{2}(\gamma^\lambda \bar\gamma^\mu C)_{\beta\gamma}(\gamma_\lambda C)_\alpha{}^{\dot\delta} - \frac{1}{2}(\gamma^\lambda \bar\gamma^\mu C)_{\alpha\gamma}(\gamma_\lambda C)_\beta{}^{\dot\delta}. \tag{3.106}$$

As a last application of triality let us discuss the correlation functions consisting of arbitrary many fermions and only two spin fields in eight dimensions. In Chapter 5.3 it is shown that results at loop-level for this large class of correlators can be derived. For tree-level scattering the findings take the form:

$$\left\langle \psi^{\mu_1}(z_1) \ldots \psi^{\mu_{2n-1}}(z_{2n-1})\, S_\alpha(z_A)\, S^{\dot\beta}(z_B) \right\rangle = \frac{1}{\sqrt{2}\, z_{AB}^{1/2} \prod_{i=1}^{2n-1}(z_{iA}\, z_{iB})^{1/2}} \sum_{l=0}^{n-1} \left(\frac{z_{AB}}{2}\right)^l$$

$$\times \sum_{\rho \in S_{2n-1}/\mathcal{P}_{n,l}} \operatorname{sgn}(\rho)\, \left(\gamma^{\mu_{\rho(1)}} \bar\gamma^{\mu_{\rho(2)}} \ldots \bar\gamma^{\mu_{\rho(2l)}} \gamma^{\mu_{\rho(2l+1)}} C\right)_\alpha{}^{\dot\beta}$$

$$\times \prod_{j=1}^{n-l-1} \frac{\eta^{\mu_{\rho(2l+2j)}\mu_{\rho(2l+2j+1)}}}{z_{\rho(2l+2j),\rho(2l+2j+1)}}\, z_{\rho(2l+2j),A}\, z_{\rho(2l+2j+1),B}, \tag{3.107}$$

$$\left\langle \psi^{\mu_1}(z_1) \ldots \psi^{\mu_{2n-2}}(z_{2n-2})\, S_\alpha(z_A)\, S_\beta(z_B) \right\rangle = \frac{1}{z_{AB} \prod_{i=1}^{2n-2}(z_{iA}\, z_{iB})^{1/2}} \sum_{l=0}^{n-1} \left(\frac{z_{AB}}{2}\right)^l$$

$$\times \sum_{\rho \in S_{2n-2}/\mathcal{Q}_{n,l}} \operatorname{sgn}(\rho)\, \left(\gamma^{\mu_{\rho(1)}} \bar\gamma^{\mu_{\rho(2)}} \ldots \gamma^{\mu_{\rho(2l-1)}} \bar\gamma^{\mu_{\rho(2l)}} C\right)_{\alpha\beta}$$

$$\times \prod_{j=1}^{n-l-1} \frac{\eta^{\mu_{\rho(2l+2j-1)}\mu_{\rho(2l+2j)}}}{z_{\rho(2l+2j-1),\rho(2l+2j)}}\, z_{\rho(2l+2j-1),A}\, z_{\rho(2l+2j),B}. \tag{3.108}$$

Via triality these correlators can be related to the unknown correlation functions

$$\left\langle \psi^{2n-1} S \dot S \right\rangle \leftrightarrow \left\langle \psi\, S^{2n-1}\, \dot S \right\rangle, \quad \left\langle \psi^{2n-2} S^2 \right\rangle \leftrightarrow \left\langle \psi^2\, S^{2n-2} \right\rangle \leftrightarrow \left\langle S^{2n-2}\, \dot S^2 \right\rangle. \tag{3.109}$$

For this purpose we need the triality covariant versions of (3.107) and (3.108). In terms of the

3.6 Techniques in Eight Dimensions

generalized fields one finds

$$\left\langle P^{i_1}(z_1)\ldots P^{i_{2n-1}}(z_{2n-1})\, Q^j(z_A)\, R^k(z_B)\right\rangle = \frac{1}{z_{AB}^{1/2}\prod_{i=1}^{2n-1}(z_{iA}\,z_{iB})^{1/2}} \sum_{l=0}^{n-1}(-z_{AB})^l$$

$$\times \sum_{\rho\in S_{2n-1}/\mathcal{P}_{n,l}} \text{sgn}(\rho)\, G^{i_{\rho(1)}jr_1}\, G^{i_{\rho(2)}q_1}{}_{r_1}\, G^{i_{\rho(3)}}{}_{q_1}{}^{r_2}\, G^{i_{\rho(4)}q_2}{}_{r_2}\ldots G^{i_{\rho(2l+1)}}{}_{q_l}{}^{k}$$

$$\times \prod_{j=1}^{n-l-1} \frac{g^{i_{\rho(2l+2j)}i_{\rho(2l+2j+1)}}}{z_{\rho(2l+2j),\rho(2l+2j+1)}}\, z_{\rho(2l+2j),A}\, z_{\rho(2l+2j+1),B}\,, \qquad (3.110)$$

$$\left\langle P^{i_1}(z_1)\ldots P^{i_{2n-2}}(z_{2n-2})\, Q^{j_1}(z_A)\, Q^{j_2}(z_B)\right\rangle = \frac{1}{z_{AB}\prod_{i=1}^{2n-2}(z_{iA}\,z_{iB})^{1/2}} \sum_{l=0}^{n-1}(-z_{AB})^l$$

$$\times \sum_{\rho\in S_{2n-2}/\mathcal{Q}_{n,l}} \text{sgn}(\rho)\, G^{i_{\rho(1)}j_1 r_1}\, G^{i_{\rho(2)}q_1}{}_{r_1}\, G^{i_{\rho(3)}}{}_{q_1}{}^{r_2}\ldots G^{i_{\rho(2l)}j_2}{}_{r_l}$$

$$\times \prod_{j=1}^{n-l-1} \frac{g^{i_{\rho(2l+2j-1)}i_{\rho(2l+2j)}}}{z_{\rho(2l+2j-1),\rho(2l+2j)}}\, z_{\rho(2l+2j-1),A}\, z_{\rho(2l+2j),B}\,. \qquad (3.111)$$

The former gives rise to a new result by setting $P = S$, $Q = \psi$, $R = \dot{S}$:

$$\left\langle \psi^\mu(z_A)\, S_{\alpha_1}(z_1)\ldots S_{\alpha_{2n-1}}(z_{2n-1})\, S^{\dot\beta}(z_B)\right\rangle = \frac{(-1)^{n-1}}{\sqrt{2}\, z_{AB}^{1/2}\prod_{i=1}^{2n-1}(z_{Ai}\,z_{iB})^{1/2}} \sum_{l=0}^{n-1}\left(\frac{z_{AB}}{2}\right)^l$$

$$\times \sum_{\rho\in S_{2n-1}/\mathcal{P}_{n,l}} \text{sgn}(\rho)\, (\gamma^\mu)_{\alpha_{\rho(1)}\dot\delta_1}\, (\gamma^{\lambda_1} C)_{\alpha_{\rho(2)}}{}^{\dot\delta_1}\, (\gamma_{\lambda_1})_{\alpha_{\rho(3)}\dot\delta_2}\, (\gamma^{\lambda_2} C)_{\alpha_{\rho(4)}}{}^{\dot\delta_2}\ldots$$

$$\times (\gamma^{\lambda_l} C)_{\alpha_{\rho(2l)}}{}^{\dot\delta_l}\, (\gamma_{\lambda_l} C)_{\alpha_{\rho(2l+1)}}{}^{\dot\beta}$$

$$\times \prod_{j=1}^{n-l-1} \frac{C_{\alpha_{\rho(2l+2j)}\alpha_{\rho(2l+2j+1)}}}{z_{\rho(2l+2j),\rho(2l+2j+1)}}\, z_{A,\rho(2l+2j)}\, z_{\rho(2l+2j+1),B}\,. \qquad (3.112)$$

The latter yields two classes of correlation functions. First, we assign $P = S$ and $Q = \psi$:

$$\left\langle \psi^\mu(z_A)\, \psi^\nu(z_B)\, S_{\alpha_1}(z_1)\ldots S_{\alpha_{2n-2}}(z_{2n-2})\right\rangle = \frac{1}{z_{AB}\prod_{i=1}^{2n-2}(z_{Ai}\,z_{Bi})^{1/2}} \sum_{l=0}^{n-1}\left(-\frac{z_{AB}}{2}\right)^l$$

$$\times \sum_{\rho\in S_{2n-2}/\mathcal{Q}_{n,l}} \text{sgn}(\rho)\, (\gamma^\mu)_{\alpha_{\rho(1)}\dot\delta_1}\, (\gamma^{\lambda_1} C)_{\alpha_{\rho(2)}}{}^{\dot\delta_1}\, (\gamma_{\lambda_1})_{\alpha_{\rho(3)}\dot\delta_2}\, (\gamma^{\lambda_2} C)_{\alpha_{\rho(4)}}{}^{\dot\delta_2}\ldots$$

$$\times (\gamma_{\lambda_{l-1}})_{\alpha_{\rho(2l-1)}\dot\delta_l}\, (\gamma^\nu C)_{\alpha_{\rho(2l)}}{}^{\dot\delta_l}$$

$$\times \prod_{j=1}^{n-l-1} \frac{C_{\alpha_{\rho(2l+2j-1)}\alpha_{\rho(2l+2j)}}}{z_{\rho(2l+2j-1),\rho(2l+2j)}} z_{A,\rho(2l+2j-1)} \, z_{B,\rho(2l+2j)} \, . \tag{3.113}$$

Secondly, one can specify $P = S$, $Q = \dot{S}$:

$$\left\langle S_{\alpha_1}(z_1) \ldots S_{\alpha_{2n-2}}(z_{2n-2}) \, S^{\dot{\beta}}(z_A) \, S^{\dot{\gamma}}(z_B) \right\rangle = \frac{1}{z_{AB} \prod_{i=1}^{2n-2}(z_{Ai} z_{Bi})^{1/2}} \sum_{l=0}^{n-1} \left(-\frac{z_{AB}}{2} \right)^l$$

$$\times \sum_{\rho \in S_{2n-2}/\mathcal{Q}_{n,l}} \mathrm{sgn}(\rho) \, (\gamma^{\lambda_1} C)_{\alpha_{\rho(1)}}{}^{\dot{\beta}} \, (\gamma_{\lambda_1})_{\alpha_{\rho(2)} \dot{\delta}_1} \, (\gamma^{\lambda_2} C)_{\alpha_{\rho(3)}}{}^{\dot{\delta}_1} \, (\gamma_{\lambda_2})_{\alpha_{\rho(4)} \dot{\delta}_2} \cdots$$

$$\times (\gamma^{\lambda_l} C)_{\alpha_{\rho(2l-1)}}{}^{\dot{\delta}_{l-1}} \, (\gamma_{\lambda_l} C)_{\alpha_{\rho(2l)}}{}^{\dot{\gamma}}$$

$$\times \prod_{j=1}^{n-l-1} \frac{C_{\alpha_{\rho(2l+2j-1)}\alpha_{\rho(2l+2j)}}}{z_{\rho(2l+2j-1),\rho(2l+2j)}} z_{\rho(2l+2j-1),A} \, z_{\rho(2l+2j),B} \, . \tag{3.114}$$

This completes our discussion of the powerful and far-reaching methods making use of $SO(8)$ triality for tree-level correlators in eight dimensions.

3.7 Techniques in Ten Dimensions

RNS correlation function in $D = 10$ are of importance whenever non-compactified string theory is considered. This might not be directly relevant for four-dimensional physics, but string amplitudes in ten dimensions are nevertheless useful and important when considering formal or mathematical aspects of string theory.

However, in ten space-time dimensions it is not possible to derive a formula for the pure spin field correlator $\langle S_{\alpha_1} \ldots S_{\alpha_{2M}} \rangle$ as has been described in Chapter 3.2. Furthermore, there are no tools like triality as in eight dimensions to relate different correlation functions. So each RNS correlator has to be calculated separately. This has been achieved for RNS correlators at loop-level in [2] up to six points. As tree-level expressions can be obtained from these we postpone showing results to Chapter 5.3.

CHAPTER 4

Results in Four Space-Time Dimensions

In this Chapter we present all tree-level RNS correlation functions in four space-time dimensions up to eight points, as well as correlators with arbitrary many fermions but only two spin fields. This work is based on [1]. The calculations have been carried out using either the iterative procedure presented in Chapter 3.2.1 or applying the factorization method from Chapter 3.4. The necessary relations between different index terms for each correlator are collected in Appendix B.1. We restrict our calculations in the following to correlators with equal or more left-handed spin fields than right-handed, i.e.

$$\langle \psi^{\mu_1} \ldots \psi^{\mu_n} S_{\alpha_1} \ldots S_{\alpha_r} S^{\dot{\beta}_1} \ldots S^{\dot{\beta}_s} \rangle \tag{4.1}$$

with $r \geq s$, because the results for the correlator with r and s interchanged can be obtained by complex conjugation. In order to transform the index terms one has to use

$$(\varepsilon_{\alpha\beta})^* = \varepsilon_{\dot{\alpha}\dot{\beta}}, \qquad (\sigma^\mu_{\alpha\dot{\beta}})^* = \sigma^\mu_{\beta\dot{\alpha}}. \tag{4.2}$$

In addition, we have to take into account a phase $(-1)^{(r-s)/2}$ that can be explained as follows. The result of (4.1) comes with a sign $(-1)^{(r+n)/2}$ due to (3.82), whereas (4.1) with interchanged spinor indices accounts for $(-1)^{(s+n)/2}$. Hence they differ by $(-1)^{(r-s)/2}$.

We want to stress again that in four dimensions the index terms as specified in (3.8) are used and that the OPE of two left-handed spin fields (3.12a) comes with a minus sign.

4.1 Review of Known Results

The four-point amplitude of one gauge field, two gauginos and one scalar has been derived in [108]. Its Lorentz structure is determined by the correlators

$$\langle \psi^\mu(z_1)\, \psi^\nu(z_2)\, S_\alpha(z_3)\, S_\beta(z_4) \rangle = \frac{-1}{(z_{13}\, z_{14}\, z_{23}\, z_{24}\, z_{34})^{1/2}} \left[\frac{\eta^{\mu\nu}}{z_{12}}\, \varepsilon_{\alpha\beta}\, z_{13}\, z_{24} + \frac{1}{2}\, (\sigma^\mu\, \bar\sigma^\nu\, \varepsilon)_{\alpha\beta}\, z_{34} \right], \tag{4.3a}$$

$$\langle \psi^\mu(z_1)\, \psi^\nu(z_2)\, S_{\dot\alpha}(z_3)\, S_{\dot\beta}(z_4) \rangle = \frac{1}{(z_{13}\, z_{14}\, z_{23}\, z_{24}\, z_{34})^{1/2}} \left[\frac{\eta^{\mu\nu}}{z_{12}}\, \varepsilon_{\dot\alpha\dot\beta}\, z_{13}\, z_{24} + \frac{1}{2}\, (\varepsilon\, \bar\sigma^\nu\, \sigma^\mu)_{\dot\alpha\dot\beta}\, z_{34} \right]. \tag{4.3b}$$

A more involved five-point amplitude involving three NS fermions and two R spin fields has been worked out in [115]:

$$\langle \psi^\mu(z_1)\, \psi^\nu(z_2)\, \psi^\lambda(z_3)\, S_\alpha(z_4)\, S_{\dot\beta}(z_5) \rangle = \frac{1}{\sqrt{2}\,(z_{14}\, z_{15}\, z_{24}\, z_{25}\, z_{34}\, z_{35})^{1/2}}$$

$$\times \left[\frac{\eta^{\mu\nu}}{z_{12}}\, \sigma^\lambda_{\alpha\dot\beta}\, z_{14}\, z_{25} - \frac{\eta^{\mu\lambda}}{z_{13}}\, \sigma^\nu_{\alpha\dot\beta}\, z_{14}\, z_{35} + \frac{\eta^{\nu\lambda}}{z_{23}}\, \sigma^\mu_{\alpha\dot\beta}\, z_{24}\, z_{35} + \frac{z_{45}}{2}\, (\sigma^\mu\, \bar\sigma^\nu\, \sigma^\lambda)_{\alpha\dot\beta} \right]. \tag{4.4}$$

This correlator enters the computation of the six-point amplitude involving four scalars and two gauginos or chiral fermions. In addition to these cases also some pure spin field correlators with four spinor indices are known. These correlators are basic ingredients of four-point amplitudes involving gauginos or chiral matter fermions [66, 108]. The correlation function consisting of two left- and right-handed spin fields each is given by

$$\langle S_\alpha(z_1)\, S_{\dot\beta}(z_2)\, S_\gamma(z_3)\, S_{\dot\delta}(z_4) \rangle = -\frac{1}{(z_{13}\, z_{24})^{1/2}}\, \varepsilon_{\alpha\gamma}\, \varepsilon_{\dot\beta\dot\delta}, \tag{4.5}$$

while for spin fields of the same chirality one obtains

$$\langle S_\alpha(z_1)\, S_\beta(z_2)\, S_\gamma(z_3)\, S_\delta(z_4) \rangle = \left(\frac{z_{12}\, z_{14}\, z_{23}\, z_{34}}{z_{13}\, z_{24}} \right)^{1/2} \left(\frac{\varepsilon_{\alpha\beta}\, \varepsilon_{\gamma\delta}}{z_{12}\, z_{34}} - \frac{\varepsilon_{\alpha\delta}\, \varepsilon_{\beta\gamma}}{z_{14}\, z_{23}} \right), \tag{4.6a}$$

$$\langle S_{\dot\alpha}(z_1)\, S_{\dot\beta}(z_2)\, S_{\dot\gamma}(z_3)\, S_{\dot\delta}(z_4) \rangle = \left(\frac{z_{12}\, z_{14}\, z_{23}\, z_{34}}{z_{13}\, z_{24}} \right)^{1/2} \left(\frac{\varepsilon_{\dot\alpha\dot\beta}\, \varepsilon_{\dot\gamma\dot\delta}}{z_{12}\, z_{34}} - \frac{\varepsilon_{\dot\alpha\dot\delta}\, \varepsilon_{\dot\beta\dot\gamma}}{z_{14}\, z_{23}} \right). \tag{4.6b}$$

With the results (4.3) and (4.6) we can explicitly check how the correlators change if the numbers of left- and right-handed spin fields are interchanged. Indeed, the index terms are replaced by their complex conjugate as stated in (4.2),

$$(\varepsilon_{\alpha\beta})^* = \varepsilon_{\dot\alpha\dot\beta}, \qquad (\sigma^\mu\, \bar\sigma^\nu\, \varepsilon)_{\alpha\beta}^* = (\varepsilon\, \bar\sigma^\nu\, \sigma^\mu)_{\dot\beta\dot\alpha} = (\varepsilon\, \bar\sigma^\mu\, \sigma^\nu)_{\dot\alpha\dot\beta}, \tag{4.7}$$

and the correlators capture a sign $(-1)^{(r-s)/2}$. From now on we only state the results for correlation functions with pre-dominantly left-handed spin fields.

4.2 Five-Point Functions

At the five-point level the only non-vanishing correlation functions are (4.4) and the correlator consisting of one fermion, three left-handed and one right-handed spin field. Either by factorization or the iterative procedure we find for this five-point function in minimal form

$$\langle \psi^\mu(z_1) \, S_\alpha(z_2) \, S_\beta(z_3) \, S_\gamma(z_4) \, S_{\dot\delta}(z_5) \rangle = \frac{-1}{\sqrt{2}\,(z_{12}\,z_{13}\,z_{14}\,z_{15}\,z_{23}\,z_{24}\,z_{34})^{1/2}} \\ \times \left(\sigma^\mu_{\alpha\dot\delta} \, \varepsilon_{\beta\gamma} \, z_{14}\,z_{23} - \sigma^\mu_{\gamma\dot\delta} \, \varepsilon_{\alpha\beta} \, z_{12}\,z_{34} \right). \tag{4.8}$$

The non-minimal expression with all three possible index terms is given by

$$\langle \psi^\mu(z_1) \, S_\alpha(z_2) \, S_\beta(z_3) \, S_\gamma(z_4) \, S_{\dot\delta}(z_5) \rangle = \frac{-1}{\sqrt{2}\,(z_{12}\,z_{13}\,z_{14}\,z_{15}\,z_{23}\,z_{24}\,z_{34})^{1/2}} \\ \times \left(\sigma^\mu_{\gamma\dot\delta} \, \varepsilon_{\alpha\beta} \, z_{12}\,z_{13} - \sigma^\mu_{\beta\dot\delta} \, \varepsilon_{\alpha\gamma} \, z_{12}\,z_{14} + \sigma^\mu_{\alpha\dot\delta} \, \varepsilon_{\beta\gamma} \, z_{13}\,z_{14} \right). \tag{4.9}$$

4.3 Six-Point Functions

The simplest six-point functions at tree-level are the ones consisting of only spin fields. In the case of four left- and two right-handed spin fields one obtains

$$\langle S_\alpha(z_1) \, S_\beta(z_2) \, S_\gamma(z_3) \, S_\delta(z_4) \, S_{\dot\epsilon}(z_5) \, S_{\dot\zeta}(z_6) \rangle \\ = \left(\frac{z_{12}\,z_{14}\,z_{23}\,z_{34}\,z_{56}}{z_{13}\,z_{24}} \right)^{1/2} \left(\frac{\varepsilon_{\alpha\beta}\,\varepsilon_{\gamma\delta}\,\varepsilon_{\dot\epsilon\dot\zeta}}{z_{12}\,z_{34}\,z_{56}} - \frac{\varepsilon_{\alpha\delta}\,\varepsilon_{\beta\gamma}\,\varepsilon_{\dot\epsilon\dot\zeta}}{z_{14}\,z_{23}\,z_{56}} \right). \tag{4.10}$$

For the six-point function involving six spin fields of the same chirality we find

$$\langle S_\alpha(z_1) \, S_\beta(z_2) \, S_\gamma(z_3) \, S_\delta(z_4) \, S_\epsilon(z_5) \, S_\zeta(z_6) \rangle = - \left(\frac{z_{12}\,z_{14}\,z_{16}\,z_{23}\,z_{25}\,z_{34}\,z_{36}\,z_{45}\,z_{56}}{z_{13}\,z_{15}\,z_{24}\,z_{26}\,z_{35}\,z_{46}} \right)^{1/2} \\ \times \left(\frac{\varepsilon_{\alpha\beta}\,\varepsilon_{\gamma\delta}\,\varepsilon_{\epsilon\zeta}}{z_{12}\,z_{34}\,z_{56}} - \frac{\varepsilon_{\alpha\beta}\,\varepsilon_{\gamma\zeta}\,\varepsilon_{\epsilon\delta}}{z_{12}\,z_{36}\,z_{54}} + \frac{\varepsilon_{\alpha\delta}\,\varepsilon_{\gamma\zeta}\,\varepsilon_{\epsilon\beta}}{z_{14}\,z_{36}\,z_{52}} - \frac{\varepsilon_{\alpha\delta}\,\varepsilon_{\gamma\beta}\,\varepsilon_{\epsilon\zeta}}{z_{14}\,z_{32}\,z_{56}} + \frac{\varepsilon_{\alpha\zeta}\,\varepsilon_{\gamma\beta}\,\varepsilon_{\epsilon\delta}}{z_{16}\,z_{32}\,z_{54}} - \frac{\varepsilon_{\alpha\zeta}\,\varepsilon_{\gamma\delta}\,\varepsilon_{\epsilon\beta}}{z_{16}\,z_{34}\,z_{52}} \right). \tag{4.11}$$

Next, we consider the case of two fermions and two left- and right-handed spin fields each for which we derive:

$$\langle \psi^\mu(z_1) \, \psi^\nu(z_2) \, S_\alpha(z_3) \, S_{\dot\beta}(z_4) \, S_\gamma(z_5) \, S_{\dot\delta}(z_6) \rangle = \frac{-1}{2 \, z_{12} \, (z_{13} \, z_{14} \, z_{15} \, z_{16} \, z_{23} \, z_{24} \, z_{25} \, z_{26} \, z_{35} \, z_{46})^{1/2}}$$
$$\times \left(\sigma^\mu_{\alpha\dot\delta} \, \sigma^\nu_{\gamma\dot\beta} \, z_{14} \, z_{15} \, z_{23} \, z_{26} - \sigma^\mu_{\alpha\dot\beta} \, \sigma^\nu_{\gamma\dot\delta} \, z_{15} \, z_{16} \, z_{23} \, z_{24} \right.$$
$$\left. + \sigma^\mu_{\gamma\dot\beta} \, \sigma^\nu_{\alpha\dot\delta} \, z_{13} \, z_{16} \, z_{24} \, z_{25} - \sigma^\mu_{\gamma\dot\delta} \, \sigma^\nu_{\alpha\dot\beta} \, z_{13} \, z_{14} \, z_{25} \, z_{26} \right). \tag{4.12}$$

This correlator is needed to compute the five-point disk amplitude of one gauge field and four chiral fermions [67]. Actually, in this reference another variant of (4.12) has been used,

$$\langle \psi^\mu(z_1) \, \psi^\nu(z_2) \, S_\alpha(z_3) \, S_{\dot\beta}(z_4) \, S_\gamma(z_5) \, S_{\dot\delta}(z_6) \rangle = \frac{-1}{(z_{13} \, z_{14} \, z_{15} \, z_{16} \, z_{23} \, z_{24} \, z_{25} \, z_{26} \, z_{35} \, z_{46})^{1/2}}$$
$$\times \left(\frac{1}{2} \sigma^\mu_{\alpha\dot\beta} \, \sigma^\nu_{\gamma\dot\delta} \, z_{15} \, z_{24} \, z_{36} - \frac{1}{2} \sigma^\mu_{\alpha\dot\delta} \, \sigma^\nu_{\gamma\dot\beta} \, z_{15} \, z_{26} \, z_{34} - \frac{1}{2} \sigma^\mu_{\gamma\dot\delta} \, \sigma^\nu_{\alpha\dot\beta} \, z_{13} \, z_{26} \, z_{45} \right.$$
$$\left. - \frac{1}{2} \sigma^\mu_{\gamma\dot\beta} \, \sigma^\nu_{\alpha\dot\delta} \, z_{13} \, z_{24} \, z_{56} + \frac{\eta^{\mu\nu}}{z_{12}} \, \varepsilon_{\alpha\gamma} \, \varepsilon_{\dot\beta\dot\delta} \, z_{13} \, z_{15} \, z_{24} \, z_{26} \right), \tag{4.13}$$

where only the index term containing $\eta^{\mu\nu}$ comes with a pole in z_{12}. This makes the calculation of the just mentioned string amplitude much simpler. For the correlator with two gauge fields and four left-handed spin fields we obtain

$$\langle \psi^\mu(z_1) \, \psi^\nu(z_2) \, S_\alpha(z_3) \, S_\beta(z_4) \, S_\gamma(z_5) \, S_\delta(z_6) \rangle$$
$$= \frac{1}{(z_{13} \, z_{14} \, z_{15} \, z_{16} \, z_{23} \, z_{24} \, z_{25} \, z_{26} \, z_{34} \, z_{35} \, z_{36} \, z_{45} \, z_{46} \, z_{56})^{1/2}}$$
$$\times \left[\frac{\eta^{\mu\nu}}{z_{12}} \, \varepsilon_{\alpha\beta} \, \varepsilon_{\gamma\delta} \, z_{13} \, z_{16} \, z_{24} \, z_{25} \, z_{36} \, z_{45} - \frac{\eta^{\mu\nu}}{z_{12}} \, \varepsilon_{\alpha\delta} \, \varepsilon_{\beta\gamma} \, z_{13} \, z_{14} \, z_{25} \, z_{26} \, z_{34} \, z_{56} \right.$$
$$+ \frac{1}{2} (\sigma^\mu \, \bar\sigma^\nu \, \varepsilon)_{\alpha\beta} \, \varepsilon_{\gamma\delta} \, z_{15} \, z_{25} \, z_{34} \, z_{36} \, z_{46} - \frac{1}{2} (\sigma^\mu \, \bar\sigma^\nu \, \varepsilon)_{\alpha\gamma} \, \varepsilon_{\beta\delta} \, z_{14} \, z_{24} \, z_{35} \, z_{36} \, z_{56}$$
$$\left. + \frac{1}{2} (\sigma^\mu \, \bar\sigma^\nu \, \varepsilon)_{\beta\gamma} \, \varepsilon_{\alpha\delta} \, z_{13} \, z_{23} \, z_{45} \, z_{46} \, z_{56} \right]. \tag{4.14}$$

This expression is relevant for the scattering amplitude of one gauge field and four chiral fermions. In a slightly more symmetric representation with respect to the spin fields this six-point function reads:

$$\langle \psi^\mu(z_1) \, \psi^\nu(z_2) \, S_\alpha(z_3) \, S_\beta(z_4) \, S_\gamma(z_5) \, S_\delta(z_6) \rangle$$
$$= \frac{1}{(z_{13} \, z_{14} \, z_{15} \, z_{16} \, z_{23} \, z_{24} \, z_{25} \, z_{26} \, z_{34} \, z_{35} \, z_{36} \, z_{45} \, z_{46} \, z_{56})^{1/2}}$$
$$\times \left[\frac{\eta^{\mu\nu}}{z_{12}} \, \varepsilon_{\alpha\beta} \, \varepsilon_{\gamma\delta} \, z_{13} \, z_{15} \, z_{24} \, z_{26} \, z_{36} \, z_{45} + \frac{\eta^{\mu\nu}}{z_{12}} \, \varepsilon_{\alpha\delta} \, \varepsilon_{\gamma\beta} \, z_{13} \, z_{15} \, z_{24} \, z_{26} \, z_{34} \, z_{56} \right.$$

4.3 Six-Point Functions

$$+ \frac{1}{2} \left(\sigma^\mu \bar{\sigma}^\nu \varepsilon \right)_{\alpha\beta} \varepsilon_{\gamma\delta} \, z_{15} \, z_{26} \, z_{34} \, z_{36} \, z_{45} - \frac{1}{2} \left(\sigma^\mu \bar{\sigma}^\nu \varepsilon \right)_{\alpha\delta} \varepsilon_{\beta\gamma} \, z_{15} \, z_{24} \, z_{34} \, z_{36} \, z_{56}$$
$$+ \frac{1}{2} \left(\sigma^\mu \bar{\sigma}^\nu \varepsilon \right)_{\gamma\delta} \varepsilon_{\alpha\beta} \, z_{13} \, z_{24} \, z_{36} \, z_{45} \, z_{56} - \frac{1}{2} \left(\sigma^\mu \bar{\sigma}^\nu \varepsilon \right)_{\gamma\beta} \varepsilon_{\alpha\delta} \, z_{13} \, z_{26} \, z_{34} \, z_{45} \, z_{56} \Bigg] . \quad (4.15)$$

Moreover, (4.15) may be cast into a form which is manifestly anti-symmetric under interchanging the two NS fermions $\psi^\mu(z_1) \leftrightarrow \psi^\nu(z_2)$:

$$\left\langle \psi^\mu(z_1) \, \psi^\nu(z_2) \, S_\alpha(z_3) \, S_\beta(z_4) \, S_\gamma(z_5) \, S_\delta(z_6) \right\rangle$$
$$= \frac{1}{2 \left(z_{13} \, z_{14} \, z_{15} \, z_{16} \, z_{23} \, z_{24} \, z_{25} \, z_{26} \, z_{34} \, z_{35} \, z_{36} \, z_{45} \, z_{46} \, z_{56} \right)^{1/2}}$$
$$\times \Bigg[\frac{\eta^{\mu\nu}}{z_{12}} \, \varepsilon_{\alpha\beta} \, \varepsilon_{\gamma\delta} \, z_{36} \, z_{45} \left(z_{13} \, z_{16} \, z_{24} \, z_{25} + z_{14} \, z_{15} \, z_{23} \, z_{26} \right)$$
$$+ \frac{\eta^{\mu\nu}}{z_{12}} \, \varepsilon_{\alpha\delta} \, \varepsilon_{\gamma\beta} \, z_{34} \, z_{56} \left(z_{13} \, z_{14} \, z_{25} \, z_{26} + z_{15} \, z_{16} \, z_{23} \, z_{24} \right)$$
$$+ \frac{1}{2} \left(\sigma^{\mu\nu} \varepsilon \right)_{\alpha\beta} \, z_{34} \, \varepsilon_{\gamma\delta} \, z_{36} \, z_{45} \left(z_{15} \, z_{26} + z_{16} \, z_{25} \right)$$
$$+ \frac{1}{2} \left(\sigma^{\mu\nu} \varepsilon \right)_{\alpha\delta} \, z_{36} \, \varepsilon_{\gamma\beta} \, z_{34} \, z_{56} \left(z_{15} \, z_{24} + z_{14} \, z_{25} \right)$$
$$+ \frac{1}{2} \left(\sigma^{\mu\nu} \varepsilon \right)_{\gamma\delta} \, z_{56} \, \varepsilon_{\alpha\beta} \, z_{36} \, z_{45} \left(z_{13} \, z_{24} + z_{14} \, z_{23} \right)$$
$$+ \frac{1}{2} \left(\sigma^{\mu\nu} \varepsilon \right)_{\gamma\beta} \, z_{54} \, \varepsilon_{\alpha\delta} \, z_{34} \, z_{56} \left(z_{13} \, z_{26} + z_{16} \, z_{23} \right) \Bigg] . \quad (4.16)$$

The last non-vanishing correlation function at six-point level contains four fermions and two left-handed spin fields. Via the factorization method and reducing it to minimal form we obtain as result

$$\left\langle \psi^\mu(z_1) \, \psi^\nu(z_2) \, \psi^\lambda(z_3) \, \psi^\rho(z_4) \, S_\alpha(z_5) \, S_\beta(z_6) \right\rangle = \frac{1}{z_{56}^{1/2} \left(z_{15} \, z_{16} \, z_{25} \, z_{26} \, z_{35} \, z_{36} \, z_{45} \, z_{46} \right)^{1/2}}$$
$$\times \Bigg[\frac{\eta^{\mu\nu} \, \eta^{\lambda\rho}}{z_{12} \, z_{34}} \, \varepsilon_{\alpha\beta} \, z_{15} \, z_{26} \, z_{35} \, z_{46} - \frac{\eta^{\mu\lambda} \, \eta^{\nu\rho}}{z_{13} \, z_{24}} \, \varepsilon_{\alpha\beta} \, z_{15} \, z_{36} \, z_{25} \, z_{46}$$
$$+ \frac{\eta^{\mu\rho} \, \eta^{\nu\lambda}}{z_{14} \, z_{23}} \, \varepsilon_{\alpha\beta} \, z_{15} \, z_{46} \, z_{25} \, z_{36} + \frac{1}{4} \left(\sigma^\mu \bar{\sigma}^\nu \sigma^\lambda \bar{\sigma}^\rho \varepsilon \right)_{\alpha\beta} z_{56}^2$$
$$+ \frac{1}{2} \frac{\eta^{\mu\nu}}{z_{12}} \left(\sigma^\lambda \bar{\sigma}^\rho \varepsilon \right)_{\alpha\beta} z_{15} \, z_{26} \, z_{56} + \frac{1}{2} \frac{\eta^{\lambda\rho}}{z_{34}} \left(\sigma^\mu \bar{\sigma}^\nu \varepsilon \right)_{\alpha\beta} z_{35} \, z_{46} \, z_{56}$$
$$- \frac{1}{2} \frac{\eta^{\mu\lambda}}{z_{13}} \left(\sigma^\nu \bar{\sigma}^\rho \varepsilon \right)_{\alpha\beta} z_{15} \, z_{36} \, z_{56} - \frac{1}{2} \frac{\eta^{\nu\rho}}{z_{24}} \left(\sigma^\mu \bar{\sigma}^\lambda \varepsilon \right)_{\alpha\beta} z_{25} \, z_{46} \, z_{56}$$
$$+ \frac{1}{2} \frac{\eta^{\mu\rho}}{z_{14}} \left(\sigma^\nu \bar{\sigma}^\lambda \varepsilon \right)_{\alpha\beta} z_{15} \, z_{46} \, z_{56} + \frac{1}{2} \frac{\eta^{\nu\lambda}}{z_{23}} \left(\sigma^\mu \bar{\sigma}^\rho \varepsilon \right)_{\alpha\beta} z_{25} \, z_{36} \, z_{56} \Bigg] . \quad (4.17)$$

4.4 Seven-Point Functions

The seven-point function built from one fermion and three spin fields of each chirality is easily evaluated using the factorization technique. We directly get the result in minimal form:

$$\langle \psi^\mu(z_1) S_\alpha(z_2) S_\beta(z_3) S_\gamma(z_4) S_{\dot\delta}(z_5) S_{\dot\epsilon}(z_6) S_{\dot\zeta}(z_7) \rangle$$
$$= \frac{-1}{\sqrt{2}\,(z_{12}\,z_{13}\,z_{14}\,z_{15}\,z_{16}\,z_{17}\,z_{23}\,z_{24}\,z_{34}\,z_{56}\,z_{57}\,z_{67})^{1/2}}$$
$$\times \Big(\sigma^\mu_{\gamma\dot\zeta}\,\varepsilon_{\alpha\beta}\,\varepsilon_{\dot\delta\dot\epsilon}\,z_{12}\,z_{15}\,z_{34}\,z_{67} - \sigma^\mu_{\gamma\dot\delta}\,\varepsilon_{\alpha\beta}\,\varepsilon_{\dot\epsilon\dot\zeta}\,z_{12}\,z_{17}\,z_{34}\,z_{56}$$
$$+ \sigma^\mu_{\alpha\dot\delta}\,\varepsilon_{\beta\gamma}\,\varepsilon_{\dot\epsilon\dot\zeta}\,z_{14}\,z_{17}\,z_{23}\,z_{56} - \sigma^\mu_{\alpha\dot\zeta}\,\varepsilon_{\beta\gamma}\,\varepsilon_{\dot\delta\dot\epsilon}\,z_{14}\,z_{15}\,z_{23}\,z_{67} \Big). \tag{4.18}$$

The spin fields can also carry different chirality. The same method yields:

$$\langle \psi^\mu(z_1) S_\alpha(z_2) S_\beta(z_3) S_\gamma(z_4) S_\delta(z_5) S_\epsilon(z_6) S_{\dot\zeta}(z_7) \rangle$$
$$= \frac{1}{\sqrt{2}\,(z_{12}\,z_{13}\,z_{14}\,z_{15}\,z_{16}\,z_{17}\,z_{23}\,z_{24}\,z_{25}\,z_{26}\,z_{34}\,z_{35}\,z_{36}\,z_{45}\,z_{46}\,z_{56})^{1/2}}$$
$$\times \Big(\sigma^\mu_{\alpha\dot\zeta}\,\varepsilon_{\beta\gamma}\,\varepsilon_{\delta\epsilon}\,z_{14}\,z_{16}\,z_{23}\,z_{25}\,z_{36}\,z_{45} - \sigma^\mu_{\alpha\dot\zeta}\,\varepsilon_{\beta\epsilon}\,\varepsilon_{\gamma\delta}\,z_{14}\,z_{16}\,z_{23}\,z_{25}\,z_{34}\,z_{56}$$
$$- \sigma^\mu_{\gamma\dot\zeta}\,\varepsilon_{\alpha\beta}\,\varepsilon_{\delta\epsilon}\,z_{12}\,z_{16}\,z_{25}\,z_{34}\,z_{36}\,z_{45} + \sigma^\mu_{\gamma\dot\zeta}\,\varepsilon_{\alpha\delta}\,\varepsilon_{\beta\epsilon}\,z_{12}\,z_{16}\,z_{23}\,z_{34}\,z_{45}\,z_{56}$$
$$+ \sigma^\mu_{\epsilon\dot\zeta}\,\varepsilon_{\alpha\beta}\,\varepsilon_{\gamma\delta}\,z_{12}\,z_{14}\,z_{25}\,z_{34}\,z_{36}\,z_{56} - \sigma^\mu_{\epsilon\dot\zeta}\,\varepsilon_{\alpha\delta}\,\varepsilon_{\beta\gamma}\,z_{12}\,z_{14}\,z_{23}\,z_{36}\,z_{45}\,z_{56} \Big). \tag{4.19}$$

By eliminating one of the Clebsch–Gordan coefficients this can be brought into minimal form,

$$\langle \psi^\mu(z_1) S_\alpha(z_2) S_\beta(z_3) S_\gamma(z_4) S_\delta(z_5) S_\epsilon(z_6) S_{\dot\zeta}(z_7) \rangle$$
$$= \frac{1}{\sqrt{2}\,(z_{12}\,z_{13}\,z_{14}\,z_{15}\,z_{16}\,z_{17}\,z_{23}\,z_{24}\,z_{25}\,z_{26}\,z_{34}\,z_{35}\,z_{36}\,z_{45}\,z_{46}\,z_{56})^{1/2}}$$
$$\times \Big[\sigma^\mu_{\alpha\dot\zeta}\,\varepsilon_{\beta\gamma}\,\varepsilon_{\delta\epsilon}\,z_{16}\,z_{23}\,z_{45}\,(z_{14}\,z_{25}\,z_{36} + z_{12}\,z_{34}\,z_{56}) - \sigma^\mu_{\alpha\dot\zeta}\,\varepsilon_{\beta\epsilon}\,\varepsilon_{\gamma\delta}\,z_{15}\,z_{16}\,z_{23}\,z_{24}\,z_{34}\,z_{56}$$
$$+ \sigma^\mu_{\epsilon\dot\zeta}\,\varepsilon_{\alpha\beta}\,\varepsilon_{\gamma\delta}\,z_{12}\,z_{34}\,z_{56}\,(z_{14}\,z_{25}\,z_{36} + z_{16}\,z_{23}\,z_{45}) - \sigma^\mu_{\epsilon\dot\zeta}\,\varepsilon_{\alpha\delta}\,\varepsilon_{\beta\gamma}\,z_{12}\,z_{13}\,z_{23}\,z_{45}\,z_{46}\,z_{56}$$
$$- \sigma^\mu_{\gamma\dot\zeta}\,\varepsilon_{\alpha\beta}\,\varepsilon_{\delta\epsilon}\,z_{12}\,z_{16}\,z_{26}\,z_{34}\,z_{35}\,z_{45} \Big]. \tag{4.20}$$

However, the z dependence is more complicated in non-minimal form. The correlation function consisting of three fermions, three left-handed and one right-handed spin field is given by the expression

$$\langle \psi^\mu(z_1)\,\psi^\nu(z_2)\,\psi^\lambda(z_3)\,S_\alpha(z_4)\,S_\beta(z_5)\,S_\gamma(z_6)\,S_{\dot\delta}(z_7) \rangle$$
$$= \frac{1}{\sqrt{2}\,z_{23}\,(z_{14}\,z_{15}\,z_{16}\,z_{17}\,z_{24}\,z_{25}\,z_{26}\,z_{27}\,z_{34}\,z_{35}\,z_{36}\,z_{37}\,z_{45}\,z_{46}\,z_{56})^{1/2}}$$

4.4 Seven-Point Functions

$$\times \left[\frac{\eta^{\mu\nu}}{z_{12}} \varepsilon_{\alpha\beta} \sigma^{\lambda}_{\gamma\dot{\delta}} z_{13} z_{14} z_{25} z_{26} z_{27} z_{34} z_{56} + \frac{\eta^{\mu\lambda}}{z_{13}} \varepsilon_{\alpha\beta} \sigma^{\nu}_{\gamma\dot{\delta}} z_{12} z_{14} z_{24} z_{35} z_{36} z_{37} z_{56} \right.$$

$$- \frac{\eta^{\mu\nu}}{z_{12}} \varepsilon_{\beta\gamma} \sigma^{\lambda}_{\alpha\dot{\delta}} z_{14} z_{15} z_{23} z_{26} z_{27} z_{36} z_{45} + \frac{\eta^{\mu\lambda}}{z_{13}} \varepsilon_{\beta\gamma} \sigma^{\nu}_{\alpha\dot{\delta}} z_{14} z_{15} z_{23} z_{26} z_{36} z_{37} z_{45}$$

$$- \frac{1}{2} (\sigma^{\mu} \bar{\sigma}^{\nu} \varepsilon)_{\alpha\beta} \sigma^{\lambda}_{\gamma\dot{\delta}} z_{16} z_{26} z_{27} z_{34} z_{35} z_{45} - \frac{1}{2} (\sigma^{\mu} \bar{\sigma}^{\lambda} \varepsilon)_{\alpha\beta} \sigma^{\nu}_{\gamma\dot{\delta}} z_{16} z_{24} z_{25} z_{36} z_{37} z_{45}$$

$$- \frac{1}{2} (\sigma^{\mu} \bar{\sigma}^{\nu} \varepsilon)_{\beta\gamma} \sigma^{\lambda}_{\alpha\dot{\delta}} z_{14} z_{24} z_{27} z_{35} z_{36} z_{56} - \frac{1}{2} (\sigma^{\mu} \bar{\sigma}^{\lambda} \varepsilon)_{\beta\gamma} \sigma^{\nu}_{\alpha\dot{\delta}} z_{14} z_{25} z_{26} z_{34} z_{37} z_{56}$$

$$\left. + \frac{1}{2} (\sigma^{\mu} \bar{\sigma}^{\nu} \varepsilon)_{\alpha\gamma} \sigma^{\lambda}_{\beta\dot{\delta}} z_{15} z_{25} z_{27} z_{34} z_{36} z_{46} + \frac{1}{2} (\sigma^{\mu} \bar{\sigma}^{\lambda} \varepsilon)_{\alpha\gamma} \sigma^{\nu}_{\beta\dot{\delta}} z_{15} z_{24} z_{26} z_{35} z_{37} z_{46} \right] . \quad (4.21)$$

This results may be also cast into a form where most of the z_{23} poles are absent:

$$\left\langle \psi^{\mu}(z_1) \, \psi^{\nu}(z_2) \, \psi^{\lambda}(z_3) \, S_{\alpha}(z_4) \, S_{\beta}(z_5) \, S_{\gamma}(z_6) \, S_{\dot{\delta}}(z_7) \right\rangle$$

$$= \frac{1}{\sqrt{2} \, (z_{14} z_{15} z_{16} z_{17} z_{24} z_{25} z_{26} z_{27} z_{34} z_{35} z_{36} z_{37} z_{45} z_{46} z_{56})^{1/2}}$$

$$\times \left[\frac{\eta^{\mu\nu}}{z_{12}} \sigma^{\lambda}_{\gamma\dot{\delta}} \varepsilon_{\alpha\beta} z_{14} z_{16} z_{25} z_{27} z_{34} z_{56} - \frac{\eta^{\mu\nu}}{z_{12}} \sigma^{\lambda}_{\alpha\dot{\delta}} \varepsilon_{\beta\gamma} z_{14} z_{16} z_{25} z_{27} z_{36} z_{45} \right.$$

$$- \frac{\eta^{\mu\lambda}}{z_{13}} \sigma^{\nu}_{\gamma\dot{\delta}} \varepsilon_{\alpha\beta} z_{14} z_{16} z_{24} z_{35} z_{37} z_{56} + \frac{\eta^{\mu\lambda}}{z_{13}} \sigma^{\nu}_{\alpha\dot{\delta}} \varepsilon_{\beta\gamma} z_{14} z_{16} z_{26} z_{35} z_{37} z_{45}$$

$$+ \frac{\eta^{\nu\lambda}}{z_{23}} \sigma^{\mu}_{\gamma\dot{\delta}} \varepsilon_{\alpha\beta} z_{14} z_{24} z_{26} z_{35} z_{37} z_{56} - \frac{\eta^{\nu\lambda}}{z_{23}} \sigma^{\mu}_{\alpha\dot{\delta}} \varepsilon_{\beta\gamma} z_{16} z_{24} z_{26} z_{35} z_{37} z_{45}$$

$$+ \frac{1}{2} (\sigma^{\mu} \bar{\sigma}^{\nu} \sigma^{\lambda})_{\gamma\dot{\delta}} \varepsilon_{\alpha\beta} z_{14} z_{25} z_{34} z_{56} z_{67} - \frac{1}{2} (\sigma^{\mu} \bar{\sigma}^{\nu} \sigma^{\lambda})_{\alpha\dot{\delta}} \varepsilon_{\beta\gamma} z_{16} z_{25} z_{36} z_{45} z_{47}$$

$$- \frac{1}{2} (\sigma^{\mu} \bar{\sigma}^{\nu} \varepsilon)_{\alpha\beta} \sigma^{\lambda}_{\gamma\dot{\delta}} z_{16} z_{23} z_{45} z_{47} z_{56} + \frac{1}{2} (\sigma^{\mu} \bar{\sigma}^{\nu} \varepsilon)_{\gamma\beta} \sigma^{\lambda}_{\alpha\dot{\delta}} z_{14} z_{23} z_{45} z_{56} z_{67}$$

$$\left. + \frac{1}{2} (\sigma^{\nu} \bar{\sigma}^{\lambda} \varepsilon)_{\alpha\beta} \sigma^{\mu}_{\gamma\dot{\delta}} z_{14} z_{26} z_{37} z_{45} z_{56} - \frac{1}{2} (\sigma^{\nu} \bar{\sigma}^{\lambda} \varepsilon)_{\gamma\beta} \sigma^{\mu}_{\alpha\dot{\delta}} z_{16} z_{24} z_{37} z_{45} z_{56} \right] . \quad (4.22)$$

The most complicated, non-vanishing seven-point function is given by five fermions and two spin fields of different chirality. Applying the iterative procedure explained before, one finds

$$\left\langle \psi^{\mu}(z_1) \, \psi^{\nu}(z_2) \, \psi^{\lambda}(z_3) \, \psi^{\rho}(z_4) \, \psi^{\tau}(z_5) \, S_{\alpha}(z_6) \, S_{\dot{\beta}}(z_7) \right\rangle$$

$$= \frac{1}{\sqrt{2} \, (z_{16} z_{17} z_{26} z_{27} z_{36} z_{37} z_{46} z_{47} z_{56} z_{57})^{1/2}}$$

$$\times \left[\frac{\eta^{\nu\lambda} \, \eta^{\rho\tau}}{z_{23} \, z_{45}} \sigma^{\mu}_{\alpha\dot{\beta}} z_{26} z_{37} z_{46} z_{57} + \frac{\eta^{\nu\tau} \, \eta^{\lambda\rho}}{z_{25} \, z_{34}} \sigma^{\mu}_{\alpha\dot{\beta}} z_{26} z_{57} z_{36} z_{47} - \frac{\eta^{\nu\rho} \, \eta^{\lambda\tau}}{z_{24} \, z_{35}} \sigma^{\mu}_{\alpha\dot{\beta}} z_{26} z_{47} z_{36} z_{57} \right.$$

$$- \frac{\eta^{\mu\lambda} \, \eta^{\rho\tau}}{z_{13} \, z_{45}} \sigma^{\nu}_{\alpha\dot{\beta}} z_{16} z_{37} z_{46} z_{57} - \frac{\eta^{\mu\tau} \, \eta^{\lambda\rho}}{z_{15} \, z_{34}} \sigma^{\nu}_{\alpha\dot{\beta}} z_{16} z_{57} z_{36} z_{47} + \frac{\eta^{\mu\rho} \, \eta^{\lambda\tau}}{z_{14} \, z_{35}} \sigma^{\nu}_{\alpha\dot{\beta}} z_{16} z_{47} z_{36} z_{57}$$

$$\left. + \frac{\eta^{\mu\nu} \, \eta^{\rho\tau}}{z_{12} \, z_{45}} \sigma^{\lambda}_{\alpha\dot{\beta}} z_{16} z_{27} z_{46} z_{57} + \frac{\eta^{\mu\tau} \, \eta^{\nu\rho}}{z_{15} \, z_{24}} \sigma^{\lambda}_{\alpha\dot{\beta}} z_{16} z_{57} z_{26} z_{47} - \frac{\eta^{\mu\rho} \, \eta^{\nu\tau}}{z_{14} \, z_{25}} \sigma^{\lambda}_{\alpha\dot{\beta}} z_{16} z_{47} z_{26} z_{57} \right.$$

$$-\frac{\eta^{\mu\nu}\eta^{\lambda\tau}}{z_{12}\,z_{35}}\sigma^{\rho}_{\alpha\dot{\beta}}z_{16}\,z_{27}\,z_{36}\,z_{57} - \frac{\eta^{\mu\tau}\eta^{\nu\lambda}}{z_{15}\,z_{23}}\sigma^{\rho}_{\alpha\dot{\beta}}z_{16}\,z_{57}\,z_{26}\,z_{37} + \frac{\eta^{\mu\lambda}\eta^{\nu\tau}}{z_{13}\,z_{25}}\sigma^{\rho}_{\alpha\dot{\beta}}z_{16}\,z_{37}\,z_{26}\,z_{57}$$

$$+\frac{\eta^{\mu\nu}\eta^{\lambda\rho}}{z_{12}\,z_{34}}\sigma^{\tau}_{\alpha\dot{\beta}}z_{16}\,z_{27}\,z_{36}\,z_{47} + \frac{\eta^{\mu\rho}\eta^{\nu\lambda}}{z_{14}\,z_{23}}\sigma^{\tau}_{\alpha\dot{\beta}}z_{16}\,z_{47}\,z_{26}\,z_{37} - \frac{\eta^{\mu\lambda}\eta^{\nu\rho}}{z_{13}\,z_{24}}\sigma^{\tau}_{\alpha\dot{\beta}}z_{16}\,z_{37}\,z_{26}\,z_{47}$$

$$+\frac{1}{2}\frac{\eta^{\mu\nu}}{z_{12}}(\sigma^{\lambda}\bar{\sigma}^{\rho}\sigma^{\tau})_{\alpha\dot{\beta}}z_{16}\,z_{27}\,z_{67} - \frac{1}{2}\frac{\eta^{\mu\lambda}}{z_{13}}(\sigma^{\nu}\bar{\sigma}^{\rho}\sigma^{\tau})_{\alpha\dot{\beta}}z_{16}\,z_{37}\,z_{67}$$

$$+\frac{1}{2}\frac{\eta^{\mu\rho}}{z_{14}}(\sigma^{\nu}\bar{\sigma}^{\lambda}\sigma^{\tau})_{\alpha\dot{\beta}}z_{16}\,z_{47}\,z_{67} - \frac{1}{2}\frac{\eta^{\mu\tau}}{z_{15}}(\sigma^{\nu}\bar{\sigma}^{\lambda}\sigma^{\rho})_{\alpha\dot{\beta}}z_{16}\,z_{57}\,z_{67}$$

$$+\frac{1}{2}\frac{\eta^{\nu\lambda}}{z_{23}}(\sigma^{\mu}\bar{\sigma}^{\rho}\sigma^{\tau})_{\alpha\dot{\beta}}z_{26}\,z_{37}\,z_{67} - \frac{1}{2}\frac{\eta^{\nu\rho}}{z_{24}}(\sigma^{\mu}\bar{\sigma}^{\lambda}\sigma^{\tau})_{\alpha\dot{\beta}}z_{26}\,z_{47}\,z_{67}$$

$$+\frac{1}{2}\frac{\eta^{\nu\tau}}{z_{25}}(\sigma^{\mu}\bar{\sigma}^{\lambda}\sigma^{\rho})_{\alpha\dot{\beta}}z_{26}\,z_{57}\,z_{67} + \frac{1}{2}\frac{\eta^{\lambda\rho}}{z_{34}}(\sigma^{\mu}\bar{\sigma}^{\nu}\sigma^{\tau})_{\alpha\dot{\beta}}z_{36}\,z_{47}\,z_{67}$$

$$-\frac{1}{2}\frac{\eta^{\lambda\tau}}{z_{35}}(\sigma^{\mu}\bar{\sigma}^{\nu}\sigma^{\rho})_{\alpha\dot{\beta}}z_{36}\,z_{57}\,z_{67} + \frac{1}{2}\frac{\eta^{\rho\tau}}{z_{45}}(\sigma^{\mu}\bar{\sigma}^{\nu}\sigma^{\lambda})_{\alpha\dot{\beta}}z_{46}\,z_{57}\,z_{67}$$

$$+\frac{1}{4}(\sigma^{\mu}\bar{\sigma}^{\nu}\sigma^{\lambda}\bar{\sigma}^{\rho}\sigma^{\tau})_{\alpha\dot{\beta}}z_{67}^{2}\Big]. \tag{4.23}$$

4.5 Eight-Point Functions

At the eight-point level all correlation functions, apart from the pure spin field correlators, have ten or more independent index terms. The expressions become rather lengthy and we therefore do not proceed to higher-point level. Let us discuss the pure spin field eight-point functions first. The expression for eight left-handed spin fields reads

$$\langle S_{\alpha}(z_1)\,S_{\beta}(z_2)\,S_{\gamma}(z_3)\,S_{\delta}(z_4)\,S_{\epsilon}(z_5)\,S_{\zeta}(z_6)\,S_{\theta}(z_7)\,S_{\iota}(z_8)\rangle$$

$$= \left(\frac{z_{12}\,z_{14}\,z_{16}\,z_{18}\,z_{23}\,z_{25}\,z_{27}\,z_{34}\,z_{36}\,z_{38}\,z_{45}\,z_{47}\,z_{56}\,z_{58}\,z_{67}\,z_{78}}{z_{13}\,z_{15}\,z_{17}\,z_{24}\,z_{26}\,z_{28}\,z_{35}\,z_{37}\,z_{46}\,z_{48}\,z_{57}\,z_{68}}\right)^{1/2}$$

$$\times \left(\frac{\varepsilon_{\alpha\beta}\,\varepsilon_{\gamma\delta}\,\varepsilon_{\epsilon\zeta}\,\varepsilon_{\theta\iota}}{z_{12}\,z_{34}\,z_{56}\,z_{78}} - \frac{\varepsilon_{\alpha\beta}\,\varepsilon_{\gamma\delta}\,\varepsilon_{\epsilon\iota}\,\varepsilon_{\theta\zeta}}{z_{12}\,z_{34}\,z_{58}\,z_{76}} + \frac{\varepsilon_{\alpha\beta}\,\varepsilon_{\gamma\zeta}\,\varepsilon_{\epsilon\iota}\,\varepsilon_{\theta\delta}}{z_{12}\,z_{36}\,z_{58}\,z_{74}} - \frac{\varepsilon_{\alpha\beta}\,\varepsilon_{\gamma\zeta}\,\varepsilon_{\epsilon\delta}\,\varepsilon_{\theta\iota}}{z_{12}\,z_{36}\,z_{54}\,z_{78}}\right.$$

$$+ \frac{\varepsilon_{\alpha\beta}\,\varepsilon_{\gamma\iota}\,\varepsilon_{\epsilon\delta}\,\varepsilon_{\theta\zeta}}{z_{12}\,z_{38}\,z_{54}\,z_{76}} - \frac{\varepsilon_{\alpha\beta}\,\varepsilon_{\gamma\iota}\,\varepsilon_{\epsilon\zeta}\,\varepsilon_{\theta\delta}}{z_{12}\,z_{38}\,z_{56}\,z_{74}} - \frac{\varepsilon_{\alpha\delta}\,\varepsilon_{\gamma\beta}\,\varepsilon_{\epsilon\zeta}\,\varepsilon_{\theta\iota}}{z_{14}\,z_{32}\,z_{56}\,z_{78}} + \frac{\varepsilon_{\alpha\delta}\,\varepsilon_{\gamma\beta}\,\varepsilon_{\epsilon\iota}\,\varepsilon_{\theta\zeta}}{z_{14}\,z_{32}\,z_{58}\,z_{76}}$$

$$- \frac{\varepsilon_{\alpha\delta}\,\varepsilon_{\gamma\zeta}\,\varepsilon_{\epsilon\iota}\,\varepsilon_{\theta\beta}}{z_{14}\,z_{36}\,z_{58}\,z_{72}} + \frac{\varepsilon_{\alpha\delta}\,\varepsilon_{\gamma\zeta}\,\varepsilon_{\epsilon\beta}\,\varepsilon_{\theta\iota}}{z_{14}\,z_{36}\,z_{52}\,z_{78}} - \frac{\varepsilon_{\alpha\delta}\,\varepsilon_{\gamma\iota}\,\varepsilon_{\epsilon\beta}\,\varepsilon_{\theta\zeta}}{z_{14}\,z_{38}\,z_{52}\,z_{76}} + \frac{\varepsilon_{\alpha\delta}\,\varepsilon_{\gamma\iota}\,\varepsilon_{\epsilon\zeta}\,\varepsilon_{\theta\beta}}{z_{14}\,z_{38}\,z_{56}\,z_{72}}$$

$$+ \frac{\varepsilon_{\alpha\zeta}\,\varepsilon_{\gamma\beta}\,\varepsilon_{\epsilon\delta}\,\varepsilon_{\theta\iota}}{z_{16}\,z_{32}\,z_{54}\,z_{78}} - \frac{\varepsilon_{\alpha\zeta}\,\varepsilon_{\gamma\beta}\,\varepsilon_{\epsilon\iota}\,\varepsilon_{\theta\delta}}{z_{16}\,z_{32}\,z_{58}\,z_{74}} + \frac{\varepsilon_{\alpha\zeta}\,\varepsilon_{\gamma\delta}\,\varepsilon_{\epsilon\iota}\,\varepsilon_{\theta\beta}}{z_{16}\,z_{34}\,z_{58}\,z_{72}} - \frac{\varepsilon_{\alpha\zeta}\,\varepsilon_{\gamma\delta}\,\varepsilon_{\epsilon\beta}\,\varepsilon_{\theta\iota}}{z_{16}\,z_{34}\,z_{52}\,z_{78}}$$

$$+ \frac{\varepsilon_{\alpha\zeta}\,\varepsilon_{\gamma\iota}\,\varepsilon_{\epsilon\beta}\,\varepsilon_{\theta\delta}}{z_{16}\,z_{38}\,z_{52}\,z_{74}} - \frac{\varepsilon_{\alpha\zeta}\,\varepsilon_{\gamma\iota}\,\varepsilon_{\epsilon\delta}\,\varepsilon_{\theta\beta}}{z_{16}\,z_{38}\,z_{54}\,z_{72}} - \frac{\varepsilon_{\alpha\iota}\,\varepsilon_{\gamma\beta}\,\varepsilon_{\epsilon\delta}\,\varepsilon_{\theta\zeta}}{z_{18}\,z_{32}\,z_{54}\,z_{76}} + \frac{\varepsilon_{\alpha\iota}\,\varepsilon_{\gamma\beta}\,\varepsilon_{\epsilon\zeta}\,\varepsilon_{\theta\delta}}{z_{18}\,z_{32}\,z_{56}\,z_{74}}$$

$$- \frac{\varepsilon_{\alpha\iota}\,\varepsilon_{\gamma\delta}\,\varepsilon_{\epsilon\zeta}\,\varepsilon_{\theta\beta}}{z_{18}\,z_{34}\,z_{56}\,z_{72}} + \frac{\varepsilon_{\alpha\iota}\,\varepsilon_{\gamma\delta}\,\varepsilon_{\epsilon\beta}\,\varepsilon_{\theta\zeta}}{z_{18}\,z_{34}\,z_{52}\,z_{76}} - \frac{\varepsilon_{\alpha\iota}\,\varepsilon_{\gamma\zeta}\,\varepsilon_{\epsilon\beta}\,\varepsilon_{\theta\delta}}{z_{18}\,z_{36}\,z_{52}\,z_{74}} + \frac{\varepsilon_{\alpha\iota}\,\varepsilon_{\gamma\zeta}\,\varepsilon_{\epsilon\delta}\,\varepsilon_{\theta\beta}}{z_{18}\,z_{36}\,z_{54}\,z_{72}}\right), \tag{4.24}$$

4.5 Eight-Point Functions

while the result for six left-handed and two-right handed spin field is given by the product of the respective six- and two-point functions:

$$\left\langle S_\alpha(z_1)\,S_\beta(z_2)\,S_\gamma(z_3)\,S_\delta(z_4)\,S_\epsilon(z_5)\,S_\zeta(z_6)\,S_{\dot\theta}(z_7)\,S_{\dot\iota}(z_8)\right\rangle$$
$$= -\left(\frac{z_{12}\,z_{14}\,z_{16}\,z_{23}\,z_{25}\,z_{34}\,z_{36}\,z_{45}\,z_{56}\,z_{78}}{z_{13}\,z_{15}\,z_{24}\,z_{26}\,z_{35}\,z_{46}}\right)^{1/2}$$
$$\times\left(\frac{\varepsilon_{\alpha\beta}\,\varepsilon_{\gamma\delta}\,\varepsilon_{\epsilon\zeta}\,\varepsilon_{\dot\theta\dot\iota}}{z_{12}\,z_{34}\,z_{56}} - \frac{\varepsilon_{\alpha\beta}\,\varepsilon_{\gamma\zeta}\,\varepsilon_{\epsilon\delta}\,\varepsilon_{\dot\theta\dot\iota}}{z_{12}\,z_{36}\,z_{54}\,z_{78}} + \frac{\varepsilon_{\alpha\delta}\,\varepsilon_{\gamma\zeta}\,\varepsilon_{\epsilon\beta}\,\varepsilon_{\dot\theta\dot\iota}}{z_{14}\,z_{36}\,z_{52}\,z_{78}}\right.$$
$$\left.- \frac{\varepsilon_{\alpha\delta}\,\varepsilon_{\gamma\beta}\,\varepsilon_{\epsilon\zeta}\,\varepsilon_{\dot\theta\dot\iota}}{z_{14}\,z_{32}\,z_{56}\,z_{78}} + \frac{\varepsilon_{\alpha\zeta}\,\varepsilon_{\gamma\beta}\,\varepsilon_{\epsilon\delta}\,\varepsilon_{\dot\theta\dot\iota}}{z_{16}\,z_{32}\,z_{54}\,z_{78}} - \frac{\varepsilon_{\alpha\zeta}\,\varepsilon_{\gamma\delta}\,\varepsilon_{\epsilon\beta}\,\varepsilon_{\dot\theta\dot\iota}}{z_{16}\,z_{34}\,z_{52}\,z_{78}}\right). \tag{4.25}$$

We obtain for the correlator consisting of four spin fields of each chirality

$$\left\langle S_\alpha(z_1)\,S_\beta(z_2)\,S_\gamma(z_3)\,S_\delta(z_4)\,S_{\dot\epsilon}(z_5)\,S_{\dot\zeta}(z_6)\,S_{\dot\theta}(z_7)\,S_{\dot\iota}(z_8)\right\rangle = \left(\frac{z_{12}\,z_{14}\,z_{23}\,z_{34}\,z_{56}\,z_{58}\,z_{67}\,z_{78}}{z_{13}\,z_{24}\,z_{57}\,z_{68}}\right)^{1/2}$$
$$\times\left(\frac{\varepsilon_{\alpha\beta}\,\varepsilon_{\gamma\delta}\,\varepsilon_{\dot\epsilon\dot\zeta}\,\varepsilon_{\dot\theta\dot\iota}}{z_{12}\,z_{34}\,z_{56}\,z_{78}} - \frac{\varepsilon_{\alpha\beta}\,\varepsilon_{\gamma\delta}\varepsilon_{\dot\epsilon\dot\iota}\,\varepsilon_{\dot\zeta\dot\theta}}{z_{12}\,z_{34}\,z_{58}\,z_{67}} - \frac{\varepsilon_{\alpha\delta}\,\varepsilon_{\gamma\beta}\,\varepsilon_{\dot\epsilon\dot\zeta}\,\varepsilon_{\dot\theta\dot\iota}}{z_{14}\,z_{23}\,z_{56}\,z_{78}} + \frac{\varepsilon_{\alpha\delta}\,\varepsilon_{\beta\gamma}\varepsilon_{\dot\epsilon\dot\iota}\,\varepsilon_{\dot\zeta\dot\theta}}{z_{14}\,z_{23}\,z_{58}\,z_{67}}\right), \tag{4.26}$$

which is simply the product of the four-point functions involving only spin fields. Now we concentrate on correlation functions involving also fermions. For the correlator given by two fermions, four left-handed and two right-handed spin fields we obtain:

$$\left\langle \psi^\mu(z_1)\,\psi^\nu(z_2)\,S_\alpha(z_3)\,S_\beta(z_4)\,S_\gamma(z_5)\,S_\delta(z_6)\,S_{\dot\epsilon}(z_7)\,S_{\dot\zeta}(z_8)\right\rangle$$
$$= \frac{1}{(z_{13}\,z_{14}\,z_{15}\,z_{16}\,z_{17}\,z_{18}\,z_{23}\,z_{24}\,z_{25}\,z_{26}\,z_{27}\,z_{28}\,z_{34}\,z_{35}\,z_{36}\,z_{45}\,z_{46}\,z_{56}\,z_{78})^{1/2}}$$
$$\times\left[\frac{\eta^{\mu\nu}}{z_{12}}\left(\varepsilon_{\alpha\beta}\,\varepsilon_{\gamma\delta}\,z_{36}\,z_{45} + \varepsilon_{\alpha\delta}\,\varepsilon_{\gamma\beta}\,z_{34}\,z_{56}\right)\varepsilon_{\dot\epsilon\dot\zeta}\,z_{13}\,z_{15}\,z_{17}\,z_{24}z_{26}\,z_{28}\right.$$
$$+ \frac{1}{2}\left(\varepsilon\,\bar\sigma^\mu\,\sigma^\nu\right)_{\dot\epsilon\dot\zeta}\left(\varepsilon_{\alpha\beta}\,\varepsilon_{\gamma\delta}\,z_{36}\,z_{45} + \varepsilon_{\alpha\delta}\,\varepsilon_{\gamma\beta}\,z_{34}\,z_{56}\right)z_{13}\,z_{15}\,z_{24}\,z_{26}\,z_{78}$$
$$+ \frac{1}{2}\left(\sigma^\mu_{\alpha\dot\epsilon}\,\sigma^\nu_{\beta\dot\zeta}\,z_{18}\,z_{27} - \sigma^\mu_{\alpha\dot\zeta}\,\sigma^\nu_{\beta\dot\epsilon}\,z_{17}\,z_{28}\right)\varepsilon_{\gamma\delta}\,z_{15}\,z_{26}\,z_{34}\,z_{36}\,z_{45}$$
$$- \frac{1}{2}\left(\sigma^\mu_{\gamma\dot\epsilon}\,\sigma^\nu_{\beta\dot\zeta}\,z_{18}\,z_{27} - \sigma^\mu_{\gamma\dot\zeta}\,\sigma^\nu_{\beta\dot\epsilon}\,z_{17}\,z_{28}\right)\varepsilon_{\alpha\delta}\,z_{13}\,z_{26}\,z_{34}\,z_{45}\,z_{56}$$
$$- \frac{1}{2}\left(\sigma^\mu_{\alpha\dot\epsilon}\,\sigma^\nu_{\delta\dot\zeta}\,z_{18}\,z_{27} - \sigma^\mu_{\alpha\dot\zeta}\,\sigma^\nu_{\delta\dot\epsilon}\,z_{17}\,z_{28}\right)\varepsilon_{\beta\gamma}\,z_{15}\,z_{24}\,z_{34}\,z_{36}\,z_{56}$$
$$\left.+ \frac{1}{2}\left(\sigma^\mu_{\gamma\dot\epsilon}\,\sigma^\nu_{\delta\dot\zeta}\,z_{18}\,z_{27} - \sigma^\mu_{\gamma\dot\zeta}\,\sigma^\nu_{\delta\dot\epsilon}\,z_{17}\,z_{28}\right)\varepsilon_{\alpha\beta}\,z_{13}\,z_{24}\,z_{36}\,z_{45}\,z_{56}\right]. \tag{4.27}$$

The spin fields can also be all of the same chirality. In this case the correlator, calculated via the factorization method, is found to be

$$\langle \psi^\mu(z_1)\,\psi^\nu(z_2)\,S_\alpha(z_3)\,S_\beta(z_4)\,S_\gamma(z_5)\,S_\delta(z_6)\,S_\epsilon(z_7)\,S_\zeta(z_8)\rangle = z_{12}^{1/2}\prod_{i<j}^{8} z_{ij}^{-1/2}$$

$$\times \Bigg[\frac{\eta^{\mu\nu}}{z_{12}}\,\varepsilon_{\alpha\beta}\big[\varepsilon_{\gamma\zeta}\,\varepsilon_{\delta\epsilon}\,z_{56}\,z_{78} - \varepsilon_{\gamma\delta}\,\varepsilon_{\epsilon\zeta}\,z_{58}\,z_{67}\big] z_{14}\,z_{16}\,z_{18}\,z_{23}\,z_{25}\,z_{27}\,z_{36}\,z_{38}\,z_{45}\,z_{47}$$

$$+\frac{\eta^{\mu\nu}}{z_{12}}\,\varepsilon_{\alpha\delta}\big[\varepsilon_{\beta\gamma}\,\varepsilon_{\epsilon\zeta}\,z_{47}\,z_{58} - \varepsilon_{\beta\epsilon}\,\varepsilon_{\gamma\zeta}\,z_{45}\,z_{78}\big] z_{14}\,z_{16}\,z_{18}\,z_{23}\,z_{25}\,z_{27}\,z_{34}\,z_{38}\,z_{56}\,z_{67}$$

$$+\frac{\eta^{\mu\nu}}{z_{12}}\,\varepsilon_{\alpha\zeta}\big[\varepsilon_{\beta\epsilon}\,\varepsilon_{\gamma\delta}\,z_{45}\,z_{67} - \varepsilon_{\beta\gamma}\,\varepsilon_{\delta\epsilon}\,z_{47}\,z_{56}\big] z_{14}\,z_{16}\,z_{18}\,z_{23}\,z_{25}\,z_{27}\,z_{34}\,z_{36}\,z_{58}\,z_{78}$$

$$+\frac{1}{2}(\sigma^\mu\bar\sigma^\nu\varepsilon)_{\beta\alpha}\big[\varepsilon_{\gamma\zeta}\,\varepsilon_{\delta\epsilon}\,z_{56}\,z_{78} - \varepsilon_{\gamma\delta}\,\varepsilon_{\epsilon\zeta}\,z_{58}\,z_{67}\big] z_{16}\,z_{18}\,z_{25}\,z_{27}\,z_{34}\,z_{36}\,z_{38}\,z_{45}\,z_{47}$$

$$+\frac{1}{2}(\sigma^\mu\bar\sigma^\nu\varepsilon)_{\beta\gamma}\big[\varepsilon_{\alpha\delta}\,\varepsilon_{\epsilon\zeta}\,z_{38}\,z_{67} - \varepsilon_{\alpha\zeta}\,\varepsilon_{\delta\epsilon}\,z_{36}\,z_{78}\big] z_{16}\,z_{18}\,z_{23}\,z_{27}\,z_{34}\,z_{45}\,z_{47}\,z_{56}\,z_{58}$$

$$+\frac{1}{2}(\sigma^\mu\bar\sigma^\nu\varepsilon)_{\beta\epsilon}\big[\varepsilon_{\alpha\zeta}\,\varepsilon_{\gamma\delta}\,z_{36}\,z_{58} - \varepsilon_{\alpha\delta}\,\varepsilon_{\gamma\zeta}\,z_{38}\,z_{56}\big] z_{16}\,z_{18}\,z_{23}\,z_{25}\,z_{34}\,z_{45}\,z_{47}\,z_{67}\,z_{78}$$

$$+\frac{1}{2}(\sigma^\mu\bar\sigma^\nu\varepsilon)_{\delta\alpha}\big[\varepsilon_{\beta\gamma}\,\varepsilon_{\epsilon\zeta}\,z_{47}\,z_{58} - \varepsilon_{\beta\epsilon}\,\varepsilon_{\gamma\zeta}\,z_{45}\,z_{78}\big] z_{14}\,z_{18}\,z_{25}\,z_{27}\,z_{34}\,z_{36}\,z_{38}\,z_{56}\,z_{67}$$

$$+\frac{1}{2}(\sigma^\mu\bar\sigma^\nu\varepsilon)_{\delta\gamma}\big[\varepsilon_{\alpha\zeta}\,\varepsilon_{\beta\epsilon}\,z_{34}\,z_{78} - \varepsilon_{\alpha\beta}\,\varepsilon_{\epsilon\zeta}\,z_{38}\,z_{47}\big] z_{14}\,z_{18}\,z_{23}\,z_{27}\,z_{36}\,z_{45}\,z_{56}\,z_{58}\,z_{67}$$

$$+\frac{1}{2}(\sigma^\mu\bar\sigma^\nu\varepsilon)_{\delta\epsilon}\big[\varepsilon_{\alpha\beta}\,\varepsilon_{\gamma\zeta}\,z_{38}\,z_{45} - \varepsilon_{\alpha\zeta}\,\varepsilon_{\beta\gamma}\,z_{34}\,z_{58}\big] z_{14}\,z_{18}\,z_{23}\,z_{25}\,z_{36}\,z_{47}\,z_{56}\,z_{67}\,z_{78}$$

$$+\frac{1}{2}(\sigma^\mu\bar\sigma^\nu\varepsilon)_{\zeta\alpha}\big[\varepsilon_{\beta\epsilon}\,\varepsilon_{\gamma\delta}\,z_{45}\,z_{67} - \varepsilon_{\beta\gamma}\,\varepsilon_{\delta\epsilon}\,z_{47}\,z_{56}\big] z_{14}\,z_{16}\,z_{25}\,z_{27}\,z_{34}\,z_{36}\,z_{38}\,z_{58}\,z_{78}$$

$$+\frac{1}{2}(\sigma^\mu\bar\sigma^\nu\varepsilon)_{\zeta\gamma}\big[\varepsilon_{\alpha\beta}\,\varepsilon_{\delta\epsilon}\,z_{36}\,z_{47} - \varepsilon_{\alpha\delta}\,\varepsilon_{\beta\epsilon}\,z_{34}\,z_{67}\big] z_{14}\,z_{16}\,z_{23}\,z_{27}\,z_{38}\,z_{45}\,z_{56}\,z_{58}\,z_{78}$$

$$+\frac{1}{2}(\sigma^\mu\bar\sigma^\nu\varepsilon)_{\zeta\epsilon}\big[\varepsilon_{\alpha\delta}\,\varepsilon_{\beta\gamma}\,z_{34}\,z_{56} - \varepsilon_{\alpha\beta}\,\varepsilon_{\gamma\delta}\,z_{36}\,z_{45}\big] z_{14}\,z_{16}\,z_{23}\,z_{25}\,z_{38}\,z_{47}\,z_{58}\,z_{67}\,z_{78}\Bigg]. \quad (4.28)$$

Next we focus on correlators entering the string scattering amplitude involving two gauge fields and four gauginos, which is discussed in Chapter 6. The eight-point function consisting of four fermions and two spin fields of each chirality is given by:

$$\langle \psi^\mu(z_1)\,\psi^\nu(z_2)\,\psi^\lambda(z_3)\,\psi^\rho(z_4)\,S_\alpha(z_5)\,S_{\dot\beta}(z_6)\,S_\gamma(z_7)\,S_{\dot\delta}(z_8)\rangle$$

$$= \frac{-1}{(z_{15}\,z_{16}\,z_{17}\,z_{18}\,z_{25}\,z_{26}\,z_{27}\,z_{28}\,z_{35}\,z_{36}\,z_{37}\,z_{38}\,z_{45}\,z_{46}\,z_{47}\,z_{48}\,z_{57}\,z_{68})^{1/2}}$$

$$\times \Bigg[\frac{\eta^{\mu\nu}\,\eta^{\lambda\rho}}{z_{12}\,z_{34}}\,\varepsilon_{\alpha\gamma}\,\varepsilon_{\dot\beta\dot\delta}\,z_{15}\,z_{16}\,z_{27}\,z_{28}\,z_{35}\,z_{36}\,z_{47}\,z_{48}$$

$$-\frac{\eta^{\mu\lambda}\,\eta^{\nu\rho}}{z_{13}\,z_{24}}\,\varepsilon_{\alpha\gamma}\,\varepsilon_{\dot\beta\dot\delta}\,z_{15}\,z_{16}\,z_{25}\,z_{26}\,z_{37}\,z_{38}\,z_{47}\,z_{48}$$

$$+\frac{\eta^{\mu\rho}\,\eta^{\nu\lambda}}{z_{14}\,z_{23}}\,\varepsilon_{\alpha\gamma}\,\varepsilon_{\dot\beta\dot\delta}\,z_{15}\,z_{16}\,z_{25}\,z_{26}\,z_{37}\,z_{38}\,z_{47}\,z_{48}$$

4.5 Eight-Point Functions

$$+\frac{1}{2}\frac{\eta^{\mu\nu}}{z_{12}}\left[(\varepsilon\,\bar{\sigma}^{\lambda}\,\sigma^{\rho})_{\dot{\beta}\dot{\delta}}\,\varepsilon_{\alpha\gamma}\,z_{35}\,z_{47}\,z_{68}+(\sigma^{\lambda}\,\bar{\sigma}^{\rho}\,\varepsilon)_{\alpha\gamma}\,\varepsilon_{\dot{\beta}\dot{\delta}}\,z_{36}\,z_{48}\,z_{57}\right]z_{15}\,z_{16}\,z_{27}\,z_{28}$$

$$-\frac{1}{2}\frac{\eta^{\mu\nu}}{z_{12}}\,\sigma^{\lambda}_{\alpha\dot{\beta}}\,\sigma^{\rho}_{\gamma\dot{\delta}}\,z_{15}\,z_{17}\,z_{26}\,z_{28}\,z_{34}\,z_{57}\,z_{68}$$

$$+\frac{1}{2}\frac{\eta^{\lambda\rho}}{z_{34}}\left[(\varepsilon\,\bar{\sigma}^{\mu}\,\sigma^{\nu})_{\dot{\beta}\dot{\delta}}\,\varepsilon_{\alpha\gamma}\,z_{15}\,z_{27}\,z_{68}+(\sigma^{\mu}\,\bar{\sigma}^{\nu}\,\varepsilon)_{\alpha\gamma}\,\varepsilon_{\dot{\beta}\dot{\delta}}\,z_{16}\,z_{28}\,z_{57}\right]z_{35}\,z_{36}\,z_{47}\,z_{48}$$

$$-\frac{1}{2}\frac{\eta^{\lambda\rho}}{z_{34}}\,\sigma^{\mu}_{\alpha\dot{\beta}}\,\sigma^{\nu}_{\gamma\dot{\delta}}\,z_{12}\,z_{35}\,z_{37}\,z_{46}\,z_{48}\,z_{57}\,z_{68}$$

$$-\frac{1}{2}\frac{\eta^{\mu\lambda}}{z_{13}}\left[(\varepsilon\,\bar{\sigma}^{\nu}\,\sigma^{\rho})_{\dot{\beta}\dot{\delta}}\,\varepsilon_{\alpha\gamma}\,z_{25}\,z_{47}\,z_{68}+(\sigma^{\nu}\,\bar{\sigma}^{\rho}\,\varepsilon)_{\alpha\gamma}\,\varepsilon_{\dot{\beta}\dot{\delta}}\,z_{26}\,z_{48}\,z_{57}\right]z_{15}\,z_{16}\,z_{37}\,z_{38}$$

$$+\frac{1}{2}\frac{\eta^{\mu\lambda}}{z_{13}}\,\sigma^{\nu}_{\alpha\dot{\beta}}\,\sigma^{\rho}_{\gamma\dot{\delta}}\,z_{15}\,z_{17}\,z_{24}\,z_{36}\,z_{38}\,z_{57}\,z_{68}$$

$$-\frac{1}{2}\frac{\eta^{\nu\rho}}{z_{24}}\left[(\varepsilon\,\bar{\sigma}^{\mu}\,\sigma^{\lambda})_{\dot{\beta}\dot{\delta}}\,\varepsilon_{\alpha\gamma}\,z_{15}\,z_{37}\,z_{68}+(\sigma^{\mu}\,\bar{\sigma}^{\lambda}\,\varepsilon)_{\alpha\gamma}\,\varepsilon_{\dot{\beta}\dot{\delta}}\,z_{16}\,z_{38}\,z_{57}\right]z_{25}\,z_{26}\,z_{47}\,z_{48}$$

$$+\frac{1}{2}\frac{\eta^{\nu\rho}}{z_{24}}\,\sigma^{\mu}_{\alpha\dot{\beta}}\,\sigma^{\lambda}_{\gamma\dot{\delta}}\,z_{13}\,z_{25}\,z_{27}\,z_{46}\,z_{48}\,z_{57}\,z_{68}$$

$$+\frac{1}{2}\frac{\eta^{\mu\rho}}{z_{14}}\left[(\varepsilon\,\bar{\sigma}^{\nu}\,\sigma^{\lambda})_{\dot{\beta}\dot{\delta}}\,\varepsilon_{\alpha\gamma}\,z_{25}\,z_{37}\,z_{68}+(\sigma^{\nu}\,\bar{\sigma}^{\lambda}\,\varepsilon)_{\alpha\gamma}\,\varepsilon_{\dot{\beta}\dot{\delta}}\,z_{26}\,z_{38}\,z_{57}\right]z_{15}\,z_{16}\,z_{47}\,z_{48}$$

$$-\frac{1}{2}\frac{\eta^{\mu\rho}}{z_{14}}\,\sigma^{\nu}_{\alpha\dot{\beta}}\,\sigma^{\lambda}_{\gamma\dot{\delta}}\,z_{15}\,z_{17}\,z_{23}\,z_{46}\,z_{48}\,z_{57}\,z_{68}$$

$$+\frac{1}{2}\frac{\eta^{\nu\lambda}}{z_{23}}\left[(\varepsilon\,\bar{\sigma}^{\mu}\,\sigma^{\rho})_{\dot{\beta}\dot{\delta}}\,\varepsilon_{\alpha\gamma}\,z_{15}\,z_{47}\,z_{68}+(\sigma^{\mu}\,\bar{\sigma}^{\rho}\,\varepsilon)_{\alpha\gamma}\,\varepsilon_{\dot{\beta}\dot{\delta}}\,z_{16}\,z_{48}\,z_{57}\right]z_{25}\,z_{26}\,z_{37}\,z_{38}$$

$$-\frac{1}{2}\frac{\eta^{\nu\lambda}}{z_{23}}\,\sigma^{\mu}_{\alpha\dot{\beta}}\,\sigma^{\rho}_{\gamma\dot{\delta}}\,z_{14}\,z_{25}\,z_{27}\,z_{36}\,z_{38}\,z_{57}\,z_{68}$$

$$+\frac{1}{2}\left[\eta^{\nu\rho}\,(\varepsilon\,\bar{\sigma}^{\mu}\,\sigma^{\lambda})_{\dot{\beta}\dot{\delta}}\,\varepsilon_{\alpha\gamma}-\eta^{\nu\lambda}\,(\varepsilon\,\bar{\sigma}^{\mu}\,\sigma^{\rho})_{\dot{\beta}\dot{\delta}}\,\varepsilon_{\alpha\gamma}+\eta^{\mu\lambda}\,\sigma^{\nu}_{\alpha\dot{\beta}}\,\sigma^{\rho}_{\gamma\dot{\delta}}-\eta^{\mu\rho}\,\sigma^{\nu}_{\alpha\dot{\beta}}\,\sigma^{\lambda}_{\gamma\dot{\delta}}\right]$$
$$\times\,z_{15}\,z_{23}\,z_{48}\,z_{57}\,z_{67}\,z_{68}$$

$$+\frac{1}{4}(\varepsilon\,\bar{\sigma}^{\mu}\,\sigma^{\nu}\,\bar{\sigma}^{\lambda}\,\sigma^{\rho})_{\dot{\beta}\dot{\delta}}\,\varepsilon_{\alpha\gamma}\,z_{68}^{2}\,z_{15}\,z_{27}\,z_{35}\,z_{47}+\frac{1}{4}(\sigma^{\mu}\,\bar{\sigma}^{\nu}\,\sigma^{\lambda}\,\bar{\sigma}^{\rho}\,\varepsilon)_{\alpha\gamma}\,\varepsilon_{\dot{\beta}\dot{\delta}}\,z_{57}^{2}\,z_{16}\,z_{28}\,z_{36}\,z_{48}$$

$$+\frac{1}{4}\sigma^{\mu}_{\alpha\dot{\beta}}\,(\sigma^{\nu}\,\bar{\sigma}^{\lambda}\,\sigma^{\rho})_{\gamma\dot{\delta}}\left[z_{15}\,z_{23}\,z_{46}\,z_{78}-z_{14}\,z_{25}\,z_{36}\,z_{78}+z_{16}\,z_{25}\,z_{37}\,z_{48}\right]z_{57}\,z_{68}$$

$$+\frac{1}{4}\sigma^{\rho}_{\gamma\dot{\delta}}\,(\sigma^{\mu}\,\bar{\sigma}^{\nu}\,\sigma^{\lambda})_{\alpha\dot{\beta}}\left[z_{15}\,z_{28}\,z_{36}\,z_{47}-z_{14}\,z_{28}\,z_{37}\,z_{56}\right]z_{57}\,z_{68}$$

$$-\frac{1}{4}\sigma^{\mu}_{\gamma\dot{\beta}}\,(\sigma^{\nu}\,\bar{\sigma}^{\lambda}\,\sigma^{\rho})_{\alpha\dot{\delta}}\left[z_{15}\,z_{28}\,z_{36}\,z_{47}+z_{15}\,z_{23}\,z_{46}\,z_{78}\right]z_{57}\,z_{68}$$

$$-\frac{1}{4}\sigma^{\nu}_{\gamma\dot{\delta}}\,(\sigma^{\lambda}\,\bar{\sigma}^{\rho}\,\sigma^{\mu})_{\alpha\dot{\beta}}\,z_{15}\,z_{23}\,z_{48}\,z_{57}\,z_{67}\,z_{68}+\frac{1}{4}\sigma^{\lambda}_{\gamma\dot{\beta}}\,(\sigma^{\mu}\,\bar{\sigma}^{\nu}\,\sigma^{\rho})_{\alpha\dot{\delta}}\,z_{16}\,z_{23}\,z_{47}\,z_{58}\,z_{57}\,z_{68}$$

$$\left.-\frac{1}{4}\sigma^{\rho}_{\gamma\dot{\beta}}\,(\sigma^{\mu}\,\bar{\sigma}^{\nu}\,\sigma^{\lambda})_{\alpha\dot{\delta}}\,z_{16}\,z_{25}\,z_{37}\,z_{48}\,z_{57}\,z_{68}\right]. \tag{4.29}$$

On the other hand, for the case where all spin fields are left-handed we compute:

$$\left\langle \psi^\mu(z_1)\, \psi^\nu(z_2)\, \psi^\lambda(z_3)\, \psi^\rho(z_4)\, S_\alpha(z_5)\, S_\beta(z_6)\, S_\gamma(z_7)\, S_\delta(z_8) \right\rangle$$
$$= \frac{1}{(z_{15}\, z_{16}\, z_{17}\, z_{18}\, z_{25}\, z_{26}\, z_{27}\, z_{28}\, z_{35}\, z_{36}\, z_{37}\, z_{38}\, z_{45}\, z_{46}\, z_{47}\, z_{48}\, z_{56}\, z_{57}\, z_{58}\, z_{67}\, z_{68}\, z_{78})^{1/2}}$$
$$\times \Bigg[\frac{\eta^{\mu\nu}\, \eta^{\lambda\rho}}{z_{12}\, z_{34}} \left(\varepsilon_{\alpha\beta}\, \varepsilon_{\gamma\delta}\, z_{58}\, z_{67} + \varepsilon_{\alpha\delta}\, \varepsilon_{\gamma\beta}\, z_{56}\, z_{78} \right) z_{15}\, z_{17}\, z_{26}\, z_{28}\, z_{35}\, z_{37}\, z_{46}\, z_{48}$$
$$- \frac{\eta^{\mu\lambda}\, \eta^{\nu\rho}}{z_{13}\, z_{24}} \left(\varepsilon_{\alpha\beta}\, \varepsilon_{\gamma\delta}\, z_{58}\, z_{67} + \varepsilon_{\alpha\delta}\, \varepsilon_{\gamma\beta}\, z_{56}\, z_{78} \right) z_{15}\, z_{17}\, z_{25}\, z_{27}\, z_{36}\, z_{38}\, z_{46}\, z_{48}$$
$$+ \frac{\eta^{\mu\rho}\, \eta^{\nu\lambda}}{z_{14}\, z_{23}} \left(\varepsilon_{\alpha\beta}\, \varepsilon_{\gamma\delta}\, z_{58}\, z_{67} + \varepsilon_{\alpha\delta}\, \varepsilon_{\gamma\beta}\, z_{56}\, z_{78} \right) z_{15}\, z_{17}\, z_{25}\, z_{27}\, z_{36}\, z_{38}\, z_{46}\, z_{48}$$
$$+ \frac{1}{2} \frac{\eta^{\mu\nu}}{z_{12}} \left[(\sigma^\lambda\, \bar\sigma^\rho\, \varepsilon)_{\alpha\beta}\, \varepsilon_{\gamma\delta}\, z_{37}\, z_{48}\, z_{56} + (\sigma^\lambda\, \bar\sigma^\rho\, \varepsilon)_{\gamma\delta}\, \varepsilon_{\alpha\beta}\, z_{35}\, z_{46}\, z_{78} \right] z_{15}\, z_{17}\, z_{26}\, z_{28}\, z_{58}\, z_{67}$$
$$- \frac{1}{2} \frac{\eta^{\mu\nu}}{z_{12}} \left[(\sigma^\lambda\, \bar\sigma^\rho\, \varepsilon)_{\alpha\delta}\, \varepsilon_{\beta\gamma}\, z_{37}\, z_{46}\, z_{58} + (\sigma^\lambda\, \bar\sigma^\rho\, \varepsilon)_{\gamma\beta}\, \varepsilon_{\alpha\delta}\, z_{35}\, z_{48}\, z_{67} \right] z_{15}\, z_{17}\, z_{26}\, z_{28}\, z_{56}\, z_{78}$$
$$+ \frac{1}{2} \frac{\eta^{\lambda\rho}}{z_{34}} \left[(\sigma^\mu\, \bar\sigma^\nu\, \varepsilon)_{\alpha\beta}\, \varepsilon_{\gamma\delta}\, z_{17}\, z_{28}\, z_{56} + (\sigma^\mu\, \bar\sigma^\nu\, \varepsilon)_{\gamma\delta}\, \varepsilon_{\alpha\beta}\, z_{15}\, z_{26}\, z_{78} \right] z_{35}\, z_{37}\, z_{46}\, z_{48}\, z_{58}\, z_{67}$$
$$- \frac{1}{2} \frac{\eta^{\lambda\rho}}{z_{34}} \left[(\sigma^\mu\, \bar\sigma^\nu\, \varepsilon)_{\alpha\delta}\, \varepsilon_{\beta\gamma}\, z_{17}\, z_{26}\, z_{58} + (\sigma^\mu\, \bar\sigma^\nu\, \varepsilon)_{\gamma\beta}\, \varepsilon_{\alpha\delta}\, z_{15}\, z_{28}\, z_{67} \right] z_{35}\, z_{37}\, z_{46}\, z_{48}\, z_{56}\, z_{78}$$
$$- \frac{1}{2} \frac{\eta^{\mu\lambda}}{z_{13}} \left[(\sigma^\nu\, \bar\sigma^\rho\, \varepsilon)_{\alpha\beta}\, \varepsilon_{\gamma\delta}\, z_{27}\, z_{48}\, z_{56} + (\sigma^\nu\, \bar\sigma^\rho\, \varepsilon)_{\gamma\delta}\, \varepsilon_{\alpha\beta}\, z_{25}\, z_{46}\, z_{78} \right] z_{15}\, z_{17}\, z_{36}\, z_{38}\, z_{58}\, z_{67}$$
$$+ \frac{1}{2} \frac{\eta^{\mu\lambda}}{z_{13}} \left[(\sigma^\nu\, \bar\sigma^\rho\, \varepsilon)_{\alpha\delta}\, \varepsilon_{\beta\gamma}\, z_{27}\, z_{46}\, z_{58} + (\sigma^\nu\, \bar\sigma^\rho\, \varepsilon)_{\gamma\beta}\, \varepsilon_{\alpha\delta}\, z_{25}\, z_{48}\, z_{67} \right] z_{15}\, z_{17}\, z_{36}\, z_{38}\, z_{56}\, z_{78}$$
$$- \frac{1}{2} \frac{\eta^{\nu\rho}}{z_{24}} \left[(\sigma^\mu\, \bar\sigma^\lambda\, \varepsilon)_{\alpha\beta}\, \varepsilon_{\gamma\delta}\, z_{17}\, z_{38}\, z_{56} + (\sigma^\mu\, \bar\sigma^\lambda\, \varepsilon)_{\gamma\delta}\, \varepsilon_{\alpha\beta}\, z_{15}\, z_{36}\, z_{78} \right] z_{25}\, z_{27}\, z_{46}\, z_{48}\, z_{58}\, z_{67}$$
$$+ \frac{1}{2} \frac{\eta^{\nu\rho}}{z_{24}} \left[(\sigma^\mu\, \bar\sigma^\lambda\, \varepsilon)_{\alpha\delta}\, \varepsilon_{\beta\gamma}\, z_{17}\, z_{36}\, z_{58} + (\sigma^\mu\, \bar\sigma^\lambda\, \varepsilon)_{\gamma\beta}\, \varepsilon_{\alpha\delta}\, z_{15}\, z_{38}\, z_{67} \right] z_{25}\, z_{27}\, z_{46}\, z_{48}\, z_{56}\, z_{78}$$
$$+ \frac{1}{2} \frac{\eta^{\mu\rho}}{z_{14}} \left[(\sigma^\nu\, \bar\sigma^\lambda\, \varepsilon)_{\alpha\beta}\, \varepsilon_{\gamma\delta}\, z_{27}\, z_{38}\, z_{56} + (\sigma^\nu\, \bar\sigma^\lambda\, \varepsilon)_{\gamma\delta}\, \varepsilon_{\alpha\beta}\, z_{25}\, z_{36}\, z_{78} \right] z_{15}\, z_{17}\, z_{46}\, z_{48}\, z_{58}\, z_{67}$$
$$- \frac{1}{2} \frac{\eta^{\mu\rho}}{z_{14}} \left[(\sigma^\nu\, \bar\sigma^\lambda\, \varepsilon)_{\alpha\delta}\, \varepsilon_{\beta\gamma}\, z_{27}\, z_{36}\, z_{58} + (\sigma^\nu\, \bar\sigma^\lambda\, \varepsilon)_{\gamma\beta}\, \varepsilon_{\alpha\delta}\, z_{25}\, z_{38}\, z_{67} \right] z_{15}\, z_{17}\, z_{46}\, z_{48}\, z_{56}\, z_{78}$$
$$+ \frac{1}{2} \frac{\eta^{\nu\lambda}}{z_{23}} \left[(\sigma^\mu\, \bar\sigma^\rho\, \varepsilon)_{\alpha\beta}\, \varepsilon_{\gamma\delta}\, z_{17}\, z_{48}\, z_{56} + (\sigma^\mu\, \bar\sigma^\rho\, \varepsilon)_{\gamma\delta}\, \varepsilon_{\alpha\beta}\, z_{15}\, z_{46}\, z_{78} \right] z_{25}\, z_{27}\, z_{36}\, z_{38}\, z_{58}\, z_{67}$$
$$- \frac{1}{2} \frac{\eta^{\nu\lambda}}{z_{23}} \left[(\sigma^\mu\, \bar\sigma^\rho\, \varepsilon)_{\alpha\delta}\, \varepsilon_{\beta\gamma}\, z_{17}\, z_{46}\, z_{58} + (\sigma^\mu\, \bar\sigma^\rho\, \varepsilon)_{\gamma\beta}\, \varepsilon_{\alpha\delta}\, z_{15}\, z_{48}\, z_{67} \right] z_{25}\, z_{27}\, z_{36}\, z_{38}\, z_{56}\, z_{78}$$
$$+ \frac{1}{4} (\sigma^\mu\, \bar\sigma^\nu\, \sigma^\lambda\, \bar\sigma^\rho\, \varepsilon)_{\alpha\beta}\, \varepsilon_{\gamma\delta}\, z_{56}^2\, z_{17}\, z_{28}\, z_{37}\, z_{48}\, z_{58}\, z_{67}$$
$$- \frac{1}{4} (\sigma^\mu\, \bar\sigma^\nu\, \sigma^\lambda\, \bar\sigma^\rho\, \varepsilon)_{\alpha\delta}\, \varepsilon_{\beta\gamma}\, z_{58}^2\, z_{17}\, z_{26}\, z_{37}\, z_{46}\, z_{56}\, z_{78}$$
$$+ \frac{1}{4} (\sigma^\mu\, \bar\sigma^\nu\, \sigma^\lambda\, \bar\sigma^\rho\, \varepsilon)_{\gamma\delta}\, \varepsilon_{\alpha\beta}\, z_{78}^2\, z_{15}\, z_{26}\, z_{35}\, z_{46}\, z_{58}\, z_{67}$$
$$+ \frac{1}{4} (\sigma^\mu\, \bar\sigma^\nu\, \sigma^\lambda\, \bar\sigma^\rho\, \varepsilon)_{\gamma\beta}\, \varepsilon_{\alpha\delta}\, z_{67}^2\, z_{15}\, z_{28}\, z_{35}\, z_{48}\, z_{56}\, z_{78}$$

$$+ \frac{1}{4} \left(\sigma^\mu \bar{\sigma}^\nu \varepsilon\right)_{\alpha\beta} \left(\sigma^\lambda \bar{\sigma}^\rho \varepsilon\right)_{\gamma\delta} \left(z_{17} z_{28} z_{35} z_{46} - z_{17} z_{25} z_{38} z_{46}\right) z_{56} z_{58} z_{67} z_{78}$$

$$- \frac{1}{4} \left(\sigma^\mu \bar{\sigma}^\nu \varepsilon\right)_{\alpha\delta} \left(\sigma^\lambda \bar{\sigma}^\rho \varepsilon\right)_{\gamma\beta} \left(z_{17} z_{26} z_{35} z_{48} - z_{17} z_{25} z_{36} z_{48}\right) z_{56} z_{58} z_{67} z_{78}$$

$$+ \frac{1}{4} \left(\sigma^\mu \bar{\sigma}^\nu \varepsilon\right)_{\gamma\delta} \left(\sigma^\lambda \bar{\sigma}^\rho \varepsilon\right)_{\alpha\beta} \left(z_{15} z_{26} z_{37} z_{48} - z_{15} z_{27} z_{36} z_{48}\right) z_{56} z_{58} z_{67} z_{78}$$

$$- \frac{1}{4} \left(\sigma^\mu \bar{\sigma}^\nu \varepsilon\right)_{\gamma\beta} \left(\sigma^\lambda \bar{\sigma}^\rho \varepsilon\right)_{\alpha\delta} \left(z_{15} z_{28} z_{37} z_{46} - z_{15} z_{27} z_{38} z_{46}\right) z_{56} z_{58} z_{67} z_{78}$$

$$+ \frac{1}{4} \left[\left(\sigma^\nu \bar{\sigma}^\lambda \varepsilon\right)_{\alpha\beta} \left(\sigma^\mu \bar{\sigma}^\rho \varepsilon\right)_{\gamma\delta} z_{38} z_{46} - \left(\sigma^\nu \bar{\sigma}^\lambda \varepsilon\right)_{\alpha\delta} \left(\sigma^\mu \bar{\sigma}^\rho \varepsilon\right)_{\gamma\beta} z_{36} z_{48}\right]$$
$$\times z_{15} z_{27} z_{56} z_{58} z_{67} z_{78}$$

$$+ \frac{1}{4} \left[\left(\sigma^\nu \bar{\sigma}^\lambda \varepsilon\right)_{\gamma\delta} \left(\sigma^\mu \bar{\sigma}^\rho \varepsilon\right)_{\alpha\beta} z_{36} z_{48} - \left(\sigma^\nu \bar{\sigma}^\lambda \varepsilon\right)_{\gamma\beta} \left(\sigma^\mu \bar{\sigma}^\rho \varepsilon\right)_{\alpha\delta} z_{38} z_{46}\right]$$
$$\left. \times z_{17} z_{25} z_{56} z_{58} z_{67} z_{78} \right]. \tag{4.30}$$

4.6 General Results

We come now to the class of tree-level correlators in four dimensions consisting of arbitrary many fermions and two spin fields,

$$\Omega_n \equiv \left\langle \psi^{\mu_1}(z_1) \ldots \psi^{\mu_{2n-1}}(z_{2n-1}) S_\alpha(z_A) S^{\dot{\beta}}(z_B) \right\rangle,$$
$$\omega_n \equiv \left\langle \psi^{\mu_1}(z_1) \ldots \psi^{\mu_{2n-2}}(z_{2n-2}) S_\alpha(z_A) S^\beta(z_B) \right\rangle, \tag{4.31}$$

for which some lower-point examples have just been presented. It is possible to derive results for these correlators for arbitrary $n \in \mathbb{N}$. Equation (3.65) implies that these correlators vanish if the numbers of NS fermions are chosen differently than stated above, namely odd for ω_n or even for Ω_n, because then no scalar representations exist. The correlation functions (4.31) enter the calculation of four-dimensional string scattering amplitudes with an arbitrary number of gluons but only two fermions or scalars. Such amplitudes are universal to any string compactification. The mathematical reason for this lies in the fact that only the fermion or scalar vertex operators (2.57) give rise to internal fields Σ or Ψ in the calculation. The arising two-point function is however completely determined by the conformal weights of the internal fields and therefore the scattering amplitude does not depend on the compactification details like geometry and topology. From a phenomenological point of view only Regge modes but not the KK/winding modes contribute to the scattering process. The RNS correlators (4.31) hence play an important role for testing perturbative string theory with a low string scale in collider experiments.

The results for the correlators (4.31) have been stated above for the cases $n = 2, 3$ in (4.3a) and (4.17) for ω_n and in (4.4) and (4.23) for Ω_n. They have striking similarities in their structure. The prefactor consists in all cases of the terms $(z_{iA} z_{iB})^{-1/2}$ stemming from the contractions between all fermions ψ^{μ_i} and the two spin fields. Additionally, every Minkowski metric $\eta^{\mu_i \mu_j}$ is accompanied by the coefficient $z_{iA} z_{iB}/z_{AB}$. In contrast, each σ-chain $\sigma^{\mu_i} \bar{\sigma}^{\mu_j}$ implies the factor $z_{AB}/2$. The sign of the separate terms depend on the ordering of the Lorentz indices, whether they appear as odd or even permutation of $\mu_1 \ldots \mu_{2n-1}$ or $\mu_1 \ldots \mu_{2n-2}$ respectively.

These properties lead us to claim the following expression for the correlator Ω_n:

$$\Omega_n = \frac{1}{\sqrt{2}} \prod_{i=1}^{2n-1} (z_{iA} z_{iB})^{-1/2} \sum_{l=0}^{n-1} \left(\frac{z_{AB}}{2}\right)^l \sum_{\rho \in S_{2n-1}/\mathcal{P}_{n,l}} \text{sgn}(\rho) \left(\sigma^{\mu_{\rho(1)}} \bar{\sigma}^{\mu_{\rho(2)}} \ldots \bar{\sigma}^{\mu_{\rho(2l)}} \sigma^{\mu_{\rho(2l+1)}}\right)_{\alpha \dot{\beta}}$$
$$\times \prod_{j=1}^{n-l-1} \frac{\eta^{\mu_{\rho(2l+2j)} \mu_{\rho(2l+2j+1)}}}{z_{\rho(2l+2j), \rho(2l+2j+1)}} z_{\rho(2l+2j), A} z_{\rho(2l+2j+1), B} \,. \tag{4.32}$$

We postulate that the other correlator ω_n is given by

$$\omega_n = \frac{-1}{z_{AB}^{1/2}} \prod_{i=1}^{2n-2} (z_{iA} z_{iB})^{-1/2} \sum_{l=0}^{n-1} \left(\frac{z_{AB}}{2}\right)^l \sum_{\rho \in S_{2n-2}/\mathcal{Q}_{n,l}} \text{sgn}(\rho) \left(\sigma^{\mu_{\rho(1)}} \bar{\sigma}^{\mu_{\rho(2)}} \ldots \sigma^{\mu_{\rho(2l-1)}} \bar{\sigma}^{\mu_{\rho(2l)}}\right)_\alpha^\beta$$
$$\times \prod_{j=1}^{n-l-1} \frac{\eta^{\mu_{\rho(2l+2j-1)} \mu_{\rho(2l+2j)}}}{z_{\rho(2l+2j-1), \rho(2l+2j)}} z_{\rho(2l+2j-1), A} z_{\rho(2l+2j), B} \,, \tag{4.33}$$

while its complex conjugate $\bar{\omega}$ with two right-handed spin fields reads

$$\bar{\omega}_n = \frac{1}{z_{AB}^{1/2}} \prod_{i=1}^{2n-2} (z_{iA} z_{iB})^{-1/2} \sum_{l=0}^{n-1} \left(\frac{z_{AB}}{2}\right)^l \sum_{\rho \in S_{2n-2}/\mathcal{Q}_{n,l}} \text{sgn}(\rho) \left(\bar{\sigma}^{\mu_{\rho(1)}} \sigma^{\mu_{\rho(2)}} \ldots \bar{\sigma}^{\mu_{\rho(2l-1)}} \sigma^{\mu_{\rho(2l)}}\right)^{\dot{\alpha}}_{\dot{\beta}}$$
$$\times \prod_{j=1}^{n-l-1} \frac{\eta^{\mu_{\rho(2l+2j-1)} \mu_{\rho(2l+2j)}}}{z_{\rho(2l+2j-1) \rho(2l+2j)}} z_{\rho(2l+2j-1), A} z_{\rho(2l+2j), B} \,. \tag{4.34}$$

These expressions can be proven using induction. We describe the idea of the proof, for a detailed account the reader should have a look at Appendix C of [1]. As in the proof of the pure spin field correlator (3.82) the expressions (4.32)-(4.34) must reduce to lower-point correlators if any two fields in Ω_n and ω_n are replaced by their OPE. As an example, consider the limit $z_{2n-1} \to z_B$ in Ω_n. Using the OPE (3.9b) it reduces to ω_n and hence the expression (4.32) in this limit has to reduce to (4.33). Examining all other possible limits gives rise to the web of limits illustrated in Figure 4.1.

The permutation set $S_{2n-1}/\mathcal{P}_{n,l}$ and $S_{2n-2}/\mathcal{Q}_{n,l}$ in (4.32)-(4.34) require some explanation. We only consider permutations ρ such that the vector indices attached to η's or σ-chains appear in ascending order, i.e. $\rho(i) < \rho(j)$ for $\eta^{\mu_{\rho(i)} \mu_{\rho(j)}}$ and $\sigma^{\mu_{\rho(i)}} \bar{\sigma}^{\mu_{\rho(j)}}$. Furthermore, products of η's are not counted several times. Once we get $\eta^{\mu_{\rho(i)} \mu_{\rho(j)}} \eta^{\mu_{\rho(k)} \mu_{\rho(l)}}$, the term $\eta^{\mu_{\rho(k)} \mu_{\rho(l)}} \eta^{\mu_{\rho(i)} \mu_{\rho(j)}}$ is not allowed to appear. These restrictions of S_{2n-1} and S_{2n-2} are summarized by the quotients $\mathcal{P}_{n,l}$

4.6 General Results

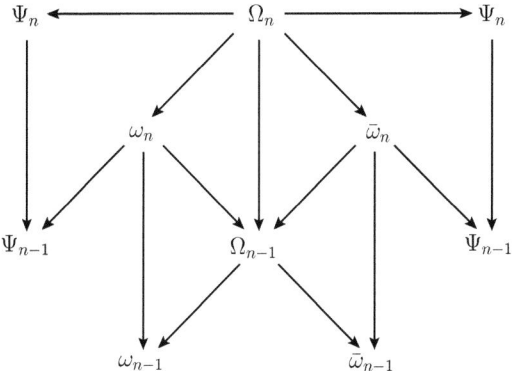

Figure 4.1: The web of limits for the correlators Ω_n and ω_n. Due to the OPEs in four dimensions the correlators reduce to lower-point correlation functions in the respective limits as indicated. Ψ_n denotes the $2n$-point function involving only NS fermions.

and $\mathcal{Q}_{n,l}$. Let us give a more formal definition:

$$S_{2n-1}/\mathcal{P}_{n,l} \equiv \big\{\rho \in S_{2n-1} : \rho(1) < \rho(2) < \ldots < \rho(2l+1),$$
$$\rho(2l+2j) < \rho(2l+2j+1) \ \forall \ j = 1, 2, \ldots, n-l-1,$$
$$\rho(2l+3) < \rho(2l+5) < \ldots < \rho(2n-1)\big\}, \tag{4.35a}$$

$$S_{2n-2}/\mathcal{Q}_{n,l} \equiv \big\{\rho \in S_{2n-2} : \rho(1) < \rho(2) < \ldots < \rho(2l),$$
$$\rho(2l+2j-1) < \rho(2l+2j) \ \forall \ j = 1, 2, \ldots, n-l-1,$$
$$\rho(2l+2) < \rho(2l+4) < \ldots < \rho(2n-2)\big\}. \tag{4.35b}$$

So the groups of permutations which are removed from S_{2n-1}, S_{2n-2} are as follows:

$$\mathcal{P}_{n,l} \leftrightarrow \begin{cases} S_{2l+1} & : \text{permute the } (2l+1) \text{ matrices } \sigma^{\mu_i}, \\ S_{n-l-1} & : \text{permute the } n-l-1 \text{ Minkowski metrics}, \\ (S_2)^{n-l-1} & : \text{exchange the indices of one of the } \eta\text{'s}, \end{cases} \tag{4.36a}$$

$$\mathcal{Q}_{n,l} \leftrightarrow \begin{cases} S_{2l} & : \text{permute the } (2l) \text{ matrices } \sigma^{\mu_i}, \\ S_{n-l-1} & : \text{permute the } n-l-1 \text{ Minkowski metrics}, \\ (S_2)^{n-l-1} & : \text{exchange the indices of one of the } \eta\text{'s}. \end{cases} \tag{4.36b}$$

Since the permutation group S_N has $N!$ elements, one can conclude from (4.36) how many terms

n	l	$\|S_{2n-2}/Q_{n,l}\|$	terms	$\|S_{2n-1}/P_{n,l}\|$	terms
1	0	$\frac{0!}{0!0!2^0} = 1$	δ	$\frac{1!}{1!0!2^0} = 1$	σ^μ
2	0	$\frac{2!}{0!1!2^1} = 1$	$\eta^{\mu\nu}$	$\frac{3!}{1!1!2^1} = 3$	$\eta^{\mu\nu}\sigma^\lambda$
	1	$\frac{2!}{2!0!2^0} = 1$	$\sigma^\mu\bar{\sigma}^\nu$	$\frac{3!}{3!0!2^0} = 1$	$\sigma^\mu\bar{\sigma}^\nu\sigma^\lambda$
3	0	$\frac{4!}{0!2!2^2} = 3$	$\eta^{\mu\nu}\eta^{\lambda\rho}$	$\frac{5!}{1!2!2^2} = 15$	$\eta^{\mu\nu}\eta^{\lambda\rho}\sigma^\tau$
	1	$\frac{4!}{2!1!2^1} = 6$	$\eta^{\mu\nu}\sigma^\lambda\bar{\sigma}^\rho$	$\frac{5!}{3!1!2^1} = 10$	$\eta^{\mu\nu}\sigma^\lambda\bar{\sigma}^\rho\sigma^\tau$
	2	$\frac{4!}{4!0!2^0} = 1$	$\sigma^\mu\bar{\sigma}^\nu\sigma^\lambda\bar{\sigma}^\rho$	$\frac{5!}{5!0!2^0} = 1$	$\sigma^\mu\bar{\sigma}^\nu\sigma^\lambda\bar{\sigma}^\rho\sigma^\tau$

Table 4.1: Number of index terms of the correlators Ω_n and ω_n.

remain in the sums over ρ in (4.32)-(4.34):

$$\left|S_{2n-1}/P_{n,l}\right| = \frac{(2n-1)!}{(2l+1)!\,(n-l-1)!\,2^{n-l-1}}, \tag{4.37a}$$

$$\left|S_{2n-2}/Q_{n,l}\right| = \frac{(2n-2)!}{(2l)!\,(n-l-1)!\,2^{n-l-1}}. \tag{4.37b}$$

Summing up these numbers for all possible values of l yields the number of index terms that appear in the results for Ω_n and ω_n. These numbers up to $n = 3$ which we have collected in Table 4.1 coincide precisely with the numbers of Clebsch–Gordan coefficients in the correlators (4.3a), (4.4), (4.17) and (4.23). However, this does not mean that all these index terms are independent. In contrast, one of the 26 terms appearing in Ω_3 can be eliminated by $\sigma^{[\mu}\bar{\sigma}^\nu\sigma^\lambda\bar{\sigma}^\rho\sigma^{\tau]} = 0$, namely the vanishing of the antisymmetric expression in more than four vector indices. However, Ω_3 in terms of a minimal set of 25 index terms would have a more complicated z dependence. From this more complicated result it would be impossible to generalize to Ω_n with arbitrary many NS fermions.

The expression for the eight point function ω_4 due to (4.33) contains 76 terms, but a group theoretic analysis determines the number of scalar representations in the tensor product to be 70, as shown in Table 3.3. This difference is explained by the six independent reduction identities $\sigma^{[\mu}\bar{\sigma}^\nu\sigma^\lambda\bar{\sigma}^\rho\sigma^\tau\bar{\sigma}^{\xi]} = 0$ and $\eta^{\mu[\nu}\varepsilon^{\lambda\rho\tau\xi]} = 0$. Similarly, for higher point examples $\Omega_{(n\geq 4)}$ and $\omega_{(n\geq 5)}$, one finds relations of both types.

CHAPTER 5

Ramond–Neveu–Schwarz Correlators at Loop-Level

In this Chapter we look at correlation functions involving the RNS fields ψ^μ and $S_\alpha, S^{\dot\beta}$ at loop-level in arbitrary even space-time dimensions $D = 2m$. First, we introduce the generalized Θ functions which capture the short distance behavior of the RNS fields as well as their periodicity along the homology cycles of the genus g Riemann surface. Then, the technique to re-express the RNS fields by m copies of an $SO(2)$ spin system is presented. In this formulations loop correlators are easy to calculate. We show how to construct Lorentz covariant expressions from these results and evaluate correlators in $D = 4, 6, 8$ and $D = 10$ space-time dimensions up to at least six-point level. In certain cases it is even possible to derive general formulas for correlators with arbitrary many external fields. The following work is based on [2].

5.1 Prerequisites

Before we show how to calculate RNS correlation functions at loop-level we have to establish some essential concepts. We comment on generalized Θ functions that add the right periodicity properties to the correlators. Additionally, $SO(2)$ spin operators and their relation to the RNS fields are introduced.

5.1.1 Generalized Θ Functions

Via the doubling trick the RNS fields of the open string can be extended to the full complex plane. For scattering at g loops these fields have therefore support on a Riemann surface of genus g. Such a surface has $2g$ one-cycles, α_I and β_I, $I = 1, \ldots, g$, which are shown in Figure 2.6. For loop scattering special attention must be paid to the change of the fields when they are shifted around these homology cycles. Fermions ψ can either satisfy periodic or antiperiodic boundary conditions. These properties are encoded in two g-dimensional vectors $\vec{a}$ and $\vec{b}$ with entries 0 or 1/2:

$$\psi(z+\alpha_I) = \exp(-\pi i\, a_I)\,\psi(z)\,, \qquad \psi(z+\beta_I) = \exp(\pi i\, b_I)\,\psi(z)\,. \tag{5.1}$$

Together, $\vec{a}$ and $\vec{b}$ are called the *spin structure*. Loop-level correlators have to respect these periodicity requirements. Further details on the periodicity properties of the RNS fields, can be found in [130] and Appendix C.1.

The unique function which imparts the above behavior to the loop correlation functions is Riemann's Θ function [131–133]:

$$\Theta(\vec{x}|\Omega) \equiv \sum_{\vec{n}\in\mathbb{Z}^g} \exp\left[2\pi i\left(\tfrac{1}{2}\vec{n}^t\,\Omega\,\vec{n} + \vec{n}^t\,\vec{x}\right)\right]. \tag{5.2}$$

In this holomorphic function $\vec{x} \in \mathbb{C}^g$ and Ω is a symmetric $g \times g$ complex matrix whose imaginary part is positive definite. The space of these matrices is called the Siegel upper-half space. The Θ function is quasi-periodic on the lattice $\mathbb{Z}^g + \Omega\mathbb{Z}^g$, i.e. periodic up to a multiplicative factor $\forall \vec{s}, \vec{t} \in \mathbb{Z}^g$:

$$\Theta(\vec{x}+\vec{s}+\Omega\vec{t}|\Omega) = \exp\left[-2\pi i\left(\tfrac{1}{2}\vec{t}^t\,\Omega\,\vec{t} + \vec{t}^t\,\vec{x}\right)\right]\Theta(\vec{x}|\Omega)\,. \tag{5.3}$$

The Θ function on $\mathbb{Z}^g$ can be generalized to Θ functions $\Theta\begin{bmatrix}\vec{a}\\\vec{b}\end{bmatrix}$ with rational characteristics. These are simply translations of Θ multiplied with an exponential factor,

$$\begin{aligned}\Theta\begin{bmatrix}\vec{a}\\\vec{b}\end{bmatrix}(\vec{x}|\Omega) &\equiv \exp\left[2\pi i\left(\tfrac{1}{2}\vec{a}^t\,\Omega\,\vec{a} + \vec{a}^t(\vec{x}+\vec{b})\right)\right]\Theta(\vec{x}+\Omega\vec{a}+\vec{b}|\Omega) \\ &= \sum_{\vec{n}\in\mathbb{Z}^g} \exp\left[2\pi i\left(\tfrac{1}{2}(\vec{n}+\vec{a})^t\,\Omega\,(\vec{n}+\vec{a}) + (\vec{n}+\vec{a})^t(\vec{x}+\vec{b})\right)\right],\end{aligned} \tag{5.4}$$

for all $\vec{a}, \vec{b} \in \mathbb{C}^g$. The original Θ then is simply $\Theta\begin{bmatrix}0\\0\end{bmatrix}$. The quasi-periodicity for the generalized Θ functions becomes

$$\begin{aligned}\Theta\begin{bmatrix}\vec{a}\\\vec{b}\end{bmatrix}(\vec{x}+\vec{s}|\Omega) &= \exp\left[2\pi i\,\vec{a}^t\,\vec{s}\right]\Theta\begin{bmatrix}\vec{a}\\\vec{b}\end{bmatrix}(\vec{x}|\Omega)\,, \\ \Theta\begin{bmatrix}\vec{a}\\\vec{b}\end{bmatrix}(\vec{x}+\Omega\vec{t}|\Omega) &= \exp\left[-2\pi i\,\vec{b}^t\,\vec{t}\right]\exp\left[-2\pi i\left(\tfrac{1}{2}\vec{t}^t\,\Omega\,\vec{t} + \vec{t}^t\,\vec{x}\right)\right]\Theta\begin{bmatrix}\vec{a}\\\vec{b}\end{bmatrix}(\vec{x}|\Omega)\,.\end{aligned} \tag{5.5}$$

From these definitions we can now specify Θ functions on the Riemann surface with genus g. In order to do so we have to lift the complex coordinate of the two-dimensional Riemann surface z to its Jacobian variety $\mathbb{C}^g/(\mathbb{Z}^g + \Omega\mathbb{Z}^g)$. This is done via the canonical map $z \mapsto \int_p^z \vec{\omega}$ with some arbitrary reference point p which will drop out in the calculations below. The integrand

5.1 Prerequisites

$\vec{\omega}$ is a g-dimensional complex vector consisting of linearly independent holomorphic one-forms. These integrals then are natural arguments for the Θ function (5.4). The matrix Ω in this context becomes the period matrix of the Riemann surface defined by

$$\int_{\alpha_I} \omega_J = \delta_{IJ}, \qquad \int_{\beta_I} \omega_J = \Omega_{IJ} \tag{5.6}$$

for a normalized basis of one forms ω_i. In addition we can define the prime form

$$E(z,w) \equiv \frac{\Theta\begin{bmatrix}\vec{a}_0\\\vec{b}_0\end{bmatrix}\left(\int_w^z \vec{\omega}|\Omega\right)}{h\begin{bmatrix}\vec{a}_0\\\vec{b}_0\end{bmatrix}(z)\, h\begin{bmatrix}\vec{a}_0\\\vec{b}_0\end{bmatrix}(w)}, \tag{5.7}$$

which is the unique holomorphic differential form with a single zero at $z = w$. In this definition $\vec{a}_0$ and $\vec{b}_0$ denote an arbitrary odd spin structure, i.e. $4\,\vec{a}_0\,\vec{b}_0$ is an odd integer. This ensures that $E(z,w) = -E(w,z)$. The half differentials in the denominator are given by

$$h\begin{bmatrix}\vec{a}_0\\\vec{b}_0\end{bmatrix}(z) \equiv \sqrt{\sum_{i=1}^{g} \omega_i\, \partial_i\, \Theta\begin{bmatrix}\vec{a}_0\\\vec{b}_0\end{bmatrix}(\vec{0}|\Omega)} \tag{5.8}$$

and assure that the E is independent of the specific choice of $\vec{a}_0, \vec{b}_0$. Given the leading behavior in the arguments z and w,

$$E(z,w) \sim (z - w) + \mathcal{O}\big((z-w)^3\big), \tag{5.9}$$

singularities in correlation functions are caused by appropriate powers of prime forms.

Tremendous simplifications occur for the case $g = 1$. In this case the Riemann surface in question is a torus and the period matrix Ω becomes the modular parameter of the torus τ. The Θ functions $\Theta\begin{bmatrix}\vec{a}\\\vec{b}\end{bmatrix}$ reduce to Jacobi's Θ functions

$$\theta_1 = \Theta\begin{bmatrix}1/2\\1/2\end{bmatrix}, \qquad \theta_2 = \Theta\begin{bmatrix}1/2\\0\end{bmatrix}, \qquad \theta_3 = \Theta\begin{bmatrix}0\\1/2\end{bmatrix}, \qquad \theta_4 = \Theta\begin{bmatrix}0\\0\end{bmatrix}, \tag{5.10}$$

and the prime form becomes

$$E(z,w)\Big|_{g=1} = \frac{\theta_1(z-w|\tau)}{\partial_z \theta_1(0|\tau)}. \tag{5.11}$$

5.1.2 $SO(2)$ Spin Systems

In Chapter 3.2 we have shown how RNS correlations functions at tree-level can be calculated directly from the OPEs of the RNS fields. At loop-level it turns out to be more effective to decompose the RNS fields into smaller building blocks and calculate correlators of these fields.

For this purpose we introduce the simple system consisting of two real Weyl fermions $\Psi^\pm$ and corresponding spin fields $s^\pm$. This system has $SO(2)$ symmetry and is hence called the $SO(2)$ spin system. The spin fields $s^\pm$ create branch cuts for the associated fermions. Both types of fields are conformal with weight $h = 1/2$ and $h = 1/8$ respectively. In addition, we assign Ramond charge ± 1 to the fermions $\Psi^\pm$ and $\pm 1/2$ to $s^\pm$. The OPEs of these fields then can be calculated by bosonization,

$$\Psi^\pm(z)\,\Psi^\mp(w) \sim \frac{1}{z-w}\,, \tag{5.12a}$$

$$s^\pm(z)\,s^\mp(w) \sim \frac{1}{(z-w)^{1/4}}\,, \tag{5.12b}$$

$$\Psi^\pm(z)\,s^\mp(w) \sim \frac{s^\pm(w)}{(z-w)^{1/2}}\,. \tag{5.12c}$$

Fields of alike charge exhibit regular behavior:

$$\Psi^\pm(z)\,\Psi^\pm(w) \sim (z-w)\,\Psi^\pm(w)\,\partial\Psi^\pm(w)\,, \tag{5.13a}$$

$$s^\pm(z)\,s^\pm(w) \sim (z-w)^{1/4}\,\Psi^\pm(w)\,, \tag{5.13b}$$

$$\Psi^\pm(z)\,s^\pm(w) \sim (z-w)^{1/2}\,\hat{s}^\pm(w)\,. \tag{5.13c}$$

Here $\hat{s}^\pm$ depicts an excited spin field of conformal weight $9/8$. From these OPEs we can construct the short-distance behavior of the RNS fields given in (3.9), (3.10) and (3.11). The Lorentz group of the RNS fields in $D = 2m$ dimensions is $SO(1, D-1)$. Therefore ψ^μ and S_α, $S^{\dot\beta}$ can be built from m independent $SO(2)$ spin systems $\{\Psi_i^\pm, s_i^\pm\}$, $i = 1, \ldots, m$. We take the convention that the $\Psi_i^\pm$ are the Cartan–Weyl representation of the $SO(1, D-1)$ vector ψ^μ,

$$\psi^{2i-2}(z) \equiv \frac{1}{\sqrt{2}}\left(\Psi_i^+(z) + \Psi_i^-(z)\right), \tag{5.14a}$$

$$\psi^{2i-1}(z) \equiv \frac{1}{\sqrt{2}\,i}\left(\Psi_i^+(z) - \Psi_i^-(z)\right), \tag{5.14b}$$

whereas the R spin fields can be written as

$$S_A(z) = \bigotimes_{i=1}^m s_i^\pm(z)\,. \tag{5.15}$$

Since each of the m Ramond charges $\pm 1/2$ can be chosen independently, there are $2^m = 2^{D/2}$ such operators. This coincides with the number of components of a Dirac spinor in $D = 2m$ dimensions. We take the convention that operators with an even number of s^- operators are left-handed, whereas those with an odd number are right-handed. Analogously, the combination

5.1 Prerequisites

of the $\Psi^\pm$'s in (5.14) results in $2m$ fields, which is the right number of degrees of freedom of a vector in $D = 2m$ dimensions.

The equations above determine the assignment between the Weyl indices α, $\dot\beta$ and the Ramond charge vectors $(\pm 1/2, \ldots, \pm 1/2)$ of the R spin fields. We explain the details in the distinct scenarios of four and six space-time dimensions which can easily be applied to eight and ten dimensions as well. For $D = 4$ we choose

$$(\alpha = 1) = (+, +), \qquad (\alpha = 2) = (-, -), \tag{5.16}$$

where for simplicity we have neglected all factors of $1/2$. Inserting these together with (5.14) into the OPE (3.9b)

$$\psi^\mu(z) \, S_\alpha(w) \sim \frac{1}{\sqrt{2}} (z-w)^{-1/2} \gamma^\mu_{\alpha\dot\beta} S^{\dot\beta}(w), \tag{5.17}$$

keeping in mind the action of $\Psi^\pm$ onto $s^\pm$ from (5.12a) and using the precise representation of the gamma matrices in four dimensions given in Appendix A.4 we find that the assignment of the dotted Weyl indices must be

$$(\dot\beta = 1) = (+, -), \qquad (\dot\beta = 2) = (-, +). \tag{5.18}$$

The same reasoning determines the ordering of the Ramond charge vectors in $D = 8$.

In $D = 6$ dimensions we encounter a different scenario due to the different chirality structure of the charge conjugation matrix. We start by diagonalizing the off-diagonal blocks of $\mathcal{C}$ which is easily achieved by permuting lines and columns of the matrix. Together with $S_\alpha = C_\alpha{}^{\dot\beta} S_{\dot\beta}$ this implies that the ordering of the Ramond charge vectors for α and $\dot\beta$ are not independent from each other, i.e. if $(\alpha = 1) = (+, +, +)$ then $(\dot\beta = 1) = (-, -, -)$. Starting with these assignments we can recursively determine all others by using (5.17) as before. We find from the representation of gamma matrices stated in Appendix A.4, where we have made a change of basis such that $C_\alpha^{\dot\beta}$ is diagonal,

$$\begin{aligned}
(\alpha = 1) &= (+, +, +), & (\dot\beta = 1) &= (-, -, -), \\
(\alpha = 2) &= (-, -, +), & (\dot\beta = 2) &= (+, +, -), \\
(\alpha = 3) &= (+, -, -), & (\dot\beta = 3) &= (-, +, +), \\
(\alpha = 4) &= (-, +, -), & (\dot\beta = 4) &= (+, -, +).
\end{aligned} \tag{5.19}$$

In exactly the same way one can determine the assignment of Ramond charge vectors to the Weyl indices for $D = 10$.

5.2 Loop Correlators

After this introduction to generalized Θ functions and the $SO(2)$ spin system we come now to the calculation of RNS loop correlators. For this purpose we express the RNS fields through $SO(2)$ spin system fields as discussed above. The $SO(2)$ spin system has been completely solved on arbitrary Riemann surfaces with genus g in a series of papers by Atick and Sen. We review their results and apply them to our RNS correlators.

5.2.1 Correlators of $SO(2)$ Spin Systems

Starting with the OPEs (5.12), (5.13) and demanding the correct periodicity conditions under shifts of the $SO(2)$ spin operators along the homology cycles of a Riemann surface with genus g Atick and Sen have been able to calculate all loop correlators of the $SO(2)$ spin system [130, 150, 151]. For the case of $2N$ spin fields they find

$$\left\langle \prod_{i=1}^{N} s^+(z_i)\, s^-(w_i) \right\rangle_{\vec{b}}^{\vec{a}} = \frac{\Theta\!\begin{bmatrix}\vec{a}\\\vec{b}\end{bmatrix}\!\left(\tfrac{1}{2}\sum_{i=1}^{N}\int_{w_i}^{z_i}\vec{\omega}\right)}{\Theta\!\begin{bmatrix}\vec{a}\\\vec{b}\end{bmatrix}\!(\vec{0})} \left(\frac{\prod_{i<j}^{N} E(z_i,z_j)\, E(w_i,w_j)}{\prod_{i,j=1}^{N} E(z_i,w_j)} \right)^{1/4}. \qquad (5.20)$$

This correlator is the key ingredient for deriving correlation functions of R spin fields. The most general form with fermions $\Psi^\pm$ and spin fields $s^\pm$, which is the starting point for calculating general RNS loop-correlators, is given by

$$\left\langle \prod_{i=1}^{N_1} s^+(y_i) \prod_{j=1}^{N_2} s^-(z_j) \prod_{k=1}^{N_3} \Psi^-(u_k) \prod_{l=1}^{N_4} \Psi^+(v_l) \right\rangle_{\vec{b}}^{\vec{a}} = \left(\frac{\prod_{r<s}^{N_1} E(y_r,y_s) \prod_{r<s}^{N_2} E(z_r,z_s)}{\prod_{i=1}^{N_1} \prod_{j=1}^{N_2} E(z_j,y_i)} \right)^{1/4}$$

$$\times \left(\frac{\prod_{r<s}^{N_3} E(u_r,u_s) \prod_{r<s}^{N_4} E(v_r,v_s)}{\prod_{k=1}^{N_3} \prod_{l=1}^{N_4} E(v_l,u_k)} \right) \left(\frac{\prod_{j=1}^{N_2} \prod_{k=1}^{N_3} E(u_k,z_j) \prod_{i=1}^{N_1} \prod_{l=1}^{N_4} E(v_l,y_i)}{\prod_{i=1}^{N_1} \prod_{k=1}^{N_3} E(u_k,y_i) \prod_{j=1}^{N_2} \prod_{l=1}^{N_4} E(v_l,z_j)} \right)^{1/2}$$

$$\times \left[\Theta\!\begin{bmatrix}\vec{a}\\\vec{b}\end{bmatrix}\!(\vec{0})\right]^{-1} \Theta\!\begin{bmatrix}\vec{a}\\\vec{b}\end{bmatrix}\!\left(\tfrac{1}{2}\sum_{i=1}^{N_1}\int_p^{y_i}\vec{\omega} - \tfrac{1}{2}\sum_{j=1}^{N_2}\int_p^{z_j}\vec{\omega} - \sum_{k=1}^{N_3}\int_p^{u_k}\vec{\omega} + \sum_{l=1}^{N_4}\int_p^{v_l}\vec{\omega} \right). \qquad (5.21)$$

Ramond charge conservation demands that $\tfrac{1}{2}(N_1 - N_2) - N_3 + N_4 = 0$. Hence, the arbitrary reference point p entering (5.21) through the canonical map drops out. RNS fields in $D = 2m$ dimension are expressed through m copies of an $SO(2)$ spin system that do not interact with each other. Hence, RNS correlation functions at loop-level factorize into m $SO(2)$ correlators of type (5.20) or (5.21).

5.2 Loop Correlators

In the following we use z_i as arguments for the $SO(2)$ spin operators and define the shorthand notation $E_{ij} \equiv E(z_i, z_j)$. Additionally, we abbreviate the generalized Θ functions by

$$\Theta\begin{bmatrix}\vec{a}\\\vec{b}\end{bmatrix}\left(\tfrac{1}{2}\left[\int_{z_l}^{z_i}\vec{\omega} + \int_{z_m}^{z_j}\vec{\omega} + \ldots + \int_{z_n}^{z_k}\vec{\omega}\right]\right) \equiv \Theta_{\vec{b}}^{\vec{a}}\begin{bmatrix}i & j & \cdots & k\\ l & m & \cdots & n\end{bmatrix}. \qquad (5.22)$$

Note in particular that the factor $1/2$ in the argument of $\Theta_{\vec{b}}^{\vec{a}}$, which is omnipresent for the spin fields, is always implicit.

Considerable simplifications occur for $g = 0$, i.e. scattering at tree-level. The spin structure dependent Θ functions trivialize, $\Theta_{\vec{b}}^{\vec{a}} \to 1$, and the prime form reduces to $E(z) - E(w) \to z - w$. In this way, by neglecting the generalized Θ functions and replacing the prime forms E_{ij} by z_{ij} one obtains the tree-level correlator from the loop result. The prime forms in the coefficients of the index terms then mimic the tree-level behavior of the correlation function.

5.2.2 Results in Lorentz Covariant Form

With the background on spin systems in mind, we can now calculate RNS correlation functions $\langle \psi^{\mu_1} \ldots \psi^{\mu_n} S_{\alpha_1} \ldots S_{\alpha_r} S^{\dot\beta_1} \ldots S^{\dot\beta_s}\rangle_{\vec{b}}^{\vec{a}}$ for specific choices of μ_i, α_i and $\dot\beta_i$ by formulating the RNS fields in $D = 2m$ dimension in terms of m spin systems via (5.14) and (5.15). The m spin systems do not interact with each other and hence the resulting correlator factorizes into m separate correlation function of a single spin system. Then, using the loop correlator formulas (5.20) and (5.21) for the individual $SO(2)$ correlation functions we find a result of the RNS correlation function for the specific choice of the Lorentz indices. However, the final goal is to express the result in covariant form, i.e. in terms of Clebsch–Gordan coefficients that are built from gamma and charge conjugation matrices. Due to the conservation of Ramond charge in (5.21) then index terms can be viewed as $SO(1, 2m-1)$ covariant Ramond charge conserving delta functions, schematically $C_{\alpha\beta} \sim \delta(\alpha + \beta)$ and $(\gamma^\mu C)_\alpha{}^{\dot\beta} \sim \delta(\mu + \alpha + \dot\beta)$ where $\mu, \alpha, \beta, \dot\beta$ are treated as Ramond charge vectors with m components such as $\mu \equiv (0, \pm 1, 0, \ldots, 0)$ and $\alpha \equiv (\pm 1/2, \ldots, \pm 1/2)$.

As a starting point we make an ansatz for the correlation function with a minimal set of Clebsch–Gordan coefficients. The cardinality of this set is determined by group theory as described in Chapter 3.3.1. Each of the index terms is accompanied by a z-dependent coefficient consisting of prime forms and generalized Θ functions. The results obtained for special choices of μ_i, α_i and $\dot\beta_i$ have to be matched with this ansatz. It is most economic to first look at configurations $(\mu_i, \alpha_i, \dot\beta_i)$ where only one tensor is non-zero. Then the loop-level result (5.21) directly yields the coefficient for the respective index term. In some cases, however, it is not possible to make all Clebsch–Gordan coefficients vanish except for one. More than one index term then contribute for every choice of $\mu_i, \alpha_i, \dot\beta_i$. In this case it can be helpful to switch to different Lorentz tensors that are (anti-)symmetric in some vector or spinor indices. Otherwise

Fay's trisecant identity [131] has to be used to determine the unknown coefficients. Sign issues can be resolved by calculating certain limits $z_i \to z_j$ at tree-level using the RNS OPEs (3.10) and (3.11) and comparing the expression to the result of the arising lower-point correlator. Alternatively, the signs can be read off from the respective tree-level correlation function. As the prime forms E_{ij} reduce to z_{ij} the signs in the loop correlator must be the same as in the tree-level version.

Let us illustrate this procedure with an easy example, the calculation of the correlation function $\langle \psi^\mu \psi^\nu \psi^\lambda S_\alpha S_\beta \rangle_{\vec{b}}^{\vec{a}}$ in $D = 6$ dimensions. Table 3.3 shows that four independent Clebsch–Gordan coefficients exist for this correlator. A convenient ansatz is

$$\langle \psi^\mu(z_1)\, \psi^\nu(z_2)\, \psi^\lambda(z_3)\, S_\alpha(z_4)\, S_\beta(z_5) \rangle_{\vec{b}}^{\vec{a}} = F_1(z)\, (\gamma^{\mu\nu\lambda} C)_{\alpha\beta}$$
$$+ F_2(z)\, \eta^{\mu\nu}\, (\gamma^\lambda C)_{\alpha\beta} + F_3(z)\, \eta^{\mu\lambda}\, (\gamma^\nu C)_{\alpha\beta} + F_4(z)\, \eta^{\nu\lambda}\, (\gamma^\mu C)_{\alpha\beta}\,. \quad (5.23)$$

The task is now to determine F_1, F_2, F_3, F_4 by making clever choices for $\mu, \nu, \lambda, \alpha, \beta$. The coefficient F_1 can easily be obtained by setting $\mu = 0, \nu = 2, \lambda = 4$. As the metric η is diagonal all index terms apart from $\gamma^{\mu\nu\lambda}$ vanish for this configuration. Then, by means of (5.14), the NS fermions in terms of the $SO(2)$ spin system fields become

$$\psi^{\mu=0}(z_1) = \frac{1}{\sqrt{2}} \left(\Psi_1^+(z_1) + \Psi_1^-(z_1) \right),$$
$$\psi^{\nu=2}(z_2) = \frac{1}{\sqrt{2}} \left(\Psi_2^+(z_2) + \Psi_2^-(z_2) \right),$$
$$\psi^{\lambda=4}(z_3) = \frac{1}{\sqrt{2}} \left(\Psi_3^+(z_3) + \Psi_3^-(z_3) \right), \quad (5.24)$$

and we choose for the spin fields

$$S_{\alpha=1}(z_4) = s_1^+(z_4)\, s_2^+(z_4)\, s_3^+(z_4)\,, \qquad S_{\beta=1}(z_5) = s_1^+(z_5)\, s_2^+(z_5)\, s_3^+(z_5)\,. \quad (5.25)$$

Hence, we have to calculate

$$\frac{1}{2\sqrt{2}} \langle (\Psi_1^+ + \Psi_1^-)(z_1)\, s_1^+(z_4)\, s_1^+(z_5) \rangle_{\vec{b}}^{\vec{a}} \langle (\Psi_2^+ + \Psi_2^-)(z_2)\, s_2^-(z_4)\, s_2^-(z_5) \rangle_{\vec{b}}^{\vec{a}}$$
$$\times \langle (\Psi_3^+ + \Psi_3^-)(z_3)\, s_3^-(z_4)\, s_3^-(z_5) \rangle_{\vec{b}}^{\vec{a}}\,. \quad (5.26)$$

Due to Ramond charge conservations in (5.21) $\Psi_1^+(z_1)$, $\Psi_2^-(z_2)$ and $\Psi_3^-(z_3)$ drop out and we obtain the coefficient F_1 up to a sign:

$$F_1 = \pm \frac{\Theta_{\vec{b}}^{\vec{a}}\begin{bmatrix}1&1\\4&5\end{bmatrix}\, \Theta_{\vec{b}}^{\vec{a}}\begin{bmatrix}2&2\\4&5\end{bmatrix}\, \Theta_{\vec{b}}^{\vec{a}}\begin{bmatrix}3&3\\4&5\end{bmatrix}\, E_{45}^{3/4}}{2\sqrt{2}\, [\Theta_{\vec{b}}^{\vec{a}}(\vec{0})]^3\, (E_{14}\, E_{15}\, E_{24}\, E_{25}\, E_{34}\, E_{35})^{1/2}}\,. \quad (5.27)$$

5.2 Loop Correlators

The coefficient F_2 can be determined in a similar way by setting $\mu = \nu = 0$, $\lambda = 2$ and $\alpha = 1$, $\beta = 4$. No other tensors than $\eta^{\mu\nu}(\gamma^\lambda C)_{\alpha\beta}$ contribute as the metric is diagonal and $\gamma^{\mu\nu\lambda}$ totally antisymmetric. The NS fermions for this index choice are expressed through

$$\psi^{\mu=0}(z_1) = \frac{1}{\sqrt{2}}\left(\Psi_1^+(z_1) + \Psi_1^-(z_1)\right),$$
$$\psi^{\nu=0}(z_2) = \frac{1}{\sqrt{2}}\left(\Psi_1^+(z_2) + \Psi_1^-(z_2)\right),$$
$$\psi^{\lambda=2}(z_3) = \frac{1}{\sqrt{2}}\left(\Psi_2^+(z_3) + \Psi_2^-(z_3)\right), \qquad (5.28)$$

while the spin fields are given by

$$S_{\alpha=1}(z_4) = s_1^+(z_4)\, s_2^+(z_4)\, s_3^+(z_4), \qquad S_{\beta=4}(z_5) = s_1^-(z_5)\, s_2^+(z_5)\, s_3^-(z_5). \qquad (5.29)$$

This time the correlator

$$\frac{1}{2\sqrt{2}}\left\langle(\Psi_1^+ + \Psi_1^-)(z_1)\,(\Psi_1^+ + \Psi_1^-)(z_2)\, s_1^+(z_4)\, s_1^-(z_5)\right\rangle_{\vec{b}}^{\vec{a}}$$
$$\times \left\langle(\Psi_2^+ + \Psi_2^-)(z_3)\, s_2^+(z_4)\, s_2^+(z_5)\right\rangle_{\vec{b}}^{\vec{a}} \left\langle s_3^+(z_4)\, s_3^-(z_5)\right\rangle_{\vec{b}}^{\vec{a}} \qquad (5.30)$$

has to be evaluated. Ψ_2^- in the second correlator drops out because of Ramond charge conservation, while in the first spin system the two in-equivalent fermion configurations $\Psi_1^+(z_1)\,\Psi_1^-(z_2)$ and $\Psi_1^-(z_1)\,\Psi_1^+(z_2)$ contribute. Consequently the total result for F_2 consists of two terms:

$$F_2 = \pm \frac{\Theta_{\vec{b}}^{\vec{a}}\left[\begin{smallmatrix}3&3\\4&5\end{smallmatrix}\right]\Theta_{\vec{b}}^{\vec{a}}\left[\begin{smallmatrix}4\\5\end{smallmatrix}\right]\left(E_{14}\,E_{25}\,\Theta_{\vec{b}}^{\vec{a}}\left[\begin{smallmatrix}1&1&4\\2&2&5\end{smallmatrix}\right] + E_{15}\,E_{24}\,\Theta_{\vec{b}}^{\vec{a}}\left[\begin{smallmatrix}1&1&5\\2&2&4\end{smallmatrix}\right]\right)}{2\sqrt{2}\left[\Theta_{\vec{b}}^{\vec{a}}(\vec{0})\right]^3 E_{12}\,(E_{14}\,E_{15}\,E_{24}\,E_{25}\,E_{34}\,E_{35})^{1/2}\,E_{45}^{1/4}}. \qquad (5.31)$$

The remaining coefficients F_3 and F_4 follow from F_2 by permutation of the vector indices and the labels $(1,2,3)$ in the prime forms and Θ functions. The signs of the individual coefficients are fixed by requiring that in the limit $z_1 \to z_2$ the expression

$$\eta^{\mu\nu}\, z_{12}^{-1} \left\langle \psi^\lambda(z_3)\, S_\alpha(z_4)\, S_\beta(z_5)\right\rangle_{\vec{b}}^{\vec{a}} \qquad (5.32)$$

must emerge. Accordingly, the limit $z_4 \to z_5$ has to give rise to

$$(\gamma^\rho C)_{\alpha\beta}\, z_{45}^{-1/4} \left\langle \psi^\mu(z_1)\, \psi^\nu(z_2)\, \psi^\lambda(z_3)\, \psi_\rho(z_5)\right\rangle_{\vec{b}}^{\vec{a}}. \qquad (5.33)$$

Next we consider a second and more complicated example, the six-point function $\langle \psi^\mu(z_1)\, \psi^\nu(z_2)\, S_\alpha(z_3)\, S_\beta(z_4)\, S_\gamma(z_5)\, S_\delta(z_6)\rangle_{\vec{b}}^{\vec{a}}$. From Chapter 3.3 it is known that six independent index terms exist for this correlator in six dimensions. We choose to work with the expression

$(\gamma^\mu C)_{\alpha\beta} (\gamma^\nu C)_{\gamma\delta}$ and its five relatives coming from permutations of the spinor indices. For this correlation function it is not possible to choose the indices for the spin system correlator in such a way that all but one tensor vanish. By either using the concrete representation of gamma matrices in Appendix A.4 or understanding the index terms as Ramond charge conserving delta functions one finds that, e.g., for

$$\psi^{\mu=0}(z_1) = \frac{1}{\sqrt{2}} \left(\Psi_1^+(z_1) + \Psi_1^-(z_1) \right),$$
$$\psi^{\nu=2}(z_2) = \frac{1}{\sqrt{2}} \left(\Psi_2^+(z_2) + \Psi_2^-(z_2) \right), \quad (5.34)$$

and

$$S_{\alpha=1}(z_3) = s_1^+(z_3) \, s_2^+(z_3) \, s_3^+(z_3), \qquad S_{\beta=3}(z_4) = s_1^+(z_4) \, s_2^-(z_4) \, s_3^-(z_4),$$
$$S_{\gamma=3}(z_5) = s_1^+(z_5) \, s_2^-(z_5) \, s_3^-(z_5), \qquad S_{\delta=2}(z_6) = s_1^-(z_6) \, s_2^-(z_6) \, s_3^+(z_6) \quad (5.35)$$

the spin system correlator contributes to both z coefficients of $(\gamma^\mu C)_{\alpha\beta} (\gamma^\nu C)_{\gamma\delta}$ and $(\gamma^\mu C)_{\alpha\gamma} (\gamma^\nu C)_{\beta\delta}$. The result of

$$\frac{1}{2\sqrt{2}} \langle (\Psi_1^+ + \Psi_1^-)(z_1) \, s_1^+(z_3) \, s_1^+(z_4) \, s_1^+(z_5) \, s_1^-(z_6) \rangle_{\vec{b}}^{\vec{a}}$$
$$\times \langle (\Psi_2^+ + \Psi_2^-)(z_2) \, s_2^+(z_3) \, s_2^-(z_4) \, s_2^-(z_5) \, s_2^-(z_6) \rangle_{\vec{b}}^{\vec{a}}$$
$$\times \langle s_3^+(z_3) \, s_3^-(z_4) \, s_3^-(z_5) \, s_3^+(z_6) \rangle_{\vec{b}}^{\vec{a}} \quad (5.36)$$

must thus be split into two parts using Fay's trisecant identity. Evaluating (5.36) we find

$$\pm \Theta_{\vec{b}}^{\vec{a}} \left[\begin{smallmatrix} 1 & 1 & 6 \\ 3 & 4 & 5 \end{smallmatrix} \right] \Theta_{\vec{b}}^{\vec{a}} \left[\begin{smallmatrix} 2 & 2 & 3 \\ 4 & 5 & 6 \end{smallmatrix} \right] \Theta_{\vec{b}}^{\vec{a}} \left[\begin{smallmatrix} 3 & 6 \\ 4 & 5 \end{smallmatrix} \right] E_{12} \, E_{16} \, E_{23} \, E_{45}, \quad (5.37)$$

where we have taken out the pre-factor

$$\frac{1}{2\sqrt{2} \, [\Theta_{\vec{b}}^{\vec{a}}(\vec{0})]^3} \frac{(E_{13} \, E_{14} \, E_{15} \, E_{16} \, E_{23} \, E_{24} \, E_{25} \, E_{26})^{-1/2}}{E_{12} \, (E_{34} \, E_{35} \, E_{36} \, E_{45} \, E_{46} \, E_{56})^{1/4}}. \quad (5.38)$$

Now, using the version (C.8) of Fay's trisecant identity with

$$\vec{\Delta} = \frac{1}{2} \int_{z_1}^{z_2} \vec{\omega} + \frac{1}{2} \int_{z_6}^{z_3} \vec{\omega} \quad (5.39)$$

this becomes

$$\mp \Theta_{\vec{b}}^{\vec{a}} \left[\begin{smallmatrix} 1 & 1 & 5 & 6 \\ 2 & 2 & 3 & 4 \end{smallmatrix} \right] \Theta_{\vec{b}}^{\vec{a}} \left[\begin{smallmatrix} 3 & 5 \\ 4 & 6 \end{smallmatrix} \right] \Theta_{\vec{b}}^{\vec{a}} \left[\begin{smallmatrix} 3 & 6 \\ 4 & 5 \end{smallmatrix} \right] E_{15} \, E_{16} \, E_{23} \, E_{24}$$

$$\pm \Theta_{\bar{b}}^{\bar{a}} \begin{bmatrix} 1 & 1 & 4 & 6 \\ 2 & 2 & 3 & 5 \end{bmatrix} \Theta_{\bar{b}}^{\bar{a}} \begin{bmatrix} 3 & 4 \\ 5 & 6 \end{bmatrix} \Theta_{\bar{b}}^{\bar{a}} \begin{bmatrix} 3 & 6 \\ 4 & 5 \end{bmatrix} E_{14} E_{16} E_{23} E_{25} \,. \tag{5.40}$$

At this point we cannot decide which of these terms belongs to the tested index terms but one has to do further calculations of spin system correlators. Making a different choice for the indices, e.g. $\mu = 0$, $\nu = 2$, $\alpha = \delta = 4$, $\beta = 3$, $\gamma = 2$, probes again $(\gamma^\mu C)_{\alpha\beta} (\gamma^\nu C)_{\gamma\delta}$ but this times in combination with $(\gamma^\mu C)_{\beta\delta} (\gamma^\nu C)_{\alpha\gamma}$. Evaluating this configuration and splitting the result as before yields

$$\mp \Theta_{\bar{b}}^{\bar{a}} \begin{bmatrix} 1 & 1 & 5 & 6 \\ 2 & 2 & 3 & 4 \end{bmatrix} \Theta_{\bar{b}}^{\bar{a}} \begin{bmatrix} 3 & 5 \\ 4 & 6 \end{bmatrix} \Theta_{\bar{b}}^{\bar{a}} \begin{bmatrix} 3 & 6 \\ 4 & 5 \end{bmatrix} E_{15} E_{16} E_{23} E_{24}$$

$$\pm \Theta_{\bar{b}}^{\bar{a}} \begin{bmatrix} 1 & 1 & 3 & 5 \\ 2 & 2 & 4 & 6 \end{bmatrix} \Theta_{\bar{b}}^{\bar{a}} \begin{bmatrix} 3 & 4 \\ 5 & 6 \end{bmatrix} \Theta_{\bar{b}}^{\bar{a}} \begin{bmatrix} 3 & 6 \\ 4 & 5 \end{bmatrix} E_{13} E_{15} E_{24} E_{26} \,. \tag{5.41}$$

From the comparison of (5.40) and (5.41) we can now conclude that the first term must be the coefficient of $(\gamma^\mu C)_{\alpha\beta} (\gamma^\nu C)_{\gamma\delta}$. The second expression in (5.40) is consequently the coefficient of $(\gamma^\mu C)_{\alpha\gamma} (\gamma^\nu C)_{\beta\delta}$, while the Glebsch–Gordan coefficient $(\gamma^\mu C)_{\beta\delta} (\gamma^\nu C)_{\alpha\gamma}$ has to come with the second term in (5.41).

Progressing in this way it is possible to evaluate all loop correlators for which one fails to separate the individual index terms in the spin system calculations by appropriate choices for the Lorentz indices. In addition, one obtains the relative signs between the different index terms. The over-all sign can again be determined by looking at certain limits or by comparing with the tree-level result.

5.3 Results of RNS Loop Correlators

The discussion of the $SO(2)$ spin operators and their correlation function at loop-level has paved the way for the calculation of RNS correlators in four, six, eight and ten space-time dimensions. We want to stress that certain methods that were available at tree-level cannot be carried over to loop-level. Bosonization cannot resolve the spin structure of the RNS fields at higher genus, but yields only averages [152–154]. Hence, the factorization of four-dimensional correlators into left- and right-handed pure spin field correlators also does not hold at loop-level. Furthermore, the uniform treatment of NS fermions and R spin fields in eight dimensions via $SO(8)$ triality does not carry over to loop correlators. This is due to the different contributions in the Θ functions of the NS fermions and R spin fields as can be seen from (5.21). Therefore, all loop-correlators in covariant form have to be calculated using the method presented above or by replacing NS fermions with two R spin fields as shown in Chapter 3.4.2 and then calculating pure spin field correlation functions at loop-level.

5.3.1 Results for $D = 4$

The calculation of four-dimensional RNS correlation functions at loop-level has been the main topic of [155] and thus constitutes the generalization to arbitrary loops of the results given in Chapter 4. By expressing the RNS fields through two copies of an $SO(2)$ spin system the author has been able to calculate correlation functions at loop-level up to eight points as well as the correlators

$$\langle S_{\alpha_1} \ldots S_{\alpha_{2M}} S_{\dot\beta_1} S_{\dot\beta_2} \rangle_{\vec b}^{\vec a}, \qquad \langle S_{\alpha_1} \ldots S_{\alpha_{2M}} S_{\dot\beta_1} S_{\dot\beta_2} S_{\dot\beta_3} S_{\dot\beta_4} \rangle_{\vec b}^{\vec a} \qquad (5.42)$$

with arbitrary many left-handed spin fields. By replacing one left- and one right-handed spin field with one NS fermions as in (3.70) we can derive from the first correlator the results for

$$\langle \psi^\mu S_{\alpha_1} \ldots S_{\alpha_{2M-1}} S_{\dot\beta_1} \rangle_{\vec b}^{\vec a}, \qquad \langle \psi^\mu \psi^\nu S_{\alpha_1} \ldots S_{\alpha_{2M-2}} \rangle_{\vec b}^{\vec a}, \qquad (5.43)$$

while the second correlator can be used to calculate

$$\langle \psi^\mu S_{\alpha_1} \ldots S_{\alpha_{2M-1}} S_{\dot\beta_1} S_{\dot\beta_2} S_{\dot\beta_3} \rangle_{\vec b}^{\vec a}, \qquad \langle \psi^\mu \psi^\nu S_{\alpha_1} \ldots S_{\alpha_{2M-2}} S_{\dot\beta_1} S_{\dot\beta_2} \rangle_{\vec b}^{\vec a},$$

$$\langle \psi^\mu \psi^\nu \psi^\lambda S_{\alpha_1} \ldots S_{\alpha_{2M-3}} S_{\dot\beta_1} \rangle_{\vec b}^{\vec a}, \qquad \langle \psi^\mu \psi^\nu \psi^\lambda \psi^\rho S_{\alpha_1} \ldots S_{\alpha_{2M-4}} \rangle_{\vec b}^{\vec a}. \qquad (5.44)$$

We refrain from quoting any results but refer the reader to [155].

5.3.2 Results for $D = 6$

In six dimensions three copies of the $SO(2)$ spin system are needed and hence three Θ functions will appear in the numerator and denominator of the following correlations functions. The simplest ones involve four spin fields:

$$\langle S_\alpha(z_1) S_\beta(z_2) S_\gamma(z_3) S_\delta(z_4) \rangle_{\vec b}^{\vec a} = \frac{\Theta_{\vec b}^{\vec a}[{}^{1\,2}_{3\,4}] \, \Theta_{\vec b}^{\vec a}[{}^{1\,3}_{2\,4}] \, \Theta_{\vec b}^{\vec a}[{}^{1\,4}_{2\,3}]}{2 \left[\Theta_{\vec b}^{\vec a}(\vec 0)\right]^3} \, \frac{(\gamma^\mu C)_{\alpha\beta} (\gamma_\mu C)_{\gamma\delta}}{(E_{12} E_{13} E_{14} E_{23} E_{24} E_{34})^{1/4}}, \quad (5.45)$$

$$\langle S_\alpha(z_1) S_\beta(z_2) S^{\dot\gamma}(z_3) S^{\dot\delta}(z_4) \rangle_{\vec b}^{\vec a} = \frac{\Theta_{\vec b}^{\vec a}[{}^{1\,2}_{3\,4}]}{\left[\Theta_{\vec b}^{\vec a}(\vec 0)\right]^3} \left(\frac{E_{13} E_{14} E_{23} E_{24}}{E_{12} E_{34}} \right)^{1/4}$$

$$\times \left[\frac{C_\alpha{}^{\dot\gamma} C_\beta{}^{\dot\delta}}{E_{13} E_{24}} \Theta_{\vec b}^{\vec a}[{}^{1\,4}_{2\,3}]^2 - \frac{C_\alpha{}^{\dot\delta} C_\beta{}^{\dot\gamma}}{E_{14} E_{23}} \Theta_{\vec b}^{\vec a}[{}^{1\,3}_{2\,4}]^2 \right]. \qquad (5.46)$$

The only non-vanishing five-point correlator with four spin fields and one fermion involves three alike chiralities:

5.3 Results of RNS Loop Correlators

$$\left\langle \psi^\mu(z_1)\, S_\alpha(z_2)\, S_\beta(z_3)\, S_\gamma(z_4)\, S^{\dot\delta}(z_5) \right\rangle^{\vec a}_{\vec b}$$

$$= \frac{1}{\sqrt{2}\,[\Theta^{\vec a}_{\vec b}(\vec 0)]^3} \frac{(E_{25}\,E_{35}\,E_{45})^{1/4}}{(E_{12}\,E_{13}\,E_{14}\,E_{15})^{1/2}(E_{23}\,E_{24}\,E_{34})^{1/4}}$$

$$\times \Bigg[(\gamma^\mu C)_{\alpha\beta}\, \frac{C_\gamma^{\dot\delta}}{E_{45}}\, E_{14}\, \Theta^{\vec a}_{\vec b}{\scriptstyle\left[\begin{smallmatrix}1&1&4\\2&3&5\end{smallmatrix}\right]}\, \Theta^{\vec a}_{\vec b}{\scriptstyle\left[\begin{smallmatrix}2&4\\3&5\end{smallmatrix}\right]}\, \Theta^{\vec a}_{\vec b}{\scriptstyle\left[\begin{smallmatrix}2&5\\3&4\end{smallmatrix}\right]}$$

$$-\, (\gamma^\mu C)_{\alpha\gamma}\, \frac{C_\beta^{\dot\delta}}{E_{35}}\, E_{13}\, \Theta^{\vec a}_{\vec b}{\scriptstyle\left[\begin{smallmatrix}1&1&3\\2&4&5\end{smallmatrix}\right]}\, \Theta^{\vec a}_{\vec b}{\scriptstyle\left[\begin{smallmatrix}2&3\\4&5\end{smallmatrix}\right]}\, \Theta^{\vec a}_{\vec b}{\scriptstyle\left[\begin{smallmatrix}2&5\\3&4\end{smallmatrix}\right]}$$

$$+\, (\gamma^\mu C)_{\beta\gamma}\, \frac{C_\alpha^{\dot\delta}}{E_{25}}\, E_{12}\, \Theta^{\vec a}_{\vec b}{\scriptstyle\left[\begin{smallmatrix}1&1&2\\3&4&5\end{smallmatrix}\right]}\, \Theta^{\vec a}_{\vec b}{\scriptstyle\left[\begin{smallmatrix}2&3\\4&5\end{smallmatrix}\right]}\, \Theta^{\vec a}_{\vec b}{\scriptstyle\left[\begin{smallmatrix}2&4\\3&5\end{smallmatrix}\right]} \Bigg]. \qquad (5.47)$$

At the six-point level there are two correlators involving only spin fields. The first one consists of five left- and one right-handed spin-field:

$$\left\langle S_\alpha(z_1)\, S_\beta(z_2)\, S_\gamma(z_3)\, S_\delta(z_4)\, S_\epsilon(z_5)\, S^{\dot\zeta}(z_6) \right\rangle^{\vec a}_{\vec b}$$

$$= \frac{1}{2\,[\Theta^{\vec a}_{\vec b}(\vec 0)]^3} \left(\frac{E_{16}\,E_{26}\,E_{36}\,E_{46}\,E_{56}}{E_{12}\,E_{13}\,E_{14}\,E_{15}\,E_{23}\,E_{24}\,E_{25}\,E_{34}\,E_{35}\,E_{45}} \right)^{1/4}$$

$$\times \Bigg[(\gamma^\mu C)_{\alpha\beta}(\gamma_\mu C)_{\gamma\epsilon}\, \frac{C_\delta^{\dot\zeta}}{E_{46}}\, \frac{E_{45}}{E_{56}}\, \Theta^{\vec a}_{\vec b}{\scriptstyle\left[\begin{smallmatrix}1&2&6\\3&4&5\end{smallmatrix}\right]}\, \Theta^{\vec a}_{\vec b}{\scriptstyle\left[\begin{smallmatrix}1&3&6\\2&4&5\end{smallmatrix}\right]}\, \Theta^{\vec a}_{\vec b}{\scriptstyle\left[\begin{smallmatrix}1&4&5\\2&3&6\end{smallmatrix}\right]}$$

$$+\, (\gamma^\mu C)_{\alpha\beta}(\gamma_\mu C)_{\epsilon\delta}\, \frac{C_\gamma^{\dot\zeta}}{E_{36}}\, \frac{E_{35}}{E_{56}}\, \Theta^{\vec a}_{\vec b}{\scriptstyle\left[\begin{smallmatrix}1&2&6\\3&4&5\end{smallmatrix}\right]}\, \Theta^{\vec a}_{\vec b}{\scriptstyle\left[\begin{smallmatrix}1&3&5\\2&4&6\end{smallmatrix}\right]}\, \Theta^{\vec a}_{\vec b}{\scriptstyle\left[\begin{smallmatrix}1&4&6\\2&3&5\end{smallmatrix}\right]}$$

$$+\, (\gamma^\mu C)_{\alpha\epsilon}(\gamma_\mu C)_{\gamma\delta}\, \frac{C_\beta^{\dot\zeta}}{E_{26}}\, \frac{E_{25}}{E_{56}}\, \Theta^{\vec a}_{\vec b}{\scriptstyle\left[\begin{smallmatrix}1&2&5\\3&4&6\end{smallmatrix}\right]}\, \Theta^{\vec a}_{\vec b}{\scriptstyle\left[\begin{smallmatrix}1&3&6\\2&4&5\end{smallmatrix}\right]}\, \Theta^{\vec a}_{\vec b}{\scriptstyle\left[\begin{smallmatrix}1&4&6\\2&3&5\end{smallmatrix}\right]}$$

$$+\, (\gamma^\mu C)_{\epsilon\beta}(\gamma_\mu C)_{\gamma\delta}\, \frac{C_\alpha^{\dot\zeta}}{E_{16}}\, \frac{E_{15}}{E_{56}}\, \Theta^{\vec a}_{\vec b}{\scriptstyle\left[\begin{smallmatrix}1&2&5\\3&4&6\end{smallmatrix}\right]}\, \Theta^{\vec a}_{\vec b}{\scriptstyle\left[\begin{smallmatrix}1&3&5\\2&4&6\end{smallmatrix}\right]}\, \Theta^{\vec a}_{\vec b}{\scriptstyle\left[\begin{smallmatrix}1&4&5\\2&3&6\end{smallmatrix}\right]} \Bigg]. \qquad (5.48)$$

In addition, we have the correlator with three left- and right-handed spin fields each:

$$\left\langle S_\alpha(z_1)\, S_\beta(z_2)\, S_\gamma(z_3)\, S^{\dot\delta}(z_4)\, S^{\dot\epsilon}(z_5)\, S^{\dot\zeta}(z_6) \right\rangle^{\vec a}_{\vec b}$$

$$= \frac{1}{[\Theta^{\vec a}_{\vec b}(\vec 0)]^3} \left(\frac{E_{14}\,E_{15}\,E_{16}\,E_{24}\,E_{25}\,E_{26}\,E_{34}\,E_{35}\,E_{36}}{E_{12}\,E_{13}\,E_{23}\,E_{45}\,E_{46}\,E_{56}} \right)^{1/4}$$

$$\times \Bigg[\frac{C_\alpha^{\dot\delta}\,C_\beta^{\dot\epsilon}\,C_\gamma^{\dot\zeta}}{E_{14}\,E_{25}\,E_{36}}\, \Theta^{\vec a}_{\vec b}{\scriptstyle\left[\begin{smallmatrix}1&2&6\\3&4&5\end{smallmatrix}\right]}\, \Theta^{\vec a}_{\vec b}{\scriptstyle\left[\begin{smallmatrix}1&3&5\\2&4&6\end{smallmatrix}\right]}\, \Theta^{\vec a}_{\vec b}{\scriptstyle\left[\begin{smallmatrix}1&5&6\\2&3&4\end{smallmatrix}\right]}$$

$$-\, \frac{C_\alpha^{\dot\delta}\,C_\beta^{\dot\zeta}\,C_\gamma^{\dot\epsilon}}{E_{14}\,E_{26}\,E_{35}}\, \Theta^{\vec a}_{\vec b}{\scriptstyle\left[\begin{smallmatrix}1&2&5\\3&4&6\end{smallmatrix}\right]}\, \Theta^{\vec a}_{\vec b}{\scriptstyle\left[\begin{smallmatrix}1&3&6\\2&4&5\end{smallmatrix}\right]}\, \Theta^{\vec a}_{\vec b}{\scriptstyle\left[\begin{smallmatrix}1&5&6\\2&3&4\end{smallmatrix}\right]}$$

$$+\, \frac{C_\alpha^{\dot\epsilon}\,C_\beta^{\dot\zeta}\,C_\gamma^{\dot\delta}}{E_{15}\,E_{26}\,E_{34}}\, \Theta^{\vec a}_{\vec b}{\scriptstyle\left[\begin{smallmatrix}1&2&4\\3&5&6\end{smallmatrix}\right]}\, \Theta^{\vec a}_{\vec b}{\scriptstyle\left[\begin{smallmatrix}1&3&6\\2&4&5\end{smallmatrix}\right]}\, \Theta^{\vec a}_{\vec b}{\scriptstyle\left[\begin{smallmatrix}1&4&6\\2&3&5\end{smallmatrix}\right]}$$

$$-\, \frac{C_\alpha^{\dot\epsilon}\,C_\beta^{\dot\delta}\,C_\gamma^{\dot\zeta}}{E_{15}\,E_{24}\,E_{36}}\, \Theta^{\vec a}_{\vec b}{\scriptstyle\left[\begin{smallmatrix}1&2&6\\3&4&5\end{smallmatrix}\right]}\, \Theta^{\vec a}_{\vec b}{\scriptstyle\left[\begin{smallmatrix}1&3&4\\2&5&6\end{smallmatrix}\right]}\, \Theta^{\vec a}_{\vec b}{\scriptstyle\left[\begin{smallmatrix}1&4&6\\2&3&5\end{smallmatrix}\right]}$$

$$+ \frac{C_\alpha{}^{\dot\zeta} C_\beta{}^{\dot\delta} C_\gamma{}^{\dot\epsilon}}{E_{16} E_{24} E_{35}} \Theta_{\vec{b}}^{\vec{a}}[\begin{smallmatrix}1&2&5\\3&4&6\end{smallmatrix}] \, \Theta_{\vec{b}}^{\vec{a}}[\begin{smallmatrix}1&3&4\\2&5&6\end{smallmatrix}] \, \Theta_{\vec{b}}^{\vec{a}}[\begin{smallmatrix}1&4&5\\2&3&6\end{smallmatrix}]$$

$$- \frac{C_\alpha{}^{\dot\zeta} C_\beta{}^{\dot\epsilon} C_\gamma{}^{\dot\delta}}{E_{16} E_{25} E_{34}} \Theta_{\vec{b}}^{\vec{a}}[\begin{smallmatrix}1&2&4\\3&5&6\end{smallmatrix}] \, \Theta_{\vec{b}}^{\vec{a}}[\begin{smallmatrix}1&3&5\\2&4&6\end{smallmatrix}] \, \Theta_{\vec{b}}^{\vec{a}}[\begin{smallmatrix}1&4&5\\2&3&6\end{smallmatrix}] \Big]. \tag{5.49}$$

Furthermore, two NS fermions can be accompanied by four spin fields, either with uniform chirality,

$$\langle \psi^\mu(z_1)\, \psi^\nu(z_2)\, S_\alpha(z_3)\, S_\beta(z_4)\, S_\gamma(z_5)\, S_\delta(z_6) \rangle^{\vec{a}}_{\vec{b}}$$
$$= -\frac{1}{2\left[\Theta_{\vec{b}}^{\vec{a}}(\vec{0})\right]^3} \frac{(E_{13} E_{14} E_{15} E_{16} E_{23} E_{24} E_{25} E_{26})^{-1/2}}{E_{12}\,(E_{34} E_{35} E_{36} E_{45} E_{46} E_{56})^{1/4}}$$
$$\times \Big[(\gamma^\mu C)_{\alpha\beta}(\gamma^\nu C)_{\gamma\delta}\, E_{15} E_{16} E_{23} E_{24}\, \Theta_{\vec{b}}^{\vec{a}}[\begin{smallmatrix}1&1&5&6\\2&2&3&4\end{smallmatrix}] \, \Theta_{\vec{b}}^{\vec{a}}[\begin{smallmatrix}3&5\\4&6\end{smallmatrix}] \, \Theta_{\vec{b}}^{\vec{a}}[\begin{smallmatrix}3&6\\4&5\end{smallmatrix}]$$
$$+ (\gamma^\mu C)_{\gamma\delta}(\gamma^\nu C)_{\alpha\beta}\, E_{13} E_{14} E_{25} E_{26}\, \Theta_{\vec{b}}^{\vec{a}}[\begin{smallmatrix}1&1&3&4\\2&2&5&6\end{smallmatrix}] \, \Theta_{\vec{b}}^{\vec{a}}[\begin{smallmatrix}3&5\\4&6\end{smallmatrix}] \, \Theta_{\vec{b}}^{\vec{a}}[\begin{smallmatrix}3&6\\4&5\end{smallmatrix}]$$
$$- (\gamma^\mu C)_{\alpha\gamma}(\gamma^\nu C)_{\beta\delta}\, E_{14} E_{16} E_{23} E_{25}\, \Theta_{\vec{b}}^{\vec{a}}[\begin{smallmatrix}1&1&4&6\\2&2&3&5\end{smallmatrix}] \, \Theta_{\vec{b}}^{\vec{a}}[\begin{smallmatrix}3&4\\5&6\end{smallmatrix}] \, \Theta_{\vec{b}}^{\vec{a}}[\begin{smallmatrix}3&6\\4&5\end{smallmatrix}]$$
$$- (\gamma^\mu C)_{\beta\delta}(\gamma^\nu C)_{\alpha\gamma}\, E_{13} E_{15} E_{24} E_{26}\, \Theta_{\vec{b}}^{\vec{a}}[\begin{smallmatrix}1&1&3&5\\2&2&4&6\end{smallmatrix}] \, \Theta_{\vec{b}}^{\vec{a}}[\begin{smallmatrix}3&4\\5&6\end{smallmatrix}] \, \Theta_{\vec{b}}^{\vec{a}}[\begin{smallmatrix}3&6\\4&5\end{smallmatrix}]$$
$$+ (\gamma^\mu C)_{\alpha\delta}(\gamma^\nu C)_{\beta\gamma}\, E_{14} E_{15} E_{23} E_{26}\, \Theta_{\vec{b}}^{\vec{a}}[\begin{smallmatrix}1&1&4&5\\2&2&3&6\end{smallmatrix}] \, \Theta_{\vec{b}}^{\vec{a}}[\begin{smallmatrix}3&4\\5&6\end{smallmatrix}] \, \Theta_{\vec{b}}^{\vec{a}}[\begin{smallmatrix}3&5\\4&6\end{smallmatrix}]$$
$$+ (\gamma^\mu C)_{\beta\gamma}(\gamma^\nu C)_{\alpha\delta}\, E_{13} E_{16} E_{24} E_{25}\, \Theta_{\vec{b}}^{\vec{a}}[\begin{smallmatrix}1&1&3&6\\2&2&4&5\end{smallmatrix}] \, \Theta_{\vec{b}}^{\vec{a}}[\begin{smallmatrix}3&4\\5&6\end{smallmatrix}] \, \Theta_{\vec{b}}^{\vec{a}}[\begin{smallmatrix}3&5\\4&6\end{smallmatrix}] \Big], \tag{5.50}$$

or with mixed chiralities:

$$\langle \psi^\mu(z_1)\, \psi^\nu(z_2)\, S_\alpha(z_3)\, S_\beta(z_4)\, S^{\dot\gamma}(z_5)\, S^{\dot\delta}(z_6) \rangle^{\vec{a}}_{\vec{b}}$$
$$= \frac{1}{\left[\Theta_{\vec{b}}^{\vec{a}}(\vec{0})\right]^3} \frac{(E_{35} E_{36} E_{45} E_{46})^{1/4}\,(E_{34} E_{56})^{-1/4}}{(E_{13} E_{14} E_{15} E_{16} E_{23} E_{24} E_{25} E_{26})^{1/2}}$$
$$\times \Bigg[\frac{\eta^{\mu\nu} C_\alpha{}^{\dot\gamma} C_\beta{}^{\dot\delta}}{E_{12} E_{35} E_{46}}\, E_{13} E_{14} E_{25} E_{26}\, \Theta_{\vec{b}}^{\vec{a}}[\begin{smallmatrix}1&1&3&4\\2&2&5&6\end{smallmatrix}] \, \Theta_{\vec{b}}^{\vec{a}}[\begin{smallmatrix}3&6\\4&5\end{smallmatrix}]^2$$
$$- \frac{\eta^{\mu\nu} C_\alpha{}^{\dot\delta} C_\beta{}^{\dot\gamma}}{E_{12} E_{36} E_{45}}\, E_{13} E_{14} E_{25} E_{26}\, \Theta_{\vec{b}}^{\vec{a}}[\begin{smallmatrix}1&1&3&4\\2&2&5&6\end{smallmatrix}] \, \Theta_{\vec{b}}^{\vec{a}}[\begin{smallmatrix}3&5\\4&6\end{smallmatrix}]^2$$
$$+ \frac{1}{2}\,(\gamma^\mu C)_{\alpha\beta}(\bar\gamma^\nu C)^{\dot\gamma\dot\delta}\, E_{12}\, \Theta_{\vec{b}}^{\vec{a}}[\begin{smallmatrix}1&1&2&2\\3&4&5&6\end{smallmatrix}] \, \Theta_{\vec{b}}^{\vec{a}}[\begin{smallmatrix}3&5\\4&6\end{smallmatrix}] \, \Theta_{\vec{b}}^{\vec{a}}[\begin{smallmatrix}3&6\\4&5\end{smallmatrix}]$$
$$+ \frac{1}{2}\,(\gamma^\mu \bar\gamma^\nu C)_\alpha{}^{\dot\gamma}\, \frac{C_\beta{}^{\dot\delta}}{E_{46}}\, E_{14} E_{26}\, \Theta_{\vec{b}}^{\vec{a}}[\begin{smallmatrix}1&1&4\\3&5&6\end{smallmatrix}] \, \Theta_{\vec{b}}^{\vec{a}}[\begin{smallmatrix}2&2&6\\3&4&5\end{smallmatrix}] \, \Theta_{\vec{b}}^{\vec{a}}[\begin{smallmatrix}3&6\\4&5\end{smallmatrix}]$$
$$- \frac{1}{2}\,(\gamma^\mu \bar\gamma^\nu C)_\alpha{}^{\dot\delta}\, \frac{C_\beta{}^{\dot\gamma}}{E_{45}}\, E_{14} E_{25}\, \Theta_{\vec{b}}^{\vec{a}}[\begin{smallmatrix}1&1&4\\3&5&6\end{smallmatrix}] \, \Theta_{\vec{b}}^{\vec{a}}[\begin{smallmatrix}2&2&5\\3&4&6\end{smallmatrix}] \, \Theta_{\vec{b}}^{\vec{a}}[\begin{smallmatrix}3&5\\4&6\end{smallmatrix}]$$
$$- \frac{1}{2}\,(\gamma^\mu \bar\gamma^\nu C)_\beta{}^{\dot\gamma}\, \frac{C_\alpha{}^{\dot\delta}}{E_{36}}\, E_{13} E_{26}\, \Theta_{\vec{b}}^{\vec{a}}[\begin{smallmatrix}1&1&3\\4&5&6\end{smallmatrix}] \, \Theta_{\vec{b}}^{\vec{a}}[\begin{smallmatrix}2&2&6\\3&4&5\end{smallmatrix}] \, \Theta_{\vec{b}}^{\vec{a}}[\begin{smallmatrix}3&5\\4&6\end{smallmatrix}]$$
$$+ \frac{1}{2}\,(\gamma^\mu \bar\gamma^\nu C)_\beta{}^{\dot\delta}\, \frac{C_\alpha{}^{\dot\gamma}}{E_{35}}\, E_{13} E_{25}\, \Theta_{\vec{b}}^{\vec{a}}[\begin{smallmatrix}1&1&3\\4&5&6\end{smallmatrix}] \, \Theta_{\vec{b}}^{\vec{a}}[\begin{smallmatrix}2&2&5\\3&4&6\end{smallmatrix}] \, \Theta_{\vec{b}}^{\vec{a}}[\begin{smallmatrix}3&6\\4&5\end{smallmatrix}] \Bigg]. \tag{5.51}$$

5.3 Results of RNS Loop Correlators

The seven-point correlator consisting of one NS fermion, four left- and two right-handed spin fields is found to be

$$\langle \psi^\mu(z_1) \, S_\alpha(z_2) \, S_\beta(z_3) \, S_\gamma(z_4) \, S_\delta(z_5) \, S^{\dot\epsilon}(z_6) \, S^{\dot\zeta}(z_7) \rangle_{\vec b}^{\vec a}$$
$$= \frac{1}{\sqrt{2} \, [\Theta_{\vec b}^{\vec a}(\vec 0)]^3} \frac{(E_{26} \, E_{27} \, E_{36} \, E_{37} \, E_{46} \, E_{47} \, E_{56} \, E_{57})^{1/4}}{(E_{12} \, E_{13} \, E_{14} \, E_{15} \, E_{16} \, E_{17})^{1/2} (E_{23} \, E_{24} \, E_{25} \, E_{34} \, E_{35} \, E_{45} \, E_{67})^{1/4}}$$
$$\times \Bigg[(\gamma^\mu C)_{\alpha\beta} \, E_{14} \, E_{15} \, \Theta_{\vec b}^{\vec a} [{}^{1\,1\,4\,5}_{2\,3\,6\,7}]$$
$$\left(\frac{C_\gamma^{\dot\epsilon} C_\delta^{\dot\zeta}}{E_{46} \, E_{57}} \Theta_{\vec b}^{\vec a} [{}^{2\,4\,7}_{3\,5\,6}] \, \Theta_{\vec b}^{\vec a} [{}^{2\,5\,6}_{3\,4\,7}] - \frac{C_\delta^{\dot\epsilon} C_\gamma^{\dot\zeta}}{E_{47} \, E_{56}} \Theta_{\vec b}^{\vec a} [{}^{2\,4\,6}_{3\,5\,7}] \, \Theta_{\vec b}^{\vec a} [{}^{2\,5\,7}_{3\,4\,6}] \right)$$
$$+ (\gamma^\mu C)_{\alpha\gamma} \, E_{13} \, E_{15} \, \Theta_{\vec b}^{\vec a} [{}^{1\,1\,3\,5}_{2\,4\,6\,7}]$$
$$\left(\frac{C_\delta^{\dot\epsilon} C_\beta^{\dot\zeta}}{E_{37} \, E_{56}} \Theta_{\vec b}^{\vec a} [{}^{2\,3\,6}_{4\,5\,7}] \, \Theta_{\vec b}^{\vec a} [{}^{2\,5\,7}_{3\,4\,6}] - \frac{C_\beta^{\dot\epsilon} C_\delta^{\dot\zeta}}{E_{36} \, E_{57}} \Theta_{\vec b}^{\vec a} [{}^{2\,3\,7}_{4\,5\,6}] \, \Theta_{\vec b}^{\vec a} [{}^{2\,5\,6}_{3\,4\,7}] \right)$$
$$+ (\gamma^\mu C)_{\alpha\delta} \, E_{13} \, E_{14} \, \Theta_{\vec b}^{\vec a} [{}^{1\,1\,3\,4}_{2\,5\,6\,7}]$$
$$\left(\frac{C_\beta^{\dot\epsilon} C_\gamma^{\dot\zeta}}{E_{36} \, E_{47}} \Theta_{\vec b}^{\vec a} [{}^{2\,3\,7}_{4\,5\,6}] \, \Theta_{\vec b}^{\vec a} [{}^{2\,4\,6}_{3\,5\,7}] - \frac{C_\gamma^{\dot\epsilon} C_\beta^{\dot\zeta}}{E_{37} \, E_{46}} \Theta_{\vec b}^{\vec a} [{}^{2\,3\,6}_{4\,5\,7}] \, \Theta_{\vec b}^{\vec a} [{}^{2\,4\,7}_{3\,5\,6}] \right)$$
$$+ (\gamma^\mu C)_{\beta\gamma} \, E_{12} \, E_{15} \, \Theta_{\vec b}^{\vec a} [{}^{1\,1\,2\,5}_{3\,4\,6\,7}]$$
$$\left(\frac{C_\alpha^{\dot\epsilon} C_\delta^{\dot\zeta}}{E_{26} \, E_{57}} \Theta_{\vec b}^{\vec a} [{}^{2\,3\,7}_{4\,5\,6}] \, \Theta_{\vec b}^{\vec a} [{}^{2\,4\,7}_{3\,5\,6}] - \frac{C_\delta^{\dot\epsilon} C_\alpha^{\dot\zeta}}{E_{27} \, E_{56}} \Theta_{\vec b}^{\vec a} [{}^{2\,3\,6}_{4\,5\,7}] \, \Theta_{\vec b}^{\vec a} [{}^{2\,4\,6}_{3\,5\,7}] \right)$$
$$+ (\gamma^\mu C)_{\beta\delta} \, E_{12} \, E_{14} \, \Theta_{\vec b}^{\vec a} [{}^{1\,1\,2\,4}_{3\,5\,6\,7}]$$
$$\left(\frac{C_\gamma^{\dot\epsilon} C_\alpha^{\dot\zeta}}{E_{27} \, E_{46}} \Theta_{\vec b}^{\vec a} [{}^{2\,3\,6}_{4\,5\,7}] \, \Theta_{\vec b}^{\vec a} [{}^{2\,5\,6}_{3\,4\,7}] - \frac{C_\alpha^{\dot\epsilon} C_\gamma^{\dot\zeta}}{E_{26} \, E_{47}} \Theta_{\vec b}^{\vec a} [{}^{2\,3\,7}_{4\,5\,6}] \, \Theta_{\vec b}^{\vec a} [{}^{2\,5\,7}_{3\,4\,6}] \right)$$
$$+ (\gamma^\mu C)_{\gamma\delta} \, E_{12} \, E_{13} \, \Theta_{\vec b}^{\vec a} [{}^{1\,1\,2\,3}_{4\,5\,6\,7}]$$
$$\left(\frac{C_\alpha^{\dot\epsilon} C_\beta^{\dot\zeta}}{E_{26} \, E_{37}} \Theta_{\vec b}^{\vec a} [{}^{2\,4\,7}_{3\,5\,6}] \, \Theta_{\vec b}^{\vec a} [{}^{2\,5\,7}_{3\,4\,6}] - \frac{C_\beta^{\dot\epsilon} C_\alpha^{\dot\zeta}}{E_{27} \, E_{36}} \Theta_{\vec b}^{\vec a} [{}^{2\,4\,6}_{3\,5\,7}] \, \Theta_{\vec b}^{\vec a} [{}^{2\,5\,6}_{3\,4\,7}] \right) \Bigg]. \quad (5.52)$$

5.3.3 Results for $D = 8$

Let us start the discussion of loop correlators in eight dimensions with the case of four spin fields, firstly

$$\langle S_\alpha(z_1) \, S_\beta(z_2) \, S^{\dot\gamma}(z_3) \, S^{\dot\delta}(z_4) \rangle_{\vec b}^{\vec a} = \frac{\Theta_{\vec b}^{\vec a} [{}^{1\,3}_{2\,4}] \, \Theta_{\vec b}^{\vec a} [{}^{1\,4}_{2\,3}]}{2 \, [\Theta_{\vec b}^{\vec a}(\vec 0)]^4 \, E_{12} \, E_{34} \, (E_{13} \, E_{14} \, E_{23} \, E_{24})^{1/2}}$$
$$\times \left[(\gamma^\mu C)_\alpha{}^{\dot\gamma} (\gamma_\mu C)_\beta{}^{\dot\delta} \, E_{14} \, E_{23} \, \Theta_{\vec b}^{\vec a} [{}^{1\,4}_{2\,3}]^2 + (\gamma^\mu C)_\alpha{}^{\dot\delta} (\gamma_\mu C)_\beta{}^{\dot\gamma} \, E_{13} \, E_{24} \, \Theta_{\vec b}^{\vec a} [{}^{1\,3}_{2\,4}]^2 \right] \quad (5.53)$$

and secondly

$$\langle S_\alpha(z_1) S_\beta(z_2) S_\gamma(z_3) S_\delta(z_4) \rangle_{\vec{b}}^{\vec{a}} = \frac{1}{[\Theta_{\vec{b}}^{\vec{a}}(\vec{0})]^4} \left[\frac{C_{\alpha\beta} C_{\gamma\delta}}{E_{12} E_{34}} \Theta_{\vec{b}}^{\vec{a}}[\begin{smallmatrix}1&3\\2&4\end{smallmatrix}]^2 \Theta_{\vec{b}}^{\vec{a}}[\begin{smallmatrix}1&4\\2&3\end{smallmatrix}]^2 \right.$$
$$\left. - \frac{C_{\alpha\gamma} C_{\beta\delta}}{E_{13} E_{24}} \Theta_{\vec{b}}^{\vec{a}}[\begin{smallmatrix}1&2\\3&4\end{smallmatrix}]^2 \Theta_{\vec{b}}^{\vec{a}}[\begin{smallmatrix}1&4\\2&3\end{smallmatrix}]^2 + \frac{C_{\alpha\delta} C_{\beta\gamma}}{E_{14} E_{23}} \Theta_{\vec{b}}^{\vec{a}}[\begin{smallmatrix}1&2\\3&4\end{smallmatrix}]^2 \Theta_{\vec{b}}^{\vec{a}}[\begin{smallmatrix}1&3\\2&4\end{smallmatrix}]^2 \right]. \quad (5.54)$$

As one sees, the four required $SO(2)$ spin systems needed to express the $SO(1,7)$ RNS fields in eight dimensions give rise to four generalized Θ functions. In the presence of an additional NS fermion, we find for the correlation functions

$$\langle \psi^\mu(z_1) S_\alpha(z_2) S_\beta(z_3) S_\gamma(z_4) S^{\dot\delta}(z_5) \rangle_{\vec{b}}^{\vec{a}} = \frac{-1}{\sqrt{2} [\Theta_{\vec{b}}^{\vec{a}}(\vec{0})]^4 (E_{12} E_{13} E_{14} E_{15} E_{25} E_{35} E_{45})^{1/2}}$$
$$\times \left[\frac{C_{\alpha\beta}}{E_{23}} (\gamma^\mu C)_\gamma{}^{\dot\delta} E_{13} E_{25} \Theta_{\vec{b}}^{\vec{a}}[\begin{smallmatrix}1&1&3\\2&4&5\end{smallmatrix}] \Theta_{\vec{b}}^{\vec{a}}[\begin{smallmatrix}2&4\\3&5\end{smallmatrix}] \Theta_{\vec{b}}^{\vec{a}}[\begin{smallmatrix}2&5\\3&4\end{smallmatrix}]^2 \right.$$
$$+ \frac{C_{\gamma\beta}}{E_{34}} (\gamma^\mu C)_\alpha{}^{\dot\delta} E_{13} E_{45} \Theta_{\vec{b}}^{\vec{a}}[\begin{smallmatrix}1&1&3\\2&4&5\end{smallmatrix}] \Theta_{\vec{b}}^{\vec{a}}[\begin{smallmatrix}2&4\\3&5\end{smallmatrix}] \Theta_{\vec{b}}^{\vec{a}}[\begin{smallmatrix}2&3\\4&5\end{smallmatrix}]^2$$
$$+ (\gamma^\lambda \bar{\gamma}^\mu C)_{\alpha\beta} (\gamma_\lambda C)_\gamma{}^{\dot\delta} \frac{E_{14} E_{25}}{2 E_{24}} \Theta_{\vec{b}}^{\vec{a}}[\begin{smallmatrix}1&1&4\\2&3&5\end{smallmatrix}] \Theta_{\vec{b}}^{\vec{a}}[\begin{smallmatrix}2&3\\4&5\end{smallmatrix}] \Theta_{\vec{b}}^{\vec{a}}[\begin{smallmatrix}2&5\\3&4\end{smallmatrix}]^2$$
$$\left. + (\gamma^\lambda \bar{\gamma}^\mu C)_{\gamma\beta} (\gamma_\lambda C)_\alpha{}^{\dot\delta} \frac{E_{12} E_{45}}{2 E_{24}} \Theta_{\vec{b}}^{\vec{a}}[\begin{smallmatrix}1&1&2\\3&4&5\end{smallmatrix}] \Theta_{\vec{b}}^{\vec{a}}[\begin{smallmatrix}2&5\\3&4\end{smallmatrix}] \Theta_{\vec{b}}^{\vec{a}}[\begin{smallmatrix}2&3\\4&5\end{smallmatrix}]^2 \right]. \quad (5.55)$$

Here we have chosen a different basis of tensors than in (3.100) in order to make antisymmetry in $S_\alpha(z_2) \leftrightarrow S_\gamma(z_4)$ manifest. The triality symmetric tree-level correlator with two fermions and two spin fields of each chirality generalizes as follows to higher genus:

$$\langle \psi^\mu(z_1) \psi^\nu(z_2) S_\alpha(z_3) S_\beta(z_4) S^{\dot\gamma}(z_5) S^{\dot\delta}(z_6) \rangle_{\vec{b}}^{\vec{a}}$$
$$= \frac{(E_{35} E_{36} E_{45} E_{46})^{-1/2}}{4 [\Theta_{\vec{b}}^{\vec{a}}(\vec{0})]^4 (E_{13} E_{14} E_{15} E_{16} E_{23} E_{24} E_{25} E_{26})^{1/2}}$$
$$\times \left[\frac{\eta^{\mu\nu} C_{\alpha\beta} C^{\dot\gamma\dot\delta}}{E_{12} E_{34} E_{56}} \Theta_{\vec{b}}^{\vec{a}}[\begin{smallmatrix}3&5\\4&6\end{smallmatrix}] \Theta_{\vec{b}}^{\vec{a}}[\begin{smallmatrix}3&6\\4&5\end{smallmatrix}] \right.$$
$$\left(E_{36} E_{45} \Theta_{\vec{b}}^{\vec{a}}[\begin{smallmatrix}3&6\\4&5\end{smallmatrix}] \left(E_{13} E_{16} E_{24} E_{25} \Theta_{\vec{b}}^{\vec{a}}[\begin{smallmatrix}1&1&3&6\\2&2&4&5\end{smallmatrix}] + E_{14} E_{15} E_{23} E_{26} \Theta_{\vec{b}}^{\vec{a}}[\begin{smallmatrix}1&1&4&5\\2&2&3&6\end{smallmatrix}] \right) \right.$$
$$+ E_{35} E_{46} \Theta_{\vec{b}}^{\vec{a}}[\begin{smallmatrix}3&5\\4&6\end{smallmatrix}] \left(E_{13} E_{15} E_{24} E_{26} \Theta_{\vec{b}}^{\vec{a}}[\begin{smallmatrix}1&1&3&5\\2&2&4&6\end{smallmatrix}] + E_{14} E_{16} E_{23} E_{25} \Theta_{\vec{b}}^{\vec{a}}[\begin{smallmatrix}1&1&4&6\\2&2&3&5\end{smallmatrix}] \right) \right)$$
$$+ (\gamma^\mu C)_\alpha{}^{\dot\gamma} (\gamma^\nu C)_\beta{}^{\dot\delta} \Theta_{\vec{b}}^{\vec{a}}[\begin{smallmatrix}3&4\\5&6\end{smallmatrix}] \Theta_{\vec{b}}^{\vec{a}}[\begin{smallmatrix}3&6\\4&5\end{smallmatrix}]$$
$$\left(E_{14} E_{25} E_{36} \Theta_{\vec{b}}^{\vec{a}}[\begin{smallmatrix}1&1&4\\3&5&6\end{smallmatrix}] \Theta_{\vec{b}}^{\vec{a}}[\begin{smallmatrix}2&2&5\\3&4&6\end{smallmatrix}] - E_{16} E_{23} E_{45} \Theta_{\vec{b}}^{\vec{a}}[\begin{smallmatrix}1&1&6\\3&4&5\end{smallmatrix}] \Theta_{\vec{b}}^{\vec{a}}[\begin{smallmatrix}2&2&3\\4&5&6\end{smallmatrix}] \right)$$
$$+ (\gamma^\mu C)_\alpha{}^{\dot\delta} (\gamma^\nu C)_\beta{}^{\dot\gamma} \Theta_{\vec{b}}^{\vec{a}}[\begin{smallmatrix}3&4\\5&6\end{smallmatrix}] \Theta_{\vec{b}}^{\vec{a}}[\begin{smallmatrix}3&5\\4&6\end{smallmatrix}]$$
$$\left. \left(E_{15} E_{23} E_{46} \Theta_{\vec{b}}^{\vec{a}}[\begin{smallmatrix}1&1&5\\3&4&6\end{smallmatrix}] \Theta_{\vec{b}}^{\vec{a}}[\begin{smallmatrix}2&2&3\\4&5&6\end{smallmatrix}] - E_{14} E_{26} E_{35} \Theta_{\vec{b}}^{\vec{a}}[\begin{smallmatrix}1&1&4\\3&5&6\end{smallmatrix}] \Theta_{\vec{b}}^{\vec{a}}[\begin{smallmatrix}2&2&6\\3&4&5\end{smallmatrix}] \right) \right]$$

5.3 Results of RNS Loop Correlators

$$+ (\gamma^\mu C)_\beta{}^{\dot\gamma} (\gamma^\nu C)_\alpha{}^{\dot\delta} \, \Theta_{\vec b}^{\vec a} \begin{bmatrix} 3 & 4 \\ 5 & 6 \end{bmatrix} \Theta_{\vec b}^{\vec a} \begin{bmatrix} 3 & 5 \\ 4 & 6 \end{bmatrix}$$
$$\left(E_{16} E_{24} E_{35} \, \Theta_{\vec b}^{\vec a} \begin{bmatrix} 1 & 1 & 6 \\ 3 & 4 & 5 \end{bmatrix} \Theta_{\vec b}^{\vec a} \begin{bmatrix} 2 & 2 & 4 \\ 3 & 5 & 6 \end{bmatrix} - E_{13} E_{25} E_{46} \, \Theta_{\vec b}^{\vec a} \begin{bmatrix} 1 & 1 & 3 \\ 4 & 5 & 6 \end{bmatrix} \Theta_{\vec b}^{\vec a} \begin{bmatrix} 2 & 2 & 5 \\ 3 & 4 & 6 \end{bmatrix} \right)$$
$$+ (\gamma^\mu C)_\beta{}^{\dot\delta} (\gamma^\nu C)_\alpha{}^{\dot\gamma} \, \Theta_{\vec b}^{\vec a} \begin{bmatrix} 3 & 4 \\ 5 & 6 \end{bmatrix} \Theta_{\vec b}^{\vec a} \begin{bmatrix} 3 & 6 \\ 4 & 5 \end{bmatrix}$$
$$\left(E_{13} E_{26} E_{45} \, \Theta_{\vec b}^{\vec a} \begin{bmatrix} 1 & 1 & 3 \\ 4 & 5 & 6 \end{bmatrix} \Theta_{\vec b}^{\vec a} \begin{bmatrix} 2 & 2 & 6 \\ 3 & 4 & 5 \end{bmatrix} - E_{15} E_{24} E_{36} \, \Theta_{\vec b}^{\vec a} \begin{bmatrix} 1 & 1 & 5 \\ 3 & 4 & 6 \end{bmatrix} \Theta_{\vec b}^{\vec a} \begin{bmatrix} 2 & 2 & 4 \\ 3 & 5 & 6 \end{bmatrix} \right)$$
$$+ \frac{C^{\dot\gamma\dot\delta}}{E_{56}} (\gamma^{\mu\nu} C)_{\alpha\beta} \, \Theta_{\vec b}^{\vec a} \begin{bmatrix} 3 & 5 \\ 4 & 6 \end{bmatrix} \Theta_{\vec b}^{\vec a} \begin{bmatrix} 3 & 6 \\ 4 & 5 \end{bmatrix}$$
$$\left(E_{15} E_{25} E_{36} E_{46} \, \Theta_{\vec b}^{\vec a} \begin{bmatrix} 1 & 1 & 5 \\ 3 & 4 & 6 \end{bmatrix} \Theta_{\vec b}^{\vec a} \begin{bmatrix} 2 & 2 & 5 \\ 3 & 4 & 6 \end{bmatrix} - E_{16} E_{26} E_{35} E_{45} \, \Theta_{\vec b}^{\vec a} \begin{bmatrix} 1 & 1 & 6 \\ 3 & 4 & 5 \end{bmatrix} \Theta_{\vec b}^{\vec a} \begin{bmatrix} 2 & 2 & 6 \\ 3 & 4 & 5 \end{bmatrix} \right)$$
$$+ \frac{C_{\alpha\beta}}{E_{34}} (\bar\gamma^{\mu\nu} C)^{\dot\gamma\dot\delta} \, \Theta_{\vec b}^{\vec a} \begin{bmatrix} 3 & 5 \\ 4 & 6 \end{bmatrix} \Theta_{\vec b}^{\vec a} \begin{bmatrix} 3 & 6 \\ 4 & 5 \end{bmatrix}$$
$$\left(E_{13} E_{23} E_{45} E_{46} \, \Theta_{\vec b}^{\vec a} \begin{bmatrix} 1 & 1 & 3 \\ 4 & 5 & 6 \end{bmatrix} \Theta_{\vec b}^{\vec a} \begin{bmatrix} 2 & 2 & 3 \\ 4 & 5 & 6 \end{bmatrix} - E_{14} E_{24} E_{35} E_{36} \, \Theta_{\vec b}^{\vec a} \begin{bmatrix} 1 & 1 & 4 \\ 3 & 5 & 6 \end{bmatrix} \Theta_{\vec b}^{\vec a} \begin{bmatrix} 2 & 2 & 4 \\ 3 & 5 & 6 \end{bmatrix} \right)$$
$$- \frac{\eta^{\mu\nu}}{E_{12}} (\gamma_\lambda C)_{[\alpha}{}^{\dot\gamma} (\gamma^\lambda C)_{\beta]}{}^{\dot\delta} \, \Theta_{\vec b}^{\vec a} \begin{bmatrix} 3 & 4 \\ 5 & 6 \end{bmatrix} \Theta_{\vec b}^{\vec a} \begin{bmatrix} 3 & 5 \\ 4 & 6 \end{bmatrix} \Theta_{\vec b}^{\vec a} \begin{bmatrix} 3 & 6 \\ 4 & 5 \end{bmatrix}$$
$$\left(E_{13} E_{14} E_{25} E_{26} \, \Theta_{\vec b}^{\vec a} \begin{bmatrix} 1 & 1 & 3 & 4 \\ 2 & 2 & 5 & 6 \end{bmatrix} + E_{15} E_{16} E_{23} E_{24} \, \Theta_{\vec b}^{\vec a} \begin{bmatrix} 1 & 1 & 5 & 6 \\ 2 & 2 & 3 & 4 \end{bmatrix} \right)$$
$$+ (\gamma^{[\mu}{}_\lambda C)_{\alpha\beta} (\bar\gamma^{\nu]\lambda} C)^{\dot\gamma\dot\delta} \, \Theta_{\vec b}^{\vec a} \begin{bmatrix} 3 & 4 \\ 5 & 6 \end{bmatrix} \Theta_{\vec b}^{\vec a} \begin{bmatrix} 3 & 6 \\ 4 & 5 \end{bmatrix}$$
$$\left. \left(E_{14} E_{25} E_{36} \, \Theta_{\vec b}^{\vec a} \begin{bmatrix} 1 & 1 & 4 \\ 3 & 5 & 6 \end{bmatrix} \Theta_{\vec b}^{\vec a} \begin{bmatrix} 2 & 2 & 5 \\ 3 & 4 & 6 \end{bmatrix} + E_{16} E_{23} E_{45} \, \Theta_{\vec b}^{\vec a} \begin{bmatrix} 1 & 1 & 6 \\ 3 & 4 & 5 \end{bmatrix} \Theta_{\vec b}^{\vec a} \begin{bmatrix} 2 & 2 & 3 \\ 4 & 5 & 6 \end{bmatrix} \right) \right]. \quad (5.56)$$

Using Fay's trisecant identity, one can alternatively write the last two lines as

$$\frac{1}{E_{12}} (\gamma^{[\mu}{}_\lambda C)_{\alpha\beta} (\bar\gamma^{\nu]\lambda} C)^{\dot\gamma\dot\delta} \, \Theta_{\vec b}^{\vec a} \begin{bmatrix} 3 & 4 \\ 5 & 6 \end{bmatrix} \Theta_{\vec b}^{\vec a} \begin{bmatrix} 3 & 5 \\ 4 & 6 \end{bmatrix} \Theta_{\vec b}^{\vec a} \begin{bmatrix} 3 & 6 \\ 4 & 5 \end{bmatrix}$$
$$\times \left(E_{13} E_{14} E_{25} E_{26} \, \Theta_{\vec b}^{\vec a} \begin{bmatrix} 1 & 1 & 3 & 4 \\ 2 & 2 & 5 & 6 \end{bmatrix} - E_{15} E_{16} E_{23} E_{24} \, \Theta_{\vec b}^{\vec a} \begin{bmatrix} 1 & 1 & 5 & 6 \\ 2 & 2 & 3 & 4 \end{bmatrix} \right). \quad (5.57)$$

The second non-vanishing six-point function with four spin fields reads:

$$\left\langle \psi^\mu(z_1) \, \psi^\nu(z_2) \, S_\alpha(z_3) \, S_\beta(z_4) \, S_\gamma(z_5) \, S_\delta(z_6) \right\rangle_{\vec b}^{\vec a}$$
$$= \frac{1}{4 \left[\Theta_{\vec b}^{\vec a}(\vec 0) \right]^4 (E_{13} E_{14} E_{15} E_{16} E_{23} E_{24} E_{25} E_{26})^{1/2}}$$
$$\times \left[\frac{\eta^{\mu\nu} C_{\alpha\beta} C_{\gamma\delta}}{E_{12} E_{34} E_{56}} \Theta_{\vec b}^{\vec a} \begin{bmatrix} 3 & 5 \\ 4 & 6 \end{bmatrix} \Theta_{\vec b}^{\vec a} \begin{bmatrix} 3 & 6 \\ 4 & 5 \end{bmatrix} \right.$$
$$\left(E_{13} E_{24} E_{15} E_{26} \, \Theta_{\vec b}^{\vec a} \begin{bmatrix} 3 & 6 \\ 4 & 5 \end{bmatrix} \Theta_{\vec b}^{\vec a} \begin{bmatrix} 1 & 1 & 3 & 5 \\ 2 & 2 & 4 & 6 \end{bmatrix} + E_{23} E_{14} E_{25} E_{16} \, \Theta_{\vec b}^{\vec a} \begin{bmatrix} 3 & 6 \\ 4 & 5 \end{bmatrix} \Theta_{\vec b}^{\vec a} \begin{bmatrix} 2 & 2 & 3 & 5 \\ 1 & 1 & 4 & 6 \end{bmatrix} \right.$$
$$\left. + E_{13} E_{24} E_{25} E_{16} \, \Theta_{\vec b}^{\vec a} \begin{bmatrix} 3 & 5 \\ 4 & 6 \end{bmatrix} \Theta_{\vec b}^{\vec a} \begin{bmatrix} 1 & 1 & 3 & 6 \\ 2 & 2 & 4 & 5 \end{bmatrix} + E_{23} E_{14} E_{15} E_{26} \, \Theta_{\vec b}^{\vec a} \begin{bmatrix} 3 & 5 \\ 4 & 6 \end{bmatrix} \Theta_{\vec b}^{\vec a} \begin{bmatrix} 2 & 2 & 3 & 6 \\ 1 & 1 & 4 & 5 \end{bmatrix} \right)$$
$$- \frac{2 \eta^{\mu\nu} C_{\alpha\gamma} C_{\beta\delta}}{E_{12} E_{35} E_{46}} \Theta_{\vec b}^{\vec a} \begin{bmatrix} 3 & 6 \\ 4 & 5 \end{bmatrix} \Theta_{\vec b}^{\vec a} \begin{bmatrix} 3 & 4 \\ 5 & 6 \end{bmatrix}^2$$
$$\left(E_{13} E_{24} E_{25} E_{16} \, \Theta_{\vec b}^{\vec a} \begin{bmatrix} 1 & 1 & 3 & 6 \\ 2 & 2 & 4 & 5 \end{bmatrix} + E_{23} E_{14} E_{15} E_{26} \, \Theta_{\vec b}^{\vec a} \begin{bmatrix} 2 & 2 & 3 & 6 \\ 1 & 1 & 4 & 5 \end{bmatrix} \right)$$

$$
\begin{aligned}
&+ \frac{2\eta^{\mu\nu} C_{\alpha\delta} C_{\beta\gamma}}{E_{12} E_{36} E_{45}} \Theta^{\vec{a}}_{\vec{b}}\begin{bmatrix}3&5\\4&6\end{bmatrix} \Theta^{\vec{a}}_{\vec{b}}\begin{bmatrix}3&4\\5&6\end{bmatrix}^2 \\
&\qquad \left(E_{13} E_{24} E_{15} E_{26} \Theta^{\vec{a}}_{\vec{b}}\begin{bmatrix}1&1&3&5\\2&2&4&6\end{bmatrix} + E_{23} E_{14} E_{25} E_{16} \Theta^{\vec{a}}_{\vec{b}}\begin{bmatrix}2&2&3&5\\1&1&4&6\end{bmatrix} \right) \\
&+ \frac{C_{\alpha\beta}(\gamma^{\mu\nu} C)_{\gamma\delta}}{E_{34}} \Theta^{\vec{a}}_{\vec{b}}\begin{bmatrix}3&5\\4&6\end{bmatrix} \Theta^{\vec{a}}_{\vec{b}}\begin{bmatrix}3&6\\4&5\end{bmatrix} \\
&\qquad \left(E_{13} E_{24} \Theta^{\vec{a}}_{\vec{b}}\begin{bmatrix}1&1&3\\4&5&6\end{bmatrix} \Theta^{\vec{a}}_{\vec{b}}\begin{bmatrix}2&2&4\\3&5&6\end{bmatrix} + E_{14} E_{23} \Theta^{\vec{a}}_{\vec{b}}\begin{bmatrix}1&1&4\\3&5&6\end{bmatrix} \Theta^{\vec{a}}_{\vec{b}}\begin{bmatrix}2&2&3\\4&5&6\end{bmatrix} \right) \\
&+ \frac{C_{\gamma\delta}(\gamma^{\mu\nu} C)_{\alpha\beta}}{E_{56}} \Theta^{\vec{a}}_{\vec{b}}\begin{bmatrix}3&5\\4&6\end{bmatrix} \Theta^{\vec{a}}_{\vec{b}}\begin{bmatrix}3&6\\4&5\end{bmatrix} \\
&\qquad \left(E_{15} E_{26} \Theta^{\vec{a}}_{\vec{b}}\begin{bmatrix}1&1&5\\3&4&6\end{bmatrix} \Theta^{\vec{a}}_{\vec{b}}\begin{bmatrix}2&2&6\\3&4&5\end{bmatrix} + E_{16} E_{25} \Theta^{\vec{a}}_{\vec{b}}\begin{bmatrix}1&1&6\\3&4&5\end{bmatrix} \Theta^{\vec{a}}_{\vec{b}}\begin{bmatrix}2&2&5\\3&4&6\end{bmatrix} \right) \\
&- \frac{C_{\alpha\gamma}(\gamma^{\mu\nu} C)_{\beta\delta}}{E_{35}} \Theta^{\vec{a}}_{\vec{b}}\begin{bmatrix}3&4\\5&6\end{bmatrix} \Theta^{\vec{a}}_{\vec{b}}\begin{bmatrix}3&6\\4&5\end{bmatrix} \\
&\qquad \left(E_{13} E_{25} \Theta^{\vec{a}}_{\vec{b}}\begin{bmatrix}1&1&3\\4&5&6\end{bmatrix} \Theta^{\vec{a}}_{\vec{b}}\begin{bmatrix}2&2&5\\3&4&6\end{bmatrix} + E_{15} E_{23} \Theta^{\vec{a}}_{\vec{b}}\begin{bmatrix}2&2&3\\4&5&6\end{bmatrix} \Theta^{\vec{a}}_{\vec{b}}\begin{bmatrix}1&1&5\\3&4&6\end{bmatrix} \right) \\
&- \frac{C_{\beta\delta}(\gamma^{\mu\nu} C)_{\alpha\gamma}}{E_{46}} \Theta^{\vec{a}}_{\vec{b}}\begin{bmatrix}3&4\\5&6\end{bmatrix} \Theta^{\vec{a}}_{\vec{b}}\begin{bmatrix}3&6\\4&5\end{bmatrix} \\
&\qquad \left(E_{14} E_{26} \Theta^{\vec{a}}_{\vec{b}}\begin{bmatrix}1&1&4\\3&5&6\end{bmatrix} \Theta^{\vec{a}}_{\vec{b}}\begin{bmatrix}2&2&6\\3&4&5\end{bmatrix} + E_{16} E_{24} \Theta^{\vec{a}}_{\vec{b}}\begin{bmatrix}2&2&4\\3&5&6\end{bmatrix} \Theta^{\vec{a}}_{\vec{b}}\begin{bmatrix}1&1&6\\3&4&5\end{bmatrix} \right) \\
&+ \frac{C_{\alpha\delta}(\gamma^{\mu\nu} C)_{\beta\gamma}}{E_{36}} \Theta^{\vec{a}}_{\vec{b}}\begin{bmatrix}3&4\\5&6\end{bmatrix} \Theta^{\vec{a}}_{\vec{b}}\begin{bmatrix}3&5\\4&6\end{bmatrix} \\
&\qquad \left(E_{13} E_{26} \Theta^{\vec{a}}_{\vec{b}}\begin{bmatrix}1&1&3\\4&5&6\end{bmatrix} \Theta^{\vec{a}}_{\vec{b}}\begin{bmatrix}2&2&6\\3&4&5\end{bmatrix} + E_{16} E_{23} \Theta^{\vec{a}}_{\vec{b}}\begin{bmatrix}2&2&3\\4&5&6\end{bmatrix} \Theta^{\vec{a}}_{\vec{b}}\begin{bmatrix}1&1&6\\3&4&5\end{bmatrix} \right) \\
&+ \frac{C_{\beta\gamma}(\gamma^{\mu\nu} C)_{\alpha\delta}}{E_{45}} \Theta^{\vec{a}}_{\vec{b}}\begin{bmatrix}3&4\\5&6\end{bmatrix} \Theta^{\vec{a}}_{\vec{b}}\begin{bmatrix}3&5\\4&6\end{bmatrix} \\
&\qquad \left(E_{14} E_{25} \Theta^{\vec{a}}_{\vec{b}}\begin{bmatrix}1&1&4\\3&5&6\end{bmatrix} \Theta^{\vec{a}}_{\vec{b}}\begin{bmatrix}2&2&5\\3&4&6\end{bmatrix} + E_{15} E_{24} \Theta^{\vec{a}}_{\vec{b}}\begin{bmatrix}2&2&4\\3&5&6\end{bmatrix} \Theta^{\vec{a}}_{\vec{b}}\begin{bmatrix}1&1&5\\3&4&6\end{bmatrix} \right) \\
&+ (\gamma^{\lambda(\mu} C)_{\alpha\beta} (\gamma^{\nu)}{}_{\lambda} C)_{\gamma\delta} E_{12} \Theta^{\vec{a}}_{\vec{b}}\begin{bmatrix}1&1&2&2\\3&4&5&6\end{bmatrix} \Theta^{\vec{a}}_{\vec{b}}\begin{bmatrix}3&4\\5&6\end{bmatrix} \Theta^{\vec{a}}_{\vec{b}}\begin{bmatrix}3&5\\4&6\end{bmatrix} \Theta^{\vec{a}}_{\vec{b}}\begin{bmatrix}3&6\\4&5\end{bmatrix} \Bigg].
\end{aligned} \quad (5.58)
$$

On the level of six spin fields, there is the rather trivial correlator

$$
\begin{aligned}
\langle S_\alpha(z_1) S_\beta(z_2) S_\gamma(z_3) S_\delta(z_4) S_\epsilon(z_5) S_\iota(z_6) \rangle^{\vec{a}}_{\vec{b}} &= \frac{1}{\left[\Theta^{\vec{a}}_{\vec{b}}(\vec{0})\right]^4} \\
\times \Bigg[&\frac{C_{\alpha\beta} C_{\gamma\delta} C_{\epsilon\iota}}{E_{12} E_{34} E_{56}} \Theta^{\vec{a}}_{\vec{b}}\begin{bmatrix}1&3&5\\2&4&6\end{bmatrix} \Theta^{\vec{a}}_{\vec{b}}\begin{bmatrix}1&3&6\\2&4&5\end{bmatrix} \Theta^{\vec{a}}_{\vec{b}}\begin{bmatrix}1&4&6\\2&3&5\end{bmatrix} \Theta^{\vec{a}}_{\vec{b}}\begin{bmatrix}1&4&5\\2&3&6\end{bmatrix} \\
&- \frac{C_{\alpha\beta} C_{\gamma\epsilon} C_{\delta\iota}}{E_{12} E_{35} E_{46}} \Theta^{\vec{a}}_{\vec{b}}\begin{bmatrix}1&3&4\\2&5&6\end{bmatrix} \Theta^{\vec{a}}_{\vec{b}}\begin{bmatrix}1&3&6\\2&5&4\end{bmatrix} \Theta^{\vec{a}}_{\vec{b}}\begin{bmatrix}1&5&6\\2&3&4\end{bmatrix} \Theta^{\vec{a}}_{\vec{b}}\begin{bmatrix}1&5&4\\2&3&6\end{bmatrix} \\
&+ \frac{C_{\alpha\beta} C_{\gamma\iota} C_{\delta\epsilon}}{E_{12} E_{36} E_{45}} \Theta^{\vec{a}}_{\vec{b}}\begin{bmatrix}1&3&4\\2&6&5\end{bmatrix} \Theta^{\vec{a}}_{\vec{b}}\begin{bmatrix}1&3&5\\2&6&4\end{bmatrix} \Theta^{\vec{a}}_{\vec{b}}\begin{bmatrix}1&6&5\\2&3&4\end{bmatrix} \Theta^{\vec{a}}_{\vec{b}}\begin{bmatrix}1&6&4\\2&3&5\end{bmatrix} \\
&+ \text{cyclic completion in } (2,\beta), (3,\gamma), (4,\delta), (5,\epsilon), (6,\iota) \Bigg].
\end{aligned} \quad (5.59)
$$

5.3 Results of RNS Loop Correlators

A more complicated case is the correlator where two left-handed spin fields are replaced by right-handed ones:

$$\left\langle S_\alpha(z_1)\, S_\beta(z_2)\, S_\gamma(z_3)\, S_\delta(z_4)\, S^{\dot\epsilon}(z_5)\, S^{\dot\zeta}(z_6) \right\rangle^{\vec a}_{\vec b}$$

$$= \frac{1}{4\left[\Theta^{\vec a}_{\vec b}(\vec 0)\right]^4 (E_{15}\, E_{16}\, E_{25}\, E_{26}\, E_{35}\, E_{36}\, E_{45}\, E_{46})^{1/2}}$$

$$\times \left[\frac{C_{\alpha\delta}\, C_{\beta\gamma}\, C^{\dot\epsilon\dot\zeta}}{E_{14}\, E_{23}\, E_{56}} \left(E_{15}\, E_{46}\, E_{25}\, E_{36}\, \Theta^{\vec a}_{\vec b}\!\left[\begin{smallmatrix}1&2&5\\3&4&6\end{smallmatrix}\right]^2 \Theta^{\vec a}_{\vec b}\!\left[\begin{smallmatrix}1&3&5\\2&4&6\end{smallmatrix}\right] \Theta^{\vec a}_{\vec b}\!\left[\begin{smallmatrix}1&3&6\\2&4&5\end{smallmatrix}\right] \right.\right.$$

$$+ E_{15}\, E_{46}\, E_{26}\, E_{35}\, \Theta^{\vec a}_{\vec b}\!\left[\begin{smallmatrix}1&3&5\\2&4&6\end{smallmatrix}\right]^2 \Theta^{\vec a}_{\vec b}\!\left[\begin{smallmatrix}1&2&5\\3&4&6\end{smallmatrix}\right] \Theta^{\vec a}_{\vec b}\!\left[\begin{smallmatrix}1&2&6\\3&4&5\end{smallmatrix}\right]$$

$$+ E_{16}\, E_{45}\, E_{25}\, E_{36}\, \Theta^{\vec a}_{\vec b}\!\left[\begin{smallmatrix}1&3&6\\2&4&5\end{smallmatrix}\right]^2 \Theta^{\vec a}_{\vec b}\!\left[\begin{smallmatrix}1&2&5\\3&4&6\end{smallmatrix}\right] \Theta^{\vec a}_{\vec b}\!\left[\begin{smallmatrix}1&2&6\\3&4&5\end{smallmatrix}\right]$$

$$\left. + E_{16}\, E_{45}\, E_{26}\, E_{35}\, \Theta^{\vec a}_{\vec b}\!\left[\begin{smallmatrix}1&2&6\\3&4&5\end{smallmatrix}\right]^2 \Theta^{\vec a}_{\vec b}\!\left[\begin{smallmatrix}1&3&5\\2&4&6\end{smallmatrix}\right] \Theta^{\vec a}_{\vec b}\!\left[\begin{smallmatrix}1&3&6\\2&4&5\end{smallmatrix}\right] \right)$$

$$+ 2 \left(\frac{C_{\alpha\beta}\, C_{\gamma\delta}\, C^{\dot\epsilon\dot\zeta}}{E_{12}\, E_{34}\, E_{56}} \Theta^{\vec a}_{\vec b}\!\left[\begin{smallmatrix}1&3&5\\2&4&6\end{smallmatrix}\right] \Theta^{\vec a}_{\vec b}\!\left[\begin{smallmatrix}1&3&6\\2&4&5\end{smallmatrix}\right] - \frac{C_{\alpha\gamma}\, C_{\beta\delta}\, C^{\dot\epsilon\dot\zeta}}{E_{13}\, E_{24}\, E_{56}} \Theta^{\vec a}_{\vec b}\!\left[\begin{smallmatrix}1&2&5\\3&4&6\end{smallmatrix}\right] \Theta^{\vec a}_{\vec b}\!\left[\begin{smallmatrix}1&2&6\\3&4&5\end{smallmatrix}\right] \right)$$

$$\left(E_{15}\, E_{26}\, E_{36}\, E_{45}\, \Theta^{\vec a}_{\vec b}\!\left[\begin{smallmatrix}1&4&5\\2&3&6\end{smallmatrix}\right]^2 + E_{16}\, E_{25}\, E_{35}\, E_{46}\, \Theta^{\vec a}_{\vec b}\!\left[\begin{smallmatrix}1&4&6\\2&3&5\end{smallmatrix}\right]^2 \right)$$

$$- \frac{C_{\alpha\beta}}{E_{12}} (\gamma_\lambda C)_{[\gamma}{}^{\dot\epsilon} (\gamma^\lambda C)_{\delta]}{}^{\dot\zeta}\, \Theta^{\vec a}_{\vec b}\!\left[\begin{smallmatrix}1&5&6\\2&3&4\end{smallmatrix}\right] \Theta^{\vec a}_{\vec b}\!\left[\begin{smallmatrix}2&5&6\\1&3&4\end{smallmatrix}\right]$$

$$\left(E_{15}\, E_{26}\, \Theta^{\vec a}_{\vec b}\!\left[\begin{smallmatrix}1&3&5\\2&4&6\end{smallmatrix}\right] \Theta^{\vec a}_{\vec b}\!\left[\begin{smallmatrix}1&4&5\\2&3&6\end{smallmatrix}\right] + E_{16}\, E_{25}\, \Theta^{\vec a}_{\vec b}\!\left[\begin{smallmatrix}1&3&6\\2&4&5\end{smallmatrix}\right] \Theta^{\vec a}_{\vec b}\!\left[\begin{smallmatrix}1&4&6\\2&3&5\end{smallmatrix}\right] \right)$$

$$+ \frac{C_{\alpha\gamma}}{E_{13}} (\gamma_\lambda C)_{[\beta}{}^{\dot\epsilon} (\gamma^\lambda C)_{\delta]}{}^{\dot\zeta}\, \Theta^{\vec a}_{\vec b}\!\left[\begin{smallmatrix}1&5&6\\2&3&4\end{smallmatrix}\right] \Theta^{\vec a}_{\vec b}\!\left[\begin{smallmatrix}3&5&6\\1&2&4\end{smallmatrix}\right]$$

$$\left(E_{15}\, E_{36}\, \Theta^{\vec a}_{\vec b}\!\left[\begin{smallmatrix}1&2&5\\3&4&6\end{smallmatrix}\right] \Theta^{\vec a}_{\vec b}\!\left[\begin{smallmatrix}1&4&5\\2&3&6\end{smallmatrix}\right] + E_{16}\, E_{35}\, \Theta^{\vec a}_{\vec b}\!\left[\begin{smallmatrix}1&2&6\\3&4&5\end{smallmatrix}\right] \Theta^{\vec a}_{\vec b}\!\left[\begin{smallmatrix}1&4&6\\2&3&5\end{smallmatrix}\right] \right)$$

$$- \frac{C_{\alpha\delta}}{E_{14}} (\gamma_\lambda C)_{[\beta}{}^{\dot\epsilon} (\gamma^\lambda C)_{\gamma]}{}^{\dot\zeta}\, \Theta^{\vec a}_{\vec b}\!\left[\begin{smallmatrix}1&5&6\\2&3&4\end{smallmatrix}\right] \Theta^{\vec a}_{\vec b}\!\left[\begin{smallmatrix}4&5&6\\1&2&3\end{smallmatrix}\right]$$

$$\left(E_{15}\, E_{46}\, \Theta^{\vec a}_{\vec b}\!\left[\begin{smallmatrix}1&2&5\\3&4&6\end{smallmatrix}\right] \Theta^{\vec a}_{\vec b}\!\left[\begin{smallmatrix}1&3&5\\2&4&6\end{smallmatrix}\right] + E_{16}\, E_{45}\, \Theta^{\vec a}_{\vec b}\!\left[\begin{smallmatrix}1&2&6\\3&4&5\end{smallmatrix}\right] \Theta^{\vec a}_{\vec b}\!\left[\begin{smallmatrix}1&3&6\\2&4&5\end{smallmatrix}\right] \right)$$

$$- \frac{C_{\beta\gamma}}{E_{23}} (\gamma_\lambda C)_{[\alpha}{}^{\dot\epsilon} (\gamma^\lambda C)_{\delta]}{}^{\dot\zeta}\, \Theta^{\vec a}_{\vec b}\!\left[\begin{smallmatrix}2&5&6\\1&3&4\end{smallmatrix}\right] \Theta^{\vec a}_{\vec b}\!\left[\begin{smallmatrix}3&5&6\\1&2&4\end{smallmatrix}\right]$$

$$\left(E_{25}\, E_{36}\, \Theta^{\vec a}_{\vec b}\!\left[\begin{smallmatrix}1&2&5\\3&4&6\end{smallmatrix}\right] \Theta^{\vec a}_{\vec b}\!\left[\begin{smallmatrix}1&3&6\\2&4&5\end{smallmatrix}\right] + E_{26}\, E_{35}\, \Theta^{\vec a}_{\vec b}\!\left[\begin{smallmatrix}1&2&6\\3&4&5\end{smallmatrix}\right] \Theta^{\vec a}_{\vec b}\!\left[\begin{smallmatrix}1&3&5\\2&4&6\end{smallmatrix}\right] \right)$$

$$+ \frac{C_{\beta\delta}}{E_{24}} (\gamma_\lambda C)_{[\alpha}{}^{\dot\epsilon} (\gamma^\lambda C)_{\gamma]}{}^{\dot\zeta}\, \Theta^{\vec a}_{\vec b}\!\left[\begin{smallmatrix}2&5&6\\1&3&4\end{smallmatrix}\right] \Theta^{\vec a}_{\vec b}\!\left[\begin{smallmatrix}4&5&6\\1&2&3\end{smallmatrix}\right]$$

$$\left(E_{25}\, E_{46}\, \Theta^{\vec a}_{\vec b}\!\left[\begin{smallmatrix}1&2&5\\3&4&6\end{smallmatrix}\right] \Theta^{\vec a}_{\vec b}\!\left[\begin{smallmatrix}1&4&6\\2&3&5\end{smallmatrix}\right] + E_{26}\, E_{45}\, \Theta^{\vec a}_{\vec b}\!\left[\begin{smallmatrix}1&2&6\\3&4&5\end{smallmatrix}\right] \Theta^{\vec a}_{\vec b}\!\left[\begin{smallmatrix}1&4&5\\2&3&6\end{smallmatrix}\right] \right)$$

$$- \frac{C_{\gamma\delta}}{E_{34}} (\gamma_\lambda C)_{[\alpha}{}^{\dot\epsilon} (\gamma^\lambda C)_{\beta]}{}^{\dot\zeta}\, \Theta^{\vec a}_{\vec b}\!\left[\begin{smallmatrix}3&5&6\\1&2&4\end{smallmatrix}\right] \Theta^{\vec a}_{\vec b}\!\left[\begin{smallmatrix}4&5&6\\1&2&3\end{smallmatrix}\right]$$

$$\left(E_{35}\, E_{46}\, \Theta^{\vec a}_{\vec b}\!\left[\begin{smallmatrix}1&3&5\\2&4&6\end{smallmatrix}\right] \Theta^{\vec a}_{\vec b}\!\left[\begin{smallmatrix}1&4&6\\2&3&5\end{smallmatrix}\right] + E_{36}\, E_{45}\, \Theta^{\vec a}_{\vec b}\!\left[\begin{smallmatrix}1&3&6\\2&4&5\end{smallmatrix}\right] \Theta^{\vec a}_{\vec b}\!\left[\begin{smallmatrix}1&4&5\\2&3&6\end{smallmatrix}\right] \right)$$

$$\left. - (\gamma_\lambda C)_\alpha{}^{(\dot\epsilon} (\gamma^{\lambda\rho} C)_{\beta\gamma} (\gamma_\rho C)_\delta{}^{\dot\zeta)}\, E_{56}\, \Theta^{\vec a}_{\vec b}\!\left[\begin{smallmatrix}1&5&6\\2&3&4\end{smallmatrix}\right] \Theta^{\vec a}_{\vec b}\!\left[\begin{smallmatrix}2&5&6\\1&3&4\end{smallmatrix}\right] \Theta^{\vec a}_{\vec b}\!\left[\begin{smallmatrix}3&5&6\\1&2&4\end{smallmatrix}\right] \Theta^{\vec a}_{\vec b}\!\left[\begin{smallmatrix}4&5&6\\1&2&3\end{smallmatrix}\right] \right]. \quad (5.60)$$

Seven-point correlators require a basis of at least 24 independent Lorentz tensors, so we refrain from computing higher-point examples without systematics.

5.3.4 Results for $D = 10$

Ten-dimensional correlators with four spin fields only are given by

$$\langle S_\alpha(z_1) S_\beta(z_2) S_\gamma(z_3) S_\delta(z_4) \rangle_{\vec{b}}^{\vec{a}} = \frac{1}{2\,(E_{12} E_{13} E_{14} E_{23} E_{24} E_{34})^{3/4}} \frac{\Theta_{\vec{b}}^{\vec{a}}[\begin{smallmatrix}1&2\\3&4\end{smallmatrix}]\,\Theta_{\vec{b}}^{\vec{a}}[\begin{smallmatrix}1&3\\2&4\end{smallmatrix}]\,\Theta_{\vec{b}}^{\vec{a}}[\begin{smallmatrix}1&4\\2&3\end{smallmatrix}]}{[\Theta_{\vec{b}}^{\vec{a}}(\vec{0})]^5}$$

$$\times \left[(\gamma^\mu C)_{\alpha\beta} (\gamma_\mu C)_{\gamma\delta} E_{14} E_{23} \Theta_{\vec{b}}^{\vec{a}}[\begin{smallmatrix}1&4\\2&3\end{smallmatrix}]^2 - (\gamma^\mu C)_{\alpha\delta} (\gamma_\mu C)_{\gamma\beta} E_{12} E_{34} \Theta_{\vec{b}}^{\vec{a}}[\begin{smallmatrix}1&2\\3&4\end{smallmatrix}]^2 \right], \tag{5.61}$$

$$\langle S_\alpha(z_1) S_\beta(z_2) S^{\dot\gamma}(z_3) S^{\dot\delta}(z_4) \rangle_{\vec{b}}^{\vec{a}} = \left(\frac{E_{12} E_{34}}{E_{13} E_{14} E_{23} E_{24}} \right)^{1/4} \frac{\Theta_{\vec{b}}^{\vec{a}}[\begin{smallmatrix}1&2\\3&4\end{smallmatrix}]}{[\Theta_{\vec{b}}^{\vec{a}}(\vec{0})]^5}$$

$$\times \left[\frac{C_\alpha^{\ \dot\delta} C_\beta^{\ \dot\gamma}}{E_{14} E_{23}} \Theta_{\vec{b}}^{\vec{a}}[\begin{smallmatrix}1&2\\3&4\end{smallmatrix}]^2 \Theta_{\vec{b}}^{\vec{a}}[\begin{smallmatrix}1&3\\2&4\end{smallmatrix}]^2 - \frac{C_\alpha^{\ \dot\gamma} C_\beta^{\ \dot\delta}}{E_{13} E_{24}} \Theta_{\vec{b}}^{\vec{a}}[\begin{smallmatrix}1&2\\3&4\end{smallmatrix}]^2 \Theta_{\vec{b}}^{\vec{a}}[\begin{smallmatrix}1&4\\2&3\end{smallmatrix}]^2 \right.$$

$$\left. + \frac{1}{2} \frac{(\gamma^\mu C)_{\alpha\beta} (\bar\gamma_\mu C)^{\dot\gamma\dot\delta}}{E_{12} E_{34}} \Theta_{\vec{b}}^{\vec{a}}[\begin{smallmatrix}1&3\\2&4\end{smallmatrix}]^2 \Theta_{\vec{b}}^{\vec{a}}[\begin{smallmatrix}1&4\\2&3\end{smallmatrix}]^2 \right]. \tag{5.62}$$

In the presence of one NS fermion, we find:

$$\langle \psi^\mu(z_1) S_\alpha(z_2) S_\beta(z_3) S_\gamma(z_4) S^{\dot\delta}(z_5) \rangle_{\vec{b}}^{\vec{a}}$$

$$= \frac{(E_{23} E_{24} E_{34})^{-3/4}}{\sqrt{2}\,[\Theta_{\vec{b}}^{\vec{a}}(\vec{0})]^5 (E_{12} E_{13} E_{14} E_{15})^{1/2} (E_{25} E_{35} E_{45})^{1/4}}$$

$$\times \left[\frac{C_\gamma^{\ \dot\delta}}{E_{45}} (\gamma^\mu C)_{\alpha\beta} E_{15} E_{24} E_{34} \Theta_{\vec{b}}^{\vec{a}}[\begin{smallmatrix}2&5\\3&4\end{smallmatrix}]^2 \Theta_{\vec{b}}^{\vec{a}}[\begin{smallmatrix}2&4\\3&5\end{smallmatrix}]^2 \Theta_{\vec{b}}^{\vec{a}}[\begin{smallmatrix}1&1&5\\2&3&4\end{smallmatrix}] \right.$$

$$+ \frac{C_\alpha^{\ \dot\delta}}{E_{25}} (\gamma^\mu C)_{\beta\gamma} E_{15} E_{23} E_{24} \Theta_{\vec{b}}^{\vec{a}}[\begin{smallmatrix}2&3\\4&5\end{smallmatrix}]^2 \Theta_{\vec{b}}^{\vec{a}}[\begin{smallmatrix}2&4\\3&5\end{smallmatrix}]^2 \Theta_{\vec{b}}^{\vec{a}}[\begin{smallmatrix}1&1&5\\2&3&4\end{smallmatrix}]$$

$$- \frac{C_\beta^{\ \dot\delta}}{E_{35}} (\gamma^\mu C)_{\alpha\gamma} E_{15} E_{23} E_{34} \Theta_{\vec{b}}^{\vec{a}}[\begin{smallmatrix}1&1&5\\2&3&4\end{smallmatrix}] \Theta_{\vec{b}}^{\vec{a}}[\begin{smallmatrix}2&3\\4&5\end{smallmatrix}]^2 \Theta_{\vec{b}}^{\vec{a}}[\begin{smallmatrix}2&5\\3&4\end{smallmatrix}]^2$$

$$- \frac{1}{2} (\gamma^\nu \bar\gamma^\mu C)_\gamma^{\ \dot\delta} (\gamma_\nu C)_{\alpha\beta} E_{12} E_{34} \Theta_{\vec{b}}^{\vec{a}}[\begin{smallmatrix}1&1&2\\3&4&5\end{smallmatrix}] \Theta_{\vec{b}}^{\vec{a}}[\begin{smallmatrix}2&3\\4&5\end{smallmatrix}] \Theta_{\vec{b}}^{\vec{a}}[\begin{smallmatrix}2&4\\3&5\end{smallmatrix}] \Theta_{\vec{b}}^{\vec{a}}[\begin{smallmatrix}2&5\\3&4\end{smallmatrix}]^2$$

$$\left. + \frac{1}{2} (\gamma^\nu \bar\gamma^\mu C)_\alpha^{\ \dot\delta} (\gamma_\nu C)_{\beta\gamma} E_{14} E_{23} \Theta_{\vec{b}}^{\vec{a}}[\begin{smallmatrix}1&1&4\\2&3&5\end{smallmatrix}] \Theta_{\vec{b}}^{\vec{a}}[\begin{smallmatrix}2&4\\3&5\end{smallmatrix}] \Theta_{\vec{b}}^{\vec{a}}[\begin{smallmatrix}2&5\\3&4\end{smallmatrix}] \Theta_{\vec{b}}^{\vec{a}}[\begin{smallmatrix}2&3\\4&5\end{smallmatrix}]^2 \right]. \tag{5.63}$$

The identity $(\gamma_\nu \bar\gamma^\mu C)_{(\alpha}^{\ \dot\delta} (\gamma^\nu C)_{\beta\gamma)} = 0$ admits to recast the last two lines into the form

$$\frac{1}{2} (\gamma^\nu \bar\gamma^\mu C)_\beta^{\ \dot\delta} (\gamma_\nu C)_{\alpha\gamma} E_{12} E_{34} \Theta_{\vec{b}}^{\vec{a}}[\begin{smallmatrix}1&1&2\\3&4&5\end{smallmatrix}] \Theta_{\vec{b}}^{\vec{a}}[\begin{smallmatrix}2&3\\4&5\end{smallmatrix}] \Theta_{\vec{b}}^{\vec{a}}[\begin{smallmatrix}2&4\\3&5\end{smallmatrix}] \Theta_{\vec{b}}^{\vec{a}}[\begin{smallmatrix}2&5\\3&4\end{smallmatrix}]^2$$

$$+ \frac{1}{2} (\gamma^\nu \bar\gamma^\mu C)_\alpha^{\ \dot\delta} (\gamma_\nu C)_{\beta\gamma} E_{13} E_{24} \Theta_{\vec{b}}^{\vec{a}}[\begin{smallmatrix}1&1&3\\2&4&5\end{smallmatrix}] \Theta_{\vec{b}}^{\vec{a}}[\begin{smallmatrix}2&4\\3&5\end{smallmatrix}]^2 \Theta_{\vec{b}}^{\vec{a}}[\begin{smallmatrix}2&5\\3&4\end{smallmatrix}] \Theta_{\vec{b}}^{\vec{a}}[\begin{smallmatrix}2&3\\4&5\end{smallmatrix}]. \tag{5.64}$$

5.3 Results of RNS Loop Correlators

The following correlator has been partially computed in [151] for the purpose of four fermion scattering at 1-loop. Let us give the complete g-loop result here:

$$\left\langle \psi^\mu(z_1)\,\psi^\nu(z_2)\,S_\alpha(z_3)\,S_\beta(z_4)\,S_\gamma(z_5)\,S_\delta(z_6) \right\rangle_{\vec{b}}^{\vec{a}}$$

$$= \frac{(E_{34}\,E_{35}\,E_{36}\,E_{45}\,E_{46}\,E_{56})^{-3/4}}{2\left[\Theta_{\vec{b}}^{\vec{a}}(\vec{0})\right]^5 (E_{13}\,E_{14}\,E_{15}\,E_{16}\,E_{23}\,E_{24}\,E_{25}\,E_{26})^{1/2}}$$

$$\times \Bigg[\frac{\eta^{\mu\nu}}{E_{12}}(\gamma^\lambda C)_{\alpha\beta}\,(\gamma_\lambda C)_{\gamma\delta}\,E_{36}\,E_{45}\,\Theta_{\vec{b}}^{\vec{a}}\!\left[\begin{smallmatrix}3&4\\5&6\end{smallmatrix}\right]\Theta_{\vec{b}}^{\vec{a}}\!\left[\begin{smallmatrix}3&5\\4&6\end{smallmatrix}\right]\Theta_{\vec{b}}^{\vec{a}}\!\left[\begin{smallmatrix}3&6\\4&5\end{smallmatrix}\right]^2$$

$$\left(E_{13}\,E_{16}\,E_{24}\,E_{25}\,\Theta_{\vec{b}}^{\vec{a}}\!\left[\begin{smallmatrix}1&1&3&6\\2&2&4&5\end{smallmatrix}\right] + E_{14}\,E_{15}\,E_{23}\,E_{26}\,\Theta_{\vec{b}}^{\vec{a}}\!\left[\begin{smallmatrix}1&1&4&5\\2&2&3&6\end{smallmatrix}\right]\right)$$

$$-\frac{\eta^{\mu\nu}}{E_{12}}(\gamma^\lambda C)_{\alpha\delta}\,(\gamma_\lambda C)_{\gamma\beta}\,E_{34}\,E_{56}\,\Theta_{\vec{b}}^{\vec{a}}\!\left[\begin{smallmatrix}3&6\\5&4\end{smallmatrix}\right]\Theta_{\vec{b}}^{\vec{a}}\!\left[\begin{smallmatrix}3&5\\6&4\end{smallmatrix}\right]\Theta_{\vec{b}}^{\vec{a}}\!\left[\begin{smallmatrix}3&4\\6&5\end{smallmatrix}\right]^2$$

$$\left(E_{13}\,E_{14}\,E_{25}\,E_{26}\,\Theta_{\vec{b}}^{\vec{a}}\!\left[\begin{smallmatrix}1&1&3&4\\2&2&6&5\end{smallmatrix}\right] + E_{15}\,E_{16}\,E_{23}\,E_{24}\,\Theta_{\vec{b}}^{\vec{a}}\!\left[\begin{smallmatrix}1&1&6&5\\2&2&3&4\end{smallmatrix}\right]\right)$$

$$+\frac{1}{2}(\gamma^\mu C)_{\gamma\beta}(\gamma^\nu C)_{\alpha\delta}\,E_{34}\,E_{56}\,\Theta_{\vec{b}}^{\vec{a}}\!\left[\begin{smallmatrix}3&4\\5&6\end{smallmatrix}\right]^2\Theta_{\vec{b}}^{\vec{a}}\!\left[\begin{smallmatrix}3&5\\4&6\end{smallmatrix}\right]$$

$$\left(E_{13}\,E_{25}\,E_{46}\,\Theta_{\vec{b}}^{\vec{a}}\!\left[\begin{smallmatrix}1&1&3\\4&5&6\end{smallmatrix}\right]\Theta_{\vec{b}}^{\vec{a}}\!\left[\begin{smallmatrix}2&2&5\\3&4&6\end{smallmatrix}\right] - E_{16}\,E_{24}\,E_{35}\,\Theta_{\vec{b}}^{\vec{a}}\!\left[\begin{smallmatrix}1&1&6\\3&4&5\end{smallmatrix}\right]\Theta_{\vec{b}}^{\vec{a}}\!\left[\begin{smallmatrix}2&2&4\\3&5&6\end{smallmatrix}\right]\right)$$

$$+\frac{1}{2}(\gamma^\mu C)_{\alpha\delta}(\gamma^\nu C)_{\gamma\beta}\,E_{34}\,E_{56}\,\Theta_{\vec{b}}^{\vec{a}}\!\left[\begin{smallmatrix}3&4\\5&6\end{smallmatrix}\right]^2\Theta_{\vec{b}}^{\vec{a}}\!\left[\begin{smallmatrix}3&5\\4&6\end{smallmatrix}\right]$$

$$\left(E_{14}\,E_{26}\,E_{35}\,\Theta_{\vec{b}}^{\vec{a}}\!\left[\begin{smallmatrix}1&1&4\\3&5&6\end{smallmatrix}\right]\Theta_{\vec{b}}^{\vec{a}}\!\left[\begin{smallmatrix}2&2&6\\3&4&5\end{smallmatrix}\right] - E_{15}\,E_{23}\,E_{46}\,\Theta_{\vec{b}}^{\vec{a}}\!\left[\begin{smallmatrix}1&1&5\\3&4&6\end{smallmatrix}\right]\Theta_{\vec{b}}^{\vec{a}}\!\left[\begin{smallmatrix}2&2&3\\4&5&6\end{smallmatrix}\right]\right)$$

$$-\frac{1}{2}(\gamma^\mu C)_{\alpha\beta}(\gamma^\nu C)_{\gamma\delta}\,E_{36}\,E_{45}\,\Theta_{\vec{b}}^{\vec{a}}\!\left[\begin{smallmatrix}3&6\\4&5\end{smallmatrix}\right]^2\Theta_{\vec{b}}^{\vec{a}}\!\left[\begin{smallmatrix}3&5\\4&6\end{smallmatrix}\right]$$

$$\left(E_{16}\,E_{24}\,E_{35}\,\Theta_{\vec{b}}^{\vec{a}}\!\left[\begin{smallmatrix}1&1&6\\3&4&5\end{smallmatrix}\right]\Theta_{\vec{b}}^{\vec{a}}\!\left[\begin{smallmatrix}2&2&4\\3&5&6\end{smallmatrix}\right] + E_{15}\,E_{23}\,E_{46}\,\Theta_{\vec{b}}^{\vec{a}}\!\left[\begin{smallmatrix}1&1&5\\3&4&6\end{smallmatrix}\right]\Theta_{\vec{b}}^{\vec{a}}\!\left[\begin{smallmatrix}2&2&3\\4&5&6\end{smallmatrix}\right]\right)$$

$$+\frac{1}{2}(\gamma^\mu C)_{\gamma\delta}(\gamma^\nu C)_{\alpha\beta}\,E_{36}\,E_{45}\,\Theta_{\vec{b}}^{\vec{a}}\!\left[\begin{smallmatrix}3&6\\4&5\end{smallmatrix}\right]^2\Theta_{\vec{b}}^{\vec{a}}\!\left[\begin{smallmatrix}3&5\\4&6\end{smallmatrix}\right]$$

$$\left(E_{13}\,E_{25}\,E_{46}\,\Theta_{\vec{b}}^{\vec{a}}\!\left[\begin{smallmatrix}1&1&3\\4&5&6\end{smallmatrix}\right]\Theta_{\vec{b}}^{\vec{a}}\!\left[\begin{smallmatrix}2&2&5\\3&4&6\end{smallmatrix}\right] + E_{14}\,E_{26}\,E_{35}\,\Theta_{\vec{b}}^{\vec{a}}\!\left[\begin{smallmatrix}1&1&4\\3&5&6\end{smallmatrix}\right]\Theta_{\vec{b}}^{\vec{a}}\!\left[\begin{smallmatrix}2&2&6\\3&4&5\end{smallmatrix}\right]\right)$$

$$+\frac{1}{4}(\gamma^{\mu\nu\lambda}C)_{\alpha\beta}(\gamma_\lambda C)_{\gamma\delta}\,E_{34}\,E_{36}\,E_{45}\,\Theta_{\vec{b}}^{\vec{a}}\!\left[\begin{smallmatrix}3&4\\5&6\end{smallmatrix}\right]\Theta_{\vec{b}}^{\vec{a}}\!\left[\begin{smallmatrix}3&6\\4&5\end{smallmatrix}\right]^2$$

$$\left(E_{15}\,E_{26}\,\Theta_{\vec{b}}^{\vec{a}}\!\left[\begin{smallmatrix}1&1&5\\3&4&6\end{smallmatrix}\right]\Theta_{\vec{b}}^{\vec{a}}\!\left[\begin{smallmatrix}2&2&6\\3&4&5\end{smallmatrix}\right] + E_{16}\,E_{25}\,\Theta_{\vec{b}}^{\vec{a}}\!\left[\begin{smallmatrix}1&1&6\\3&4&5\end{smallmatrix}\right]\Theta_{\vec{b}}^{\vec{a}}\!\left[\begin{smallmatrix}2&2&5\\3&4&6\end{smallmatrix}\right]\right)$$

$$+\frac{1}{4}(\gamma^{\mu\nu\lambda}C)_{\gamma\delta}(\gamma_\lambda C)_{\alpha\beta}\,E_{56}\,E_{36}\,E_{45}\,\Theta_{\vec{b}}^{\vec{a}}\!\left[\begin{smallmatrix}3&4\\5&6\end{smallmatrix}\right]\Theta_{\vec{b}}^{\vec{a}}\!\left[\begin{smallmatrix}3&6\\4&5\end{smallmatrix}\right]^2$$

$$\left(E_{13}\,E_{24}\,\Theta_{\vec{b}}^{\vec{a}}\!\left[\begin{smallmatrix}1&1&3\\4&5&6\end{smallmatrix}\right]\Theta_{\vec{b}}^{\vec{a}}\!\left[\begin{smallmatrix}2&2&4\\3&5&6\end{smallmatrix}\right] + E_{14}\,E_{23}\,\Theta_{\vec{b}}^{\vec{a}}\!\left[\begin{smallmatrix}1&1&4\\3&5&6\end{smallmatrix}\right]\Theta_{\vec{b}}^{\vec{a}}\!\left[\begin{smallmatrix}2&2&3\\4&5&6\end{smallmatrix}\right]\right)$$

$$-\frac{1}{4}(\gamma^{\mu\nu\lambda}C)_{\alpha\delta}(\gamma_\lambda C)_{\gamma\beta}\,E_{36}\,E_{34}\,E_{56}\,\Theta_{\vec{b}}^{\vec{a}}\!\left[\begin{smallmatrix}3&4\\5&6\end{smallmatrix}\right]^2\Theta_{\vec{b}}^{\vec{a}}\!\left[\begin{smallmatrix}3&6\\4&5\end{smallmatrix}\right]$$

$$\left(E_{15}\,E_{24}\,\Theta_{\vec{b}}^{\vec{a}}\!\left[\begin{smallmatrix}1&1&5\\3&4&6\end{smallmatrix}\right]\Theta_{\vec{b}}^{\vec{a}}\!\left[\begin{smallmatrix}2&2&4\\3&5&6\end{smallmatrix}\right] + E_{14}\,E_{25}\,\Theta_{\vec{b}}^{\vec{a}}\!\left[\begin{smallmatrix}1&1&4\\3&5&6\end{smallmatrix}\right]\Theta_{\vec{b}}^{\vec{a}}\!\left[\begin{smallmatrix}2&2&5\\3&4&6\end{smallmatrix}\right]\right)$$

$$-\frac{1}{4}(\gamma^{\mu\nu\lambda}C)_{\gamma\beta}(\gamma_\lambda C)_{\alpha\delta}\,E_{54}\,E_{34}\,E_{56}\,\Theta_{\vec{b}}^{\vec{a}}\!\left[\begin{smallmatrix}3&4\\5&6\end{smallmatrix}\right]^2\Theta_{\vec{b}}^{\vec{a}}\!\left[\begin{smallmatrix}3&6\\4&5\end{smallmatrix}\right]$$

$$\left(E_{13}\,E_{26}\,\Theta_{\vec{b}}^{\vec{a}}\!\left[\begin{smallmatrix}1&1&3\\4&5&6\end{smallmatrix}\right]\Theta_{\vec{b}}^{\vec{a}}\!\left[\begin{smallmatrix}2&2&6\\3&4&5\end{smallmatrix}\right] + E_{16}\,E_{23}\,\Theta_{\vec{b}}^{\vec{a}}\!\left[\begin{smallmatrix}1&1&6\\3&4&5\end{smallmatrix}\right]\Theta_{\vec{b}}^{\vec{a}}\!\left[\begin{smallmatrix}2&2&3\\4&5&6\end{smallmatrix}\right]\right)$$

$$- (\gamma^{(\mu} C)_{\alpha\gamma} (\gamma^{\nu)} C)_{\beta\delta} E_{12} E_{34} E_{36} E_{45} E_{56} \Theta_{\vec{b}}^{\vec{a}} [{}^{1\,1\,2\,2}_{3\,4\,5\,6}] \, \Theta_{\vec{b}}^{\vec{a}} [{}^{3\,4}_{5\,6}]^2 \, \Theta_{\vec{b}}^{\vec{a}} [{}^{3\,6}_{4\,5}]^2 \Big]. \tag{5.65}$$

This representation in terms of antisymmetric products $\gamma^{\mu\nu\lambda}$ rather than $\gamma^\mu \bar\gamma^\nu \gamma^\lambda$ was chosen in order to make the antisymmetry under the exchange of two spin fields $S_{\alpha_i}(z_i)$, and $S_{\alpha_j}(z_j)$ up to a prefactor $E_{ij}^{1/4}$ manifest. If the four spin fields have mixed chirality, one has

$$\begin{aligned}
&\langle \psi^\mu(z_1)\, \psi^\nu(z_2)\, S_\alpha(z_3)\, S_\beta(z_4)\, S^{\dot\gamma}(z_5)\, S^{\dot\delta}(z_6) \rangle_{\vec{b}}^{\vec{a}} \\
&= \frac{1}{[\Theta_{\vec{b}}^{\vec{a}}(\vec{0})]^5} \, \frac{(E_{34}\, E_{56})^{-3/4}\, (E_{35}\, E_{36}\, E_{45}\, E_{46})^{-1/4}}{(E_{13}\, E_{14}\, E_{15}\, E_{16}\, E_{23}\, E_{24}\, E_{25}\, E_{26})^{1/2}} \\
&\times \Bigg[\frac{1}{2} \frac{\eta^{\mu\nu}}{E_{12}} (\gamma^\lambda C)_{\alpha\beta} (\bar\gamma_\lambda C)^{\dot\gamma\dot\delta} E_{13}\, E_{15}\, E_{24}\, E_{26}\, \Theta_{\vec{b}}^{\vec{a}}[{}^{1\,1\,3\,5}_{2\,2\,4\,6}] \, \Theta_{\vec{b}}^{\vec{a}}[{}^{3\,4}_{5\,6}] \, \Theta_{\vec{b}}^{\vec{a}}[{}^{3\,5}_{4\,6}] \, \Theta_{\vec{b}}^{\vec{a}}[{}^{3\,6}_{4\,5}]^2 \\
&+ \frac{\eta^{\mu\nu}\, C_\alpha{}^{\dot\delta}\, C_\beta{}^{\dot\gamma}}{E_{12}\, E_{36}\, E_{45}} E_{13}\, E_{14}\, E_{25}\, E_{26}\, E_{34}\, E_{56}\, \Theta_{\vec{b}}^{\vec{a}}[{}^{1\,1\,3\,4}_{2\,2\,5\,6}] \, \Theta_{\vec{b}}^{\vec{a}}[{}^{3\,4}_{5\,6}]^2 \, \Theta_{\vec{b}}^{\vec{a}}[{}^{3\,5}_{4\,6}]^2 \\
&- \frac{\eta^{\mu\nu}\, C_\alpha{}^{\dot\gamma}\, C_\beta{}^{\dot\delta}}{E_{12}\, E_{35}\, E_{46}} E_{13}\, E_{14}\, E_{24}\, E_{26}\, E_{34}\, E_{56}\, \Theta_{\vec{b}}^{\vec{a}}[{}^{1\,1\,3\,4}_{2\,2\,5\,6}] \, \Theta_{\vec{b}}^{\vec{a}}[{}^{3\,4}_{5\,6}]^2 \, \Theta_{\vec{b}}^{\vec{a}}[{}^{3\,6}_{4\,5}]^2 \\
&- \frac{1}{2} (\gamma^\mu \bar\gamma^\nu C)_\alpha{}^{\dot\gamma} \frac{C_\beta{}^{\dot\delta}}{E_{46}} E_{16}\, E_{24}\, E_{34}\, E_{56}\, \Theta_{\vec{b}}^{\vec{a}}[{}^{1\,1\,6}_{3\,4\,5}] \, \Theta_{\vec{b}}^{\vec{a}}[{}^{2\,2\,4}_{3\,5\,6}] \, \Theta_{\vec{b}}^{\vec{a}}[{}^{3\,6}_{4\,5}] \, \Theta_{\vec{b}}^{\vec{a}}[{}^{3\,4}_{5\,6}]^2 \\
&+ \frac{1}{2} (\gamma^\mu \bar\gamma^\nu C)_\alpha{}^{\dot\delta} \frac{C_\beta{}^{\dot\gamma}}{E_{45}} E_{15}\, E_{24}\, E_{34}\, E_{56}\, \Theta_{\vec{b}}^{\vec{a}}[{}^{1\,1\,5}_{3\,4\,6}] \, \Theta_{\vec{b}}^{\vec{a}}[{}^{2\,2\,4}_{3\,5\,6}] \, \Theta_{\vec{b}}^{\vec{a}}[{}^{3\,5}_{4\,6}] \, \Theta_{\vec{b}}^{\vec{a}}[{}^{3\,4}_{5\,6}]^2 \\
&+ \frac{1}{2} (\gamma^\mu \bar\gamma^\nu C)_\beta{}^{\dot\gamma} \frac{C_\alpha{}^{\dot\delta}}{E_{36}} E_{13}\, E_{26}\, E_{34}\, E_{56}\, \Theta_{\vec{b}}^{\vec{a}}[{}^{1\,1\,3}_{4\,5\,6}] \, \Theta_{\vec{b}}^{\vec{a}}[{}^{2\,2\,6}_{3\,4\,5}] \, \Theta_{\vec{b}}^{\vec{a}}[{}^{3\,5}_{4\,6}] \, \Theta_{\vec{b}}^{\vec{a}}[{}^{3\,4}_{5\,6}]^2 \\
&- \frac{1}{2} (\gamma^\mu \bar\gamma^\nu C)_\beta{}^{\dot\delta} \frac{C_\alpha{}^{\dot\gamma}}{E_{35}} E_{13}\, E_{25}\, E_{34}\, E_{56}\, \Theta_{\vec{b}}^{\vec{a}}[{}^{1\,1\,3}_{4\,5\,6}] \, \Theta_{\vec{b}}^{\vec{a}}[{}^{2\,2\,5}_{3\,4\,6}] \, \Theta_{\vec{b}}^{\vec{a}}[{}^{3\,6}_{4\,5}] \, \Theta_{\vec{b}}^{\vec{a}}[{}^{3\,4}_{5\,6}]^2 \\
&- \frac{1}{2} (\gamma^\mu C)_{\alpha\beta} (\bar\gamma^\nu C)^{\dot\gamma\dot\delta} E_{15}\, E_{24}\, E_{36}\, \Theta_{\vec{b}}^{\vec{a}}[{}^{1\,1\,5}_{3\,4\,6}] \, \Theta_{\vec{b}}^{\vec{a}}[{}^{2\,2\,4}_{3\,5\,6}] \, \Theta_{\vec{b}}^{\vec{a}}[{}^{3\,5}_{4\,6}]^2 \, \Theta_{\vec{b}}^{\vec{a}}[{}^{3\,6}_{4\,5}]^2 \\
&+ \frac{1}{2} (\gamma^\nu C)_{\alpha\beta} (\bar\gamma^\mu C)^{\dot\gamma\dot\delta} E_{13}\, E_{25}\, E_{46}\, \Theta_{\vec{b}}^{\vec{a}}[{}^{1\,1\,3}_{4\,5\,6}] \, \Theta_{\vec{b}}^{\vec{a}}[{}^{2\,2\,5}_{3\,4\,6}] \, \Theta_{\vec{b}}^{\vec{a}}[{}^{3\,6}_{4\,5}]^2 \, \Theta_{\vec{b}}^{\vec{a}}[{}^{3\,5}_{4\,6}]^2 \\
&- \frac{1}{4} (\gamma^\mu \bar\gamma^\lambda C)_\alpha{}^{\dot\gamma} (\gamma^\nu \bar\gamma_\lambda C)_\beta{}^{\dot\delta} E_{16}\, E_{25}\, E_{34}\, \Theta_{\vec{b}}^{\vec{a}}[{}^{1\,1\,6}_{3\,4\,5}] \, \Theta_{\vec{b}}^{\vec{a}}[{}^{2\,2\,5}_{3\,4\,6}] \, \Theta_{\vec{b}}^{\vec{a}}[{}^{3\,4}_{5\,6}] \, \Theta_{\vec{b}}^{\vec{a}}[{}^{3\,5}_{4\,6}] \, \Theta_{\vec{b}}^{\vec{a}}[{}^{3\,6}_{4\,5}] \\
&- \frac{1}{4} (\gamma^\mu \bar\gamma^\lambda C)_\alpha{}^{\dot\delta} (\gamma^\nu \bar\gamma_\lambda C)_\beta{}^{\dot\gamma} E_{15}\, E_{26}\, E_{34}\, \Theta_{\vec{b}}^{\vec{a}}[{}^{1\,1\,5}_{3\,4\,6}] \, \Theta_{\vec{b}}^{\vec{a}}[{}^{2\,2\,6}_{3\,4\,5}] \, \Theta_{\vec{b}}^{\vec{a}}[{}^{3\,4}_{5\,6}] \, \Theta_{\vec{b}}^{\vec{a}}[{}^{3\,5}_{4\,6}] \, \Theta_{\vec{b}}^{\vec{a}}[{}^{3\,6}_{4\,5}] \\
&+ \frac{1}{4} (\bar\gamma^\mu \gamma^\nu \bar\gamma^\lambda C)^{\dot\gamma\dot\delta} (\gamma_\lambda C)_{\alpha\beta} E_{13}\, E_{24}\, E_{56}\, \Theta_{\vec{b}}^{\vec{a}}[{}^{1\,1\,3}_{4\,5\,6}] \, \Theta_{\vec{b}}^{\vec{a}}[{}^{2\,2\,4}_{3\,5\,6}] \, \Theta_{\vec{b}}^{\vec{a}}[{}^{3\,4}_{5\,6}] \, \Theta_{\vec{b}}^{\vec{a}}[{}^{3\,5}_{4\,6}] \, \Theta_{\vec{b}}^{\vec{a}}[{}^{3\,6}_{4\,5}] \Bigg].
\end{aligned} \tag{5.66}$$

The correlator with five left-handed spin fields and one right-handed spin field has appeared in the literature before, namely in [141] at tree-level for the purpose of six fermion scattering. Let us give its loop generalization here:

5.3 Results of RNS Loop Correlators

$$\left\langle S_\alpha(z_1)\, S_\beta(z_2)\, S_\gamma(z_3)\, S_\delta(z_4)\, S_\epsilon(z_5)\, S^{\dot\zeta}(z_6)\right\rangle_{\vec b}^{\vec a}$$

$$= \frac{1}{2\left[\Theta_{\vec b}^{\vec a}(\vec 0)\right]^5} \frac{(E_{12}\,E_{13}\,E_{14}\,E_{15}\,E_{23}\,E_{24}\,E_{25}\,E_{34}\,E_{35}\,E_{45})^{-3/4}}{(E_{16}\,E_{26}\,E_{36}\,E_{46}\,E_{56})^{1/4}}$$

$$\times \Bigg[(\gamma^\mu C)_{\alpha\beta}(\gamma_\mu C)_{\gamma\delta}\, \frac{C_\epsilon{}^{\dot\zeta}}{E_{56}}\, E_{14}\,E_{15}\,E_{23}\,E_{25}\,E_{35}\,E_{46}$$

$$\Theta_{\vec b}^{\vec a}\!\left[\begin{smallmatrix}1&2&5\\3&4&6\end{smallmatrix}\right]\Theta_{\vec b}^{\vec a}\!\left[\begin{smallmatrix}1&3&5\\2&4&6\end{smallmatrix}\right]\Theta_{\vec b}^{\vec a}\!\left[\begin{smallmatrix}1&4&5\\2&3&6\end{smallmatrix}\right]\Theta_{\vec b}^{\vec a}\!\left[\begin{smallmatrix}1&4&6\\2&3&5\end{smallmatrix}\right]^2$$

$$- (\gamma^\mu C)_{\alpha\delta}(\gamma_\mu C)_{\beta\gamma}\, \frac{C_\epsilon{}^{\dot\zeta}}{E_{56}}\, E_{12}\,E_{15}\,E_{25}\,E_{34}\,E_{35}\,E_{46}$$

$$\Theta_{\vec b}^{\vec a}\!\left[\begin{smallmatrix}1&2&6\\3&4&5\end{smallmatrix}\right]\Theta_{\vec b}^{\vec a}\!\left[\begin{smallmatrix}1&3&5\\2&4&6\end{smallmatrix}\right]\Theta_{\vec b}^{\vec a}\!\left[\begin{smallmatrix}1&4&6\\2&3&5\end{smallmatrix}\right]\Theta_{\vec b}^{\vec a}\!\left[\begin{smallmatrix}1&2&5\\3&4&6\end{smallmatrix}\right]^2$$

$$+ (\gamma^\mu C)_{\alpha\beta}(\gamma_\mu C)_{\gamma\epsilon}\, \frac{C_\delta{}^{\dot\zeta}}{E_{46}}\, E_{14}\,E_{15}\,E_{23}\,E_{24}\,E_{34}\,E_{56}$$

$$\Theta_{\vec b}^{\vec a}\!\left[\begin{smallmatrix}1&2&4\\3&5&6\end{smallmatrix}\right]\Theta_{\vec b}^{\vec a}\!\left[\begin{smallmatrix}1&3&4\\2&5&6\end{smallmatrix}\right]\Theta_{\vec b}^{\vec a}\!\left[\begin{smallmatrix}1&4&5\\2&3&6\end{smallmatrix}\right]\Theta_{\vec b}^{\vec a}\!\left[\begin{smallmatrix}1&5&6\\2&3&4\end{smallmatrix}\right]^2$$

$$- (\gamma^\mu C)_{\alpha\epsilon}(\gamma_\mu C)_{\beta\gamma}\, \frac{C_\delta{}^{\dot\zeta}}{E_{46}}\, E_{12}\,E_{14}\,E_{24}\,E_{34}\,E_{35}\,E_{56}$$

$$\Theta_{\vec b}^{\vec a}\!\left[\begin{smallmatrix}1&2&6\\3&4&5\end{smallmatrix}\right]\Theta_{\vec b}^{\vec a}\!\left[\begin{smallmatrix}1&3&4\\2&5&6\end{smallmatrix}\right]\Theta_{\vec b}^{\vec a}\!\left[\begin{smallmatrix}1&5&6\\2&3&4\end{smallmatrix}\right]\Theta_{\vec b}^{\vec a}\!\left[\begin{smallmatrix}1&2&4\\3&5&6\end{smallmatrix}\right]^2$$

$$+ (\gamma^\mu C)_{\alpha\beta}(\gamma_\mu C)_{\delta\epsilon}\, \frac{C_\gamma{}^{\dot\zeta}}{E_{36}}\, E_{13}\,E_{15}\,E_{24}\,E_{26}\,E_{34}\,E_{35}$$

$$\Theta_{\vec b}^{\vec a}\!\left[\begin{smallmatrix}1&2&6\\3&4&5\end{smallmatrix}\right]\Theta_{\vec b}^{\vec a}\!\left[\begin{smallmatrix}1&3&4\\2&5&6\end{smallmatrix}\right]\Theta_{\vec b}^{\vec a}\!\left[\begin{smallmatrix}1&5&6\\2&3&4\end{smallmatrix}\right]\Theta_{\vec b}^{\vec a}\!\left[\begin{smallmatrix}1&3&5\\2&4&6\end{smallmatrix}\right]^2$$

$$- (\gamma^\mu C)_{\alpha\epsilon}(\gamma_\mu C)_{\beta\delta}\, \frac{C_\gamma{}^{\dot\zeta}}{E_{36}}\, E_{12}\,E_{13}\,E_{26}\,E_{34}\,E_{35}\,E_{45}$$

$$\Theta_{\vec b}^{\vec a}\!\left[\begin{smallmatrix}1&2&3\\4&5&6\end{smallmatrix}\right]\Theta_{\vec b}^{\vec a}\!\left[\begin{smallmatrix}1&3&4\\2&5&6\end{smallmatrix}\right]\Theta_{\vec b}^{\vec a}\!\left[\begin{smallmatrix}1&3&5\\2&4&6\end{smallmatrix}\right]\Theta_{\vec b}^{\vec a}\!\left[\begin{smallmatrix}1&2&6\\3&4&5\end{smallmatrix}\right]^2$$

$$+ (\gamma^\mu C)_{\alpha\gamma}(\gamma_\mu C)_{\delta\epsilon}\, \frac{C_\beta{}^{\dot\zeta}}{E_{26}}\, E_{12}\,E_{15}\,E_{24}\,E_{25}\,E_{34}\,E_{36}$$

$$\Theta_{\vec b}^{\vec a}\!\left[\begin{smallmatrix}1&2&4\\3&5&6\end{smallmatrix}\right]\Theta_{\vec b}^{\vec a}\!\left[\begin{smallmatrix}1&3&6\\2&4&5\end{smallmatrix}\right]\Theta_{\vec b}^{\vec a}\!\left[\begin{smallmatrix}1&5&6\\2&3&4\end{smallmatrix}\right]\Theta_{\vec b}^{\vec a}\!\left[\begin{smallmatrix}1&2&5\\3&4&6\end{smallmatrix}\right]^2$$

$$- (\gamma^\mu C)_{\alpha\epsilon}(\gamma_\mu C)_{\gamma\delta}\, \frac{C_\beta{}^{\dot\zeta}}{E_{26}}\, E_{12}\,E_{13}\,E_{24}\,E_{25}\,E_{36}\,E_{45}$$

$$\Theta_{\vec b}^{\vec a}\!\left[\begin{smallmatrix}1&2&3\\4&5&6\end{smallmatrix}\right]\Theta_{\vec b}^{\vec a}\!\left[\begin{smallmatrix}1&2&4\\3&5&6\end{smallmatrix}\right]\Theta_{\vec b}^{\vec a}\!\left[\begin{smallmatrix}1&2&5\\3&4&6\end{smallmatrix}\right]\Theta_{\vec b}^{\vec a}\!\left[\begin{smallmatrix}1&3&6\\2&4&5\end{smallmatrix}\right]^2$$

$$+ (\gamma^\mu C)_{\beta\epsilon}(\gamma_\mu C)_{\gamma\delta}\, \frac{C_\alpha{}^{\dot\zeta}}{E_{16}}\, E_{12}\,E_{14}\,E_{15}\,E_{23}\,E_{36}\,E_{45}$$

$$\Theta_{\vec b}^{\vec a}\!\left[\begin{smallmatrix}1&2&3\\4&5&6\end{smallmatrix}\right]\Theta_{\vec b}^{\vec a}\!\left[\begin{smallmatrix}1&2&4\\3&5&6\end{smallmatrix}\right]\Theta_{\vec b}^{\vec a}\!\left[\begin{smallmatrix}1&2&5\\3&4&6\end{smallmatrix}\right]\Theta_{\vec b}^{\vec a}\!\left[\begin{smallmatrix}1&4&5\\2&3&6\end{smallmatrix}\right]^2$$

$$- (\gamma^\mu C)_{\beta\gamma}(\gamma_\mu C)_{\delta\epsilon}\, \frac{C_\alpha{}^{\dot\zeta}}{E_{16}}\, E_{12}\,E_{14}\,E_{15}\,E_{25}\,E_{34}\,E_{36}$$

$$\Theta_{\vec b}^{\vec a}\!\left[\begin{smallmatrix}1&2&4\\3&5&6\end{smallmatrix}\right]\Theta_{\vec b}^{\vec a}\!\left[\begin{smallmatrix}1&3&4\\2&5&6\end{smallmatrix}\right]\Theta_{\vec b}^{\vec a}\!\left[\begin{smallmatrix}1&4&5\\2&3&6\end{smallmatrix}\right]\Theta_{\vec b}^{\vec a}\!\left[\begin{smallmatrix}1&2&5\\3&4&6\end{smallmatrix}\right]^2$$

$$- \frac{1}{2}(\gamma^\mu \bar\gamma^\nu C)_\beta{}^{\dot\zeta}(\gamma_\mu C)_{\alpha\epsilon}(\gamma_\nu C)_{\gamma\delta}\, E_{12}\,E_{14}\,E_{24}\,E_{35}\,E_{35}$$

$$\Theta_{\vec b}^{\vec a}\!\left[\begin{smallmatrix}1&2&6\\3&4&5\end{smallmatrix}\right]\Theta_{\vec b}^{\vec a}\!\left[\begin{smallmatrix}1&3&5\\2&4&6\end{smallmatrix}\right]\Theta_{\vec b}^{\vec a}\!\left[\begin{smallmatrix}1&4&6\\2&3&5\end{smallmatrix}\right]\Theta_{\vec b}^{\vec a}\!\left[\begin{smallmatrix}1&2&4\\3&5&6\end{smallmatrix}\right]^2$$

$$+ \frac{1}{2} (\gamma^\mu \bar\gamma^\nu C)_\alpha{}^{\dot\zeta} (\gamma_\mu C)_{\beta\delta} (\gamma_\nu C)_{\gamma\epsilon} \, E_{12} \, E_{15} \, E_{25} \, E_{34} \, E_{34}$$
$$\Theta^{\vec a}_{\vec b} {\scriptstyle [\begin{smallmatrix}1 & 2 & 6\\3 & 4 & 5\end{smallmatrix}]} \, \Theta^{\vec a}_{\vec b} {\scriptstyle [\begin{smallmatrix}1 & 3 & 4\\2 & 5 & 6\end{smallmatrix}]} \, \Theta^{\vec a}_{\vec b} {\scriptstyle [\begin{smallmatrix}1 & 5 & 6\\2 & 3 & 4\end{smallmatrix}]} \, \Theta^{\vec a}_{\vec b} {\scriptstyle [\begin{smallmatrix}1 & 2 & 5\\3 & 4 & 6\end{smallmatrix}]}^2$$
$$+ \frac{1}{2} (\gamma^\mu \bar\gamma^\nu C)_\epsilon{}^{\dot\zeta} (\gamma_\mu C)_{\alpha\beta} (\gamma_\nu C)_{\gamma\delta} \, E_{14} \, E_{15} \, E_{23} \, E_{24} \, E_{35}$$
$$\Theta^{\vec a}_{\vec b} {\scriptstyle [\begin{smallmatrix}1 & 2 & 4\\3 & 5 & 6\end{smallmatrix}]} \, \Theta^{\vec a}_{\vec b} {\scriptstyle [\begin{smallmatrix}1 & 3 & 5\\2 & 4 & 6\end{smallmatrix}]} \, \Theta^{\vec a}_{\vec b} {\scriptstyle [\begin{smallmatrix}1 & 4 & 5\\2 & 3 & 6\end{smallmatrix}]} \, \Theta^{\vec a}_{\vec b} {\scriptstyle [\begin{smallmatrix}1 & 4 & 6\\2 & 3 & 5\end{smallmatrix}]} \, \Theta^{\vec a}_{\vec b} {\scriptstyle [\begin{smallmatrix}1 & 5 & 6\\2 & 3 & 4\end{smallmatrix}]}$$
$$+ \frac{1}{2} (\gamma^\mu \bar\gamma^\nu C)_\delta{}^{\dot\zeta} (\gamma_\mu C)_{\alpha\beta} (\gamma_\nu C)_{\gamma\epsilon} \, E_{13} \, E_{15} \, E_{24} \, E_{25} \, E_{34}$$
$$\Theta^{\vec a}_{\vec b} {\scriptstyle [\begin{smallmatrix}1 & 2 & 5\\3 & 4 & 6\end{smallmatrix}]} \, \Theta^{\vec a}_{\vec b} {\scriptstyle [\begin{smallmatrix}1 & 3 & 4\\2 & 5 & 6\end{smallmatrix}]} \, \Theta^{\vec a}_{\vec b} {\scriptstyle [\begin{smallmatrix}1 & 3 & 5\\2 & 4 & 6\end{smallmatrix}]} \, \Theta^{\vec a}_{\vec b} {\scriptstyle [\begin{smallmatrix}1 & 3 & 6\\2 & 4 & 5\end{smallmatrix}]} \, \Theta^{\vec a}_{\vec b} {\scriptstyle [\begin{smallmatrix}1 & 5 & 6\\2 & 3 & 4\end{smallmatrix}]}$$
$$- \frac{1}{2} (\gamma^\mu \bar\gamma^\nu C)_\gamma{}^{\dot\zeta} (\gamma_\mu C)_{\alpha\delta} (\gamma_\nu C)_{\beta\epsilon} \, E_{12} \, E_{15} \, E_{24} \, E_{34} \, E_{35}$$
$$\Theta^{\vec a}_{\vec b} {\scriptstyle [\begin{smallmatrix}1 & 2 & 4\\3 & 5 & 6\end{smallmatrix}]} \, \Theta^{\vec a}_{\vec b} {\scriptstyle [\begin{smallmatrix}1 & 2 & 5\\3 & 4 & 6\end{smallmatrix}]} \, \Theta^{\vec a}_{\vec b} {\scriptstyle [\begin{smallmatrix}1 & 2 & 6\\3 & 4 & 5\end{smallmatrix}]} \, \Theta^{\vec a}_{\vec b} {\scriptstyle [\begin{smallmatrix}1 & 3 & 5\\2 & 4 & 6\end{smallmatrix}]} \, \Theta^{\vec a}_{\vec b} {\scriptstyle [\begin{smallmatrix}1 & 5 & 6\\2 & 3 & 4\end{smallmatrix}]}$$
$$- \frac{1}{2} (\gamma^\mu \bar\gamma^\nu C)_\gamma{}^{\dot\zeta} (\gamma_\mu C)_{\alpha\epsilon} (\gamma_\nu C)_{\beta\delta} \, E_{12} \, E_{14} \, E_{25} \, E_{34} \, E_{35}$$
$$\left. \Theta^{\vec a}_{\vec b} {\scriptstyle [\begin{smallmatrix}1 & 2 & 4\\3 & 5 & 6\end{smallmatrix}]} \, \Theta^{\vec a}_{\vec b} {\scriptstyle [\begin{smallmatrix}1 & 2 & 5\\3 & 4 & 6\end{smallmatrix}]} \, \Theta^{\vec a}_{\vec b} {\scriptstyle [\begin{smallmatrix}1 & 2 & 6\\3 & 4 & 5\end{smallmatrix}]} \, \Theta^{\vec a}_{\vec b} {\scriptstyle [\begin{smallmatrix}1 & 3 & 4\\2 & 5 & 6\end{smallmatrix}]} \, \Theta^{\vec a}_{\vec b} {\scriptstyle [\begin{smallmatrix}1 & 4 & 6\\2 & 3 & 5\end{smallmatrix}]} \right] . \quad (5.67)$$

There is also a non-vanishing correlator with three left- and right-handed spin fields each:

$$\langle S_\alpha(z_1) \, S_\beta(z_2) \, S_\gamma(z_3) \, S^{\dot\delta}(z_4) \, S^{\dot\epsilon}(z_5) \, S^{\dot\zeta}(z_6) \rangle^{\vec a}_{\vec b}$$
$$= - \frac{1}{[\Theta^{\vec a}_{\vec b}(\vec 0)]^5} \left(\frac{E_{12} \, E_{13} \, E_{23} \, E_{45} \, E_{46} \, E_{56}}{E_{14} \, E_{15} \, E_{16} \, E_{24} \, E_{25} \, E_{26} \, E_{34} \, E_{35} \, E_{36}} \right)^{1/4}$$
$$\times \left[\frac{C_\alpha{}^{\dot\delta} \, C_\beta{}^{\dot\epsilon} \, C_\gamma{}^{\dot\zeta}}{E_{14} \, E_{25} \, E_{36}} \, \Theta^{\vec a}_{\vec b} {\scriptstyle [\begin{smallmatrix}1 & 2 & 3\\4 & 5 & 6\end{smallmatrix}]}^2 \, \Theta^{\vec a}_{\vec b} {\scriptstyle [\begin{smallmatrix}1 & 5 & 6\\2 & 3 & 4\end{smallmatrix}]} \, \Theta^{\vec a}_{\vec b} {\scriptstyle [\begin{smallmatrix}1 & 3 & 5\\2 & 4 & 6\end{smallmatrix}]} \, \Theta^{\vec a}_{\vec b} {\scriptstyle [\begin{smallmatrix}1 & 2 & 6\\3 & 4 & 5\end{smallmatrix}]} \right.$$
$$- \frac{C_\alpha{}^{\dot\delta} \, C_\beta{}^{\dot\zeta} \, C_\gamma{}^{\dot\epsilon}}{E_{14} \, E_{26} \, E_{35}} \, \Theta^{\vec a}_{\vec b} {\scriptstyle [\begin{smallmatrix}1 & 2 & 3\\4 & 5 & 6\end{smallmatrix}]}^2 \, \Theta^{\vec a}_{\vec b} {\scriptstyle [\begin{smallmatrix}1 & 5 & 6\\2 & 3 & 4\end{smallmatrix}]} \, \Theta^{\vec a}_{\vec b} {\scriptstyle [\begin{smallmatrix}1 & 3 & 6\\2 & 4 & 5\end{smallmatrix}]} \, \Theta^{\vec a}_{\vec b} {\scriptstyle [\begin{smallmatrix}1 & 2 & 5\\3 & 4 & 6\end{smallmatrix}]}$$
$$+ \frac{C_\alpha{}^{\dot\epsilon} \, C_\beta{}^{\dot\zeta} \, C_\gamma{}^{\dot\delta}}{E_{15} \, E_{26} \, E_{34}} \, \Theta^{\vec a}_{\vec b} {\scriptstyle [\begin{smallmatrix}1 & 2 & 3\\4 & 5 & 6\end{smallmatrix}]}^2 \, \Theta^{\vec a}_{\vec b} {\scriptstyle [\begin{smallmatrix}1 & 4 & 6\\2 & 3 & 5\end{smallmatrix}]} \, \Theta^{\vec a}_{\vec b} {\scriptstyle [\begin{smallmatrix}1 & 3 & 6\\2 & 4 & 5\end{smallmatrix}]} \, \Theta^{\vec a}_{\vec b} {\scriptstyle [\begin{smallmatrix}1 & 2 & 4\\3 & 5 & 6\end{smallmatrix}]}$$
$$- \frac{C_\alpha{}^{\dot\epsilon} \, C_\beta{}^{\dot\delta} \, C_\gamma{}^{\dot\zeta}}{E_{15} \, E_{24} \, E_{36}} \, \Theta^{\vec a}_{\vec b} {\scriptstyle [\begin{smallmatrix}1 & 2 & 3\\4 & 5 & 6\end{smallmatrix}]}^2 \, \Theta^{\vec a}_{\vec b} {\scriptstyle [\begin{smallmatrix}1 & 4 & 6\\2 & 3 & 5\end{smallmatrix}]} \, \Theta^{\vec a}_{\vec b} {\scriptstyle [\begin{smallmatrix}1 & 3 & 4\\2 & 5 & 6\end{smallmatrix}]} \, \Theta^{\vec a}_{\vec b} {\scriptstyle [\begin{smallmatrix}1 & 2 & 6\\3 & 4 & 5\end{smallmatrix}]}$$
$$+ \frac{C_\alpha{}^{\dot\zeta} \, C_\beta{}^{\dot\delta} \, C_\gamma{}^{\dot\epsilon}}{E_{16} \, E_{24} \, E_{35}} \, \Theta^{\vec a}_{\vec b} {\scriptstyle [\begin{smallmatrix}1 & 2 & 3\\4 & 5 & 6\end{smallmatrix}]}^2 \, \Theta^{\vec a}_{\vec b} {\scriptstyle [\begin{smallmatrix}1 & 4 & 5\\2 & 3 & 6\end{smallmatrix}]} \, \Theta^{\vec a}_{\vec b} {\scriptstyle [\begin{smallmatrix}1 & 3 & 4\\2 & 5 & 6\end{smallmatrix}]} \, \Theta^{\vec a}_{\vec b} {\scriptstyle [\begin{smallmatrix}1 & 2 & 5\\3 & 4 & 6\end{smallmatrix}]}$$
$$- \frac{C_\alpha{}^{\dot\zeta} \, C_\beta{}^{\dot\epsilon} \, C_\gamma{}^{\dot\delta}}{E_{16} \, E_{25} \, E_{34}} \, \Theta^{\vec a}_{\vec b} {\scriptstyle [\begin{smallmatrix}1 & 2 & 3\\4 & 5 & 6\end{smallmatrix}]}^2 \, \Theta^{\vec a}_{\vec b} {\scriptstyle [\begin{smallmatrix}1 & 4 & 5\\2 & 3 & 6\end{smallmatrix}]} \, \Theta^{\vec a}_{\vec b} {\scriptstyle [\begin{smallmatrix}1 & 3 & 5\\2 & 4 & 6\end{smallmatrix}]} \, \Theta^{\vec a}_{\vec b} {\scriptstyle [\begin{smallmatrix}1 & 2 & 4\\3 & 5 & 6\end{smallmatrix}]}$$
$$- \frac{1}{2} \frac{(\gamma^\mu C)_{\alpha\beta} (\bar\gamma_\mu C)^{\dot\delta\dot\epsilon} C_\gamma{}^{\dot\zeta}}{E_{12} \, E_{36} \, E_{45}} \, \Theta^{\vec a}_{\vec b} {\scriptstyle [\begin{smallmatrix}1 & 2 & 3\\4 & 5 & 6\end{smallmatrix}]} \, \Theta^{\vec a}_{\vec b} {\scriptstyle [\begin{smallmatrix}1 & 3 & 4\\2 & 5 & 6\end{smallmatrix}]} \, \Theta^{\vec a}_{\vec b} {\scriptstyle [\begin{smallmatrix}1 & 3 & 5\\2 & 4 & 6\end{smallmatrix}]} \, \Theta^{\vec a}_{\vec b} {\scriptstyle [\begin{smallmatrix}1 & 4 & 6\\2 & 3 & 5\end{smallmatrix}]} \, \Theta^{\vec a}_{\vec b} {\scriptstyle [\begin{smallmatrix}1 & 5 & 6\\2 & 3 & 4\end{smallmatrix}]}$$
$$+ \frac{1}{2} \frac{(\gamma^\mu C)_{\alpha\beta} (\bar\gamma_\mu C)^{\dot\delta\dot\zeta} C_\gamma{}^{\dot\epsilon}}{E_{12} \, E_{35} \, E_{46}} \, \Theta^{\vec a}_{\vec b} {\scriptstyle [\begin{smallmatrix}1 & 2 & 3\\4 & 5 & 6\end{smallmatrix}]} \, \Theta^{\vec a}_{\vec b} {\scriptstyle [\begin{smallmatrix}1 & 3 & 4\\2 & 5 & 6\end{smallmatrix}]} \, \Theta^{\vec a}_{\vec b} {\scriptstyle [\begin{smallmatrix}1 & 3 & 6\\2 & 4 & 5\end{smallmatrix}]} \, \Theta^{\vec a}_{\vec b} {\scriptstyle [\begin{smallmatrix}1 & 4 & 5\\2 & 3 & 6\end{smallmatrix}]} \, \Theta^{\vec a}_{\vec b} {\scriptstyle [\begin{smallmatrix}1 & 5 & 6\\2 & 3 & 4\end{smallmatrix}]}$$
$$- \frac{1}{2} \frac{(\gamma^\mu C)_{\alpha\beta} (\bar\gamma_\mu C)^{\dot\epsilon\dot\zeta} C_\gamma{}^{\dot\delta}}{E_{12} \, E_{34} \, E_{56}} \, \Theta^{\vec a}_{\vec b} {\scriptstyle [\begin{smallmatrix}1 & 2 & 3\\4 & 5 & 6\end{smallmatrix}]} \, \Theta^{\vec a}_{\vec b} {\scriptstyle [\begin{smallmatrix}1 & 3 & 5\\2 & 4 & 6\end{smallmatrix}]} \, \Theta^{\vec a}_{\vec b} {\scriptstyle [\begin{smallmatrix}1 & 3 & 6\\2 & 4 & 5\end{smallmatrix}]} \, \Theta^{\vec a}_{\vec b} {\scriptstyle [\begin{smallmatrix}1 & 4 & 5\\2 & 3 & 6\end{smallmatrix}]} \, \Theta^{\vec a}_{\vec b} {\scriptstyle [\begin{smallmatrix}1 & 4 & 6\\2 & 3 & 5\end{smallmatrix}]}$$

5.3 Results of RNS Loop Correlators

$$
\begin{aligned}
&+ \frac{1}{2} \frac{(\gamma^\mu C)_{\alpha\gamma} (\bar\gamma_\mu C)^{\dot\delta\dot\epsilon} C_\beta{}^{\dot\zeta}}{E_{13} E_{26} E_{45}} \Theta_{\vec b}^{\vec a}[\begin{smallmatrix}1&2&3\\4&5&6\end{smallmatrix}] \Theta_{\vec b}^{\vec a}[\begin{smallmatrix}1&2&4\\3&5&6\end{smallmatrix}] \Theta_{\vec b}^{\vec a}[\begin{smallmatrix}1&2&5\\3&4&6\end{smallmatrix}] \Theta_{\vec b}^{\vec a}[\begin{smallmatrix}1&4&6\\2&3&5\end{smallmatrix}] \Theta_{\vec b}^{\vec a}[\begin{smallmatrix}1&5&6\\2&3&4\end{smallmatrix}] \\
&- \frac{1}{2} \frac{(\gamma^\mu C)_{\alpha\gamma} (\bar\gamma_\mu C)^{\dot\delta\dot\zeta} C_\beta{}^{\dot\epsilon}}{E_{13} E_{25} E_{46}} \Theta_{\vec b}^{\vec a}[\begin{smallmatrix}1&2&3\\4&5&6\end{smallmatrix}] \Theta_{\vec b}^{\vec a}[\begin{smallmatrix}1&2&4\\3&5&6\end{smallmatrix}] \Theta_{\vec b}^{\vec a}[\begin{smallmatrix}1&2&6\\3&4&5\end{smallmatrix}] \Theta_{\vec b}^{\vec a}[\begin{smallmatrix}1&4&5\\2&3&6\end{smallmatrix}] \Theta_{\vec b}^{\vec a}[\begin{smallmatrix}1&5&6\\2&3&4\end{smallmatrix}] \\
&+ \frac{1}{2} \frac{(\gamma^\mu C)_{\alpha\gamma} (\bar\gamma_\mu C)^{\dot\epsilon\dot\zeta} C_\beta{}^{\dot\delta}}{E_{13} E_{24} E_{56}} \Theta_{\vec b}^{\vec a}[\begin{smallmatrix}1&2&3\\4&5&6\end{smallmatrix}] \Theta_{\vec b}^{\vec a}[\begin{smallmatrix}1&2&5\\3&4&6\end{smallmatrix}] \Theta_{\vec b}^{\vec a}[\begin{smallmatrix}1&2&6\\3&4&5\end{smallmatrix}] \Theta_{\vec b}^{\vec a}[\begin{smallmatrix}1&4&5\\2&3&6\end{smallmatrix}] \Theta_{\vec b}^{\vec a}[\begin{smallmatrix}1&4&6\\2&3&5\end{smallmatrix}] \\
&- \frac{1}{2} \frac{(\gamma^\mu C)_{\beta\gamma} (\bar\gamma_\mu C)^{\dot\delta\dot\epsilon} C_\alpha{}^{\dot\zeta}}{E_{16} E_{23} E_{45}} \Theta_{\vec b}^{\vec a}[\begin{smallmatrix}1&2&3\\4&5&6\end{smallmatrix}] \Theta_{\vec b}^{\vec a}[\begin{smallmatrix}1&2&4\\3&5&6\end{smallmatrix}] \Theta_{\vec b}^{\vec a}[\begin{smallmatrix}1&2&5\\3&4&6\end{smallmatrix}] \Theta_{\vec b}^{\vec a}[\begin{smallmatrix}1&3&4\\2&5&6\end{smallmatrix}] \Theta_{\vec b}^{\vec a}[\begin{smallmatrix}1&3&5\\2&4&6\end{smallmatrix}] \\
&+ \frac{1}{2} \frac{(\gamma^\mu C)_{\beta\gamma} (\bar\gamma_\mu C)^{\dot\delta\dot\zeta} C_\alpha{}^{\dot\epsilon}}{E_{15} E_{23} E_{46}} \Theta_{\vec b}^{\vec a}[\begin{smallmatrix}1&2&3\\4&5&6\end{smallmatrix}] \Theta_{\vec b}^{\vec a}[\begin{smallmatrix}1&2&4\\3&5&6\end{smallmatrix}] \Theta_{\vec b}^{\vec a}[\begin{smallmatrix}1&2&6\\3&4&5\end{smallmatrix}] \Theta_{\vec b}^{\vec a}[\begin{smallmatrix}1&3&4\\2&5&6\end{smallmatrix}] \Theta_{\vec b}^{\vec a}[\begin{smallmatrix}1&3&6\\2&4&5\end{smallmatrix}] \\
&- \frac{1}{2} \frac{(\gamma^\mu C)_{\beta\gamma} (\bar\gamma_\mu C)^{\dot\epsilon\dot\zeta} C_\alpha{}^{\dot\delta}}{E_{14} E_{23} E_{56}} \Theta_{\vec b}^{\vec a}[\begin{smallmatrix}1&2&3\\4&5&6\end{smallmatrix}] \Theta_{\vec b}^{\vec a}[\begin{smallmatrix}1&2&5\\3&4&6\end{smallmatrix}] \Theta_{\vec b}^{\vec a}[\begin{smallmatrix}1&2&6\\3&4&5\end{smallmatrix}] \Theta_{\vec b}^{\vec a}[\begin{smallmatrix}1&3&5\\2&4&6\end{smallmatrix}] \Theta_{\vec b}^{\vec a}[\begin{smallmatrix}1&3&6\\2&4&5\end{smallmatrix}] \\
&+ \frac{1}{4} (\gamma^\mu \bar\gamma^\nu C)_\alpha{}^{\dot\epsilon} (\gamma_\mu C)_{\beta\gamma} (\bar\gamma_\nu C)^{\dot\delta\dot\zeta} \frac{E_{36}}{E_{13} E_{23} E_{46} E_{56}} \\
&\qquad \Theta_{\vec b}^{\vec a}[\begin{smallmatrix}1&2&4\\3&5&6\end{smallmatrix}] \Theta_{\vec b}^{\vec a}[\begin{smallmatrix}1&2&5\\3&4&6\end{smallmatrix}] \Theta_{\vec b}^{\vec a}[\begin{smallmatrix}1&2&6\\3&4&5\end{smallmatrix}] \Theta_{\vec b}^{\vec a}[\begin{smallmatrix}1&3&6\\2&4&5\end{smallmatrix}] \Theta_{\vec b}^{\vec a}[\begin{smallmatrix}1&4&5\\2&3&6\end{smallmatrix}] \\
&- \frac{1}{4} (\gamma^\mu \bar\gamma^\nu C)_\gamma{}^{\dot\epsilon} (\gamma_\mu C)_{\alpha\beta} (\bar\gamma_\nu C)^{\dot\delta\dot\zeta} \frac{E_{16}}{E_{12} E_{13} E_{46} E_{56}} \\
&\qquad \Theta_{\vec b}^{\vec a}[\begin{smallmatrix}1&2&6\\3&4&5\end{smallmatrix}] \Theta_{\vec b}^{\vec a}[\begin{smallmatrix}1&3&6\\2&4&5\end{smallmatrix}] \Theta_{\vec b}^{\vec a}[\begin{smallmatrix}1&4&5\\2&3&6\end{smallmatrix}] \Theta_{\vec b}^{\vec a}[\begin{smallmatrix}1&4&6\\2&3&5\end{smallmatrix}] \Theta_{\vec b}^{\vec a}[\begin{smallmatrix}1&5&6\\2&3&4\end{smallmatrix}] \\
&+ \frac{1}{4} (\gamma^\mu \bar\gamma^\nu C)_\alpha{}^{\dot\zeta} (\gamma_\mu C)_{\beta\gamma} (\bar\gamma_\nu C)^{\dot\delta\dot\epsilon} \frac{E_{35}}{E_{13} E_{23} E_{45} E_{56}} \\
&\qquad \Theta_{\vec b}^{\vec a}[\begin{smallmatrix}1&2&4\\3&5&6\end{smallmatrix}] \Theta_{\vec b}^{\vec a}[\begin{smallmatrix}1&2&5\\3&4&6\end{smallmatrix}] \Theta_{\vec b}^{\vec a}[\begin{smallmatrix}1&2&6\\3&4&5\end{smallmatrix}] \Theta_{\vec b}^{\vec a}[\begin{smallmatrix}1&3&5\\2&4&6\end{smallmatrix}] \Theta_{\vec b}^{\vec a}[\begin{smallmatrix}1&4&6\\2&3&5\end{smallmatrix}] \\
&- \frac{1}{4} (\gamma^\mu \bar\gamma^\nu C)_\gamma{}^{\dot\zeta} (\gamma_\mu C)_{\alpha\beta} (\bar\gamma_\nu C)^{\dot\delta\dot\epsilon} \frac{E_{15}}{E_{12} E_{13} E_{45} E_{56}} \\
&\qquad \Theta_{\vec b}^{\vec a}[\begin{smallmatrix}1&2&5\\3&4&6\end{smallmatrix}] \Theta_{\vec b}^{\vec a}[\begin{smallmatrix}1&4&5\\2&3&6\end{smallmatrix}] \Theta_{\vec b}^{\vec a}[\begin{smallmatrix}1&3&5\\2&4&6\end{smallmatrix}] \Theta_{\vec b}^{\vec a}[\begin{smallmatrix}1&5&6\\2&3&4\end{smallmatrix}] \Theta_{\vec b}^{\vec a}[\begin{smallmatrix}1&4&6\\2&3&5\end{smallmatrix}] \bigg]. \quad (5.68)
\end{aligned}
$$

Note that also this result exhibits manifest antisymmetry when exchanging $S_\alpha(z_1) \leftrightarrow S_\gamma(z_3)$ and $S^{\dot\epsilon}(z_5) \leftrightarrow S^{\dot\zeta}(z_6)$ up to the powers $(E_{13} E_{56})^{1/4}$.

5.3.5 Pure Spin Field Correlators

In Chapter 3.4 and 3.5 it has been shown that certain pure spin field correlator in four and six space-time dimensions can be solved for arbitrary many fields. This is due to the fact that only index terms built from $\varepsilon_{\alpha\beta}$ and $C_\alpha{}^{\dot\beta}$ are necessary. The tree-level results (3.82) and (3.90) can also be generalized to loop-level. It has been shown in [155] that for $D = 4$

$$
\left\langle \prod_{i=1}^{2M} S_{\alpha_i}(z_i) \right\rangle_{\vec b}^{\vec a} = (-1)^M \frac{\left[\Theta_{\vec b}^{\vec a}\left(\frac{1}{2} \sum_{i=1}^M \int_{z_{2i}}^{z_{2i-1}} \vec\omega\right) \right]^{2-M}}{\left[\Theta_{\vec b}^{\vec a}(\vec 0) \right]^2} \left(\frac{\prod_{i \leq j}^M E_{2i-1, 2j} \prod_{\bar i < \bar j}^M E_{2\bar i, 2\bar j - 1}}{\prod_{k < l}^M E_{2k-1, 2l-1} E_{2k, 2l}} \right)^{1/2}
$$

$$\times \sum_{\rho \in S_M} \text{sgn}(\rho) \prod_{m=1}^{M} \frac{\varepsilon_{\alpha_{2m-1}\alpha_{\rho(2m)}}}{E_{2m-1,\rho(2m)}} \Theta_{\vec{b}}^{\vec{a}}\left(\frac{1}{2}\sum_{i=1}^{M}\int_{z_{2i}}^{z_{2i-1}}\vec{\omega} - \int_{z_{\rho(2m)}}^{z_{2m-1}}\vec{\omega}\right). \quad (5.69)$$

The proof of this formula by induction progresses in the same way as (3.83), but now also the generalized Θ functions have to be taken care of. Details can be found in Appendix B of [155].

The relative of (5.69) in $D = 6$ dimensions is given by

$$\left\langle \prod_{i=1}^{M} S_{\alpha_i}(z_{2i-1}) S^{\dot{\beta}_i}(z_{2i}) \right\rangle_{\vec{b}}^{\vec{a}} = \frac{\left[\Theta_{\vec{b}}^{\vec{a}}\left(\frac{1}{2}\sum_{i=1}^{M}\int_{z_{2i}}^{z_{2i-1}}\vec{\omega}\right)\right]^{3-M}}{[\Theta_{\vec{b}}^{\vec{a}}(\vec{0})]^3} \left(\frac{\prod_{i,j=1}^{M} E_{2i-1,2j}}{\prod_{i<j}^{M} E_{2i-1,2j-1} E_{2i,2j}}\right)^{1/4}$$

$$\times \sum_{\rho \in S_M} \text{sgn}(\rho) \prod_{m=1}^{M} \frac{C_{\alpha_{2m-1}}{}^{\dot{\beta}_{\rho(2m)}}}{E_{2m-1,\rho(2m)}} \Theta_{\vec{b}}^{\vec{a}}\left(\frac{1}{2}\sum_{i=1}^{M}\int_{z_{2i-1}}^{z_{2i}}\vec{\omega} - \int_{z_{\rho(2m)}}^{z_{2m-1}}\vec{\omega}\right), \quad (5.70)$$

which is the generalization of the four- and six-point functions (5.46) and (5.49). In order to check that the eight-point function calculated by $SO(2)$ spin operators is consistent with (5.70) for $M = 4$ one has to make heavy use of Fay's trisecant identity.

5.3.6 General Results

We come now to a second class of RNS correlation functions for which one can derive general results at loop-level. This are correlators consisting of arbitrary many NS fermions but only two spin field:

$$\left\langle \psi^{\mu_1}(z_1) \ldots \psi^{\mu_n}(z_n) S_\alpha(z_A) S^{\dot{\beta}}(z_B) \right\rangle_{\vec{b}}^{\vec{a}},$$
$$\left\langle \psi^{\mu_1}(z_1) \ldots \psi^{\mu_n}(z_n) S_\alpha(z_A) S^{\beta}(z_B) \right\rangle_{\vec{b}}^{\vec{a}}. \quad (5.71)$$

In four dimensions at tree-level these have been discussed in Chapter 4.6. We generalize these results to loop-level and furthermore evaluate the loop correlators also in six, eight and ten dimensions.

The different chirality structure of the charge conjugation matrix in $D = 4, 8$ and $D = 6, 10$ dimensions imposes constraints on the number of NS fermions in (5.71). The possible index terms for the correlators consist of products of η's, γ-chains and one C. In four and eight dimensions the charge conjugation matrix has the structure $C_{\alpha\beta}$. Therefore, the number of fermions has to be odd for Ω and even for ω, while in six and ten dimensions with the charge conjugation matrix $C_\alpha{}^{\dot\beta}$ it has to be the other way round. In the following we thus examine the correlation functions

5.3 Results of RNS Loop Correlators

- $D = 0 \mod 4$:

$$\Omega_{n,D} \equiv \left\langle \psi^{\mu_1}(z_1) \ldots \psi^{\mu_{2n-1}}(z_{2n-1}) S_\alpha(z_A) S^{\dot\beta}(z_B) \right\rangle_{\vec b}^{\vec a}\bigg|_D,$$
$$\omega_{n,D} \equiv \left\langle \psi^{\mu_1}(z_1) \ldots \psi^{\mu_{2n-2}}(z_{2n-2}) S_\alpha(z_A) S_\beta(z_B) \right\rangle_{\vec b}^{\vec a}\bigg|_D, \qquad (5.72)$$

- $D = 2 \mod 4$:

$$\Omega_{n,D} \equiv \left\langle \psi^{\mu_1}(z_1) \ldots \psi^{\mu_{2n-2}}(z_{2n-2}) S_\alpha(z_A) S^{\dot\beta}(z_B) \right\rangle_{\vec b}^{\vec a}\bigg|_D,$$
$$\omega_{n,D} \equiv \left\langle \psi^{\mu_1}(z_1) \ldots \psi^{\mu_{2n-1}}(z_{2n-1}) S_\alpha(z_A) S_\beta(z_B) \right\rangle_{\vec b}^{\vec a}\bigg|_D. \qquad (5.73)$$

If the number of fermions is chosen differently no scalar representations exist and the correlators hence vanish.

The correlation functions (5.72) in four dimensions have been determined in [155]. They are given as follows:

$$\Omega_{n,D=4} = \frac{\left[\Theta_{\vec b}^{\vec a}\!\left(\tfrac12 \int_{z_B}^{z_A}\vec\omega\right)\right]^{2-n}}{\sqrt{2}\,[\Theta_{\vec b}^{\vec a}(\vec 0)]^2 \prod_{i=1}^{2n-1}(E_{iA}E_{iB})^{1/2}} \sum_{l=0}^{n-1} \left(\frac{E_{AB}}{2\,\Theta_{\vec b}^{\vec a}\!\left(\tfrac12 \int_{z_B}^{z_A}\vec\omega\right)}\right)^l$$
$$\times \sum_{\rho\in S_{2n-1}/\mathcal P_{n,l}} \mathrm{sgn}(\rho)\left(\sigma^{\mu_{\rho(1)}}\bar\sigma^{\mu_{\rho(2)}}\ldots\bar\sigma^{\mu_{\rho(2l)}}\sigma^{\mu_{\rho(2l+1)}}\varepsilon\right)_\alpha^{\dot\beta} \prod_{k=1}^{2l+1}\Theta_{\vec b}^{\vec a}\!\left(\tfrac12\!\!\int_{z_{\rho(k)}}^{z_A}\!\!\vec\omega+\tfrac12\!\!\int_{z_{\rho(k)}}^{z_B}\!\!\vec\omega\right)$$
$$\times \prod_{j=1}^{n-l-1}\frac{\eta^{\mu_{\rho(2l+2j)}\mu_{\rho(2l+2j+1)}}}{E_{\rho(2l+2j),\rho(2l+2j+1)}}\,E_{\rho(2l+2j),A}\,E_{\rho(2l+2j+1),B}\,\Theta_{\vec b}^{\vec a}\!\left(\!\int_{z_{\rho(2l+2j+1)}}^{z_{\rho(2l+2j)}}\!\!\vec\omega+\tfrac12\!\!\int_{z_B}^{z_A}\!\!\vec\omega\right). \quad (5.74)$$

Its relative with even number of NS fermions and two alike spin fields reads

$$\omega_{n,D=4} = \frac{-\left[\Theta_{\vec b}^{\vec a}\!\left(\tfrac12 \int_{z_B}^{z_A}\vec\omega\right)\right]^{3-n}}{\Theta_{\vec b}^{\vec a}(\vec 0)\,\Theta_{\vec b}^{\vec a}(\vec 0)\,E_{AB}^{1/2}\prod_{i=1}^{2n-2}(E_{iA}E_{iB})^{1/2}} \sum_{l=0}^{n-1}\left(\frac{E_{AB}}{2\,\Theta_{\vec b}^{\vec a}\!\left(\tfrac12\int_{z_B}^{z_A}\vec\omega\right)}\right)^l$$
$$\times \sum_{\rho\in S_{2n-2}/\mathcal Q_{n,l}} \mathrm{sgn}(\rho)\left(\sigma^{\mu_{\rho(1)}}\bar\sigma^{\mu_{\rho(2)}}\ldots\sigma^{\mu_{\rho(2l-1)}}\bar\sigma^{\mu_{\rho(2l)}}\varepsilon\right)_{\alpha\beta} \prod_{k=1}^{2l}\Theta_{\vec b}^{\vec a}\!\left(\tfrac12\!\!\int_{z_{\rho(k)}}^{z_A}\!\!\vec\omega+\tfrac12\!\!\int_{z_{\rho(k)}}^{z_B}\!\!\vec\omega\right)$$
$$\times \prod_{j=1}^{n-l-1}\frac{\eta^{\mu_{\rho(2l+2j-1)}\mu_{\rho(2l+2j)}}}{E_{\rho(2l+2j-1),\rho(2l+2j)}}\,E_{\rho(2l+2j-1),A}\,E_{\rho(2l+2j),B}\,\Theta_{\vec b}^{\vec a}\!\left(\!\int_{z_{\rho(2l+2j)}}^{z_{\rho(2l+2j-1)}}\!\!\vec\omega+\tfrac12\!\!\int_{z_B}^{z_A}\!\!\vec\omega\right). \quad (5.75)$$

These results are the generalizations of the tree-level outcome (4.32) and (4.33). Here we make use of the same notation regarding the summation over the permutations ρ as in Chapter 4.6. The proof of (5.74) and (5.75) by induction proceeds in the same way as the tree-level findings following the web of limits illustrated in Figure 4.1. However, now one also has to take into account the generalized Θ functions. Details on the proof can be found in Appendix B of [155].

The generalization of these results to higher dimensions requires only minor modifications. As one goes to higher dimensions the number of $SO(2)$ spin systems increases. This does not play a role for the NS fermions because we still can choose the Lorentz indices for the calculation such that only the spin systems which were already present in the lower dimensions, enter the calculation. The additional spin systems contribute nevertheless through the R spin fields and hence Ω and ω in various dimensions differ by powers of the minimal spin system correlator:

$$\langle s^+(z_A)\, s^-(z_B)\rangle_{\vec{b}}^{\vec{a}} = \frac{1}{\Theta_{\vec{b}}^{\vec{a}}(\vec{0})\, E_{AB}^{1/4}}\, \Theta_{\vec{b}}^{\vec{a}}\!\left(\tfrac{1}{2}\int_{z_B}^{z_A}\vec{\omega}\right). \tag{5.76}$$

Following this argument Ω_n and ω_n in $D=8$ acquire two powers of (5.76) in contrast to (5.74) and (5.75):

$$\Omega_{n,D=8} = \frac{\left[\Theta_{\vec{b}}^{\vec{a}}\!\left(\tfrac{1}{2}\int_{z_B}^{z_A}\vec{\omega}\right)\right]^{4-n}}{\sqrt{2}\,\left[\Theta_{\vec{b}}^{\vec{a}}(\vec{0})\right]^4 E_{AB}^{1/2}\, \prod_{i=1}^{2n-1}(E_{iA}\,E_{iB})^{1/2}}\, \sum_{l=0}^{n-1}\left(\frac{E_{AB}}{2\,\Theta_{\vec{b}}^{\vec{a}}\!\left(\tfrac{1}{2}\int_{z_B}^{z_A}\vec{\omega}\right)}\right)^l$$

$$\times \sum_{\rho\in S_{2n-1}/\mathcal{P}_{n,l}} \mathrm{sgn}(\rho)\,\left(\gamma^{\mu_{\rho(1)}}\bar{\gamma}^{\mu_{\rho(2)}}\ldots\gamma^{\mu_{\rho(2l)}}\gamma^{\mu_{\rho(2l+1)}}\, C\right)_\alpha^{\ \dot{\beta}}\, \prod_{k=1}^{2l+1}\Theta_{\vec{b}}^{\vec{a}}\!\left(\tfrac{1}{2}\int_{z_{\rho(k)}}^{z_A}\vec{\omega}+\tfrac{1}{2}\int_{z_{\rho(k)}}^{z_B}\vec{\omega}\right)$$

$$\times \prod_{j=1}^{n-l-1}\frac{\eta^{\mu_{\rho(2l+2j)}\mu_{\rho(2l+2j+1)}}}{E_{\rho(2l+2j),\rho(2l+2j+1)}}\, E_{\rho(2l+2j),A}\, E_{\rho(2l+2j+1),B}\, \Theta_{\vec{b}}^{\vec{a}}\!\left(\int_{z_{\rho(2l+2j+1)}}^{z_{\rho(2l+2j)}}\vec{\omega}+\tfrac{1}{2}\int_{z_B}^{z_A}\vec{\omega}\right), \tag{5.77}$$

$$\omega_{n,D=8} = \frac{\left[\Theta_{\vec{b}}^{\vec{a}}\!\left(\tfrac{1}{2}\int_{z_B}^{z_A}\vec{\omega}\right)\right]^{5-n}}{\left[\Theta_{\vec{b}}^{\vec{a}}(\vec{0})\right]^4 E_{AB}\, \prod_{i=1}^{2n-2}(E_{iA}\,E_{iB})^{1/2}}\, \sum_{l=0}^{n-1}\left(\frac{E_{AB}}{2\,\Theta_{\vec{b}}^{\vec{a}}\!\left(\tfrac{1}{2}\int_{z_B}^{z_A}\vec{\omega}\right)}\right)^l$$

$$\times \sum_{\rho\in S_{2n-2}/\mathcal{Q}_{n,l}} \mathrm{sgn}(\rho)\,\left(\gamma^{\mu_{\rho(1)}}\bar{\gamma}^{\mu_{\rho(2)}}\ldots\gamma^{\mu_{\rho(2l-1)}}\bar{\gamma}^{\mu_{\rho(2l)}}\, C\right)_{\alpha\dot{\beta}}\, \prod_{k=1}^{2l}\Theta_{\vec{b}}^{\vec{a}}\!\left(\tfrac{1}{2}\int_{z_{\rho(k)}}^{z_A}\vec{\omega}+\tfrac{1}{2}\int_{z_{\rho(k)}}^{z_B}\vec{\omega}\right)$$

$$\times \prod_{j=1}^{n-l-1}\frac{\eta^{\mu_{\rho(2l+2j-1)}\mu_{\rho(2l+2j)}}}{E_{\rho(2l+2j-1),\rho(2l+2j)}}\, E_{\rho(2l+2j-1),A}\, E_{\rho(2l+2j),B}\, \Theta_{\vec{b}}^{\vec{a}}\!\left(\int_{z_{\rho(2l+2j)}}^{z_{\rho(2l+2j-1)}}\vec{\omega}+\tfrac{1}{2}\int_{z_B}^{z_A}\vec{\omega}\right). \tag{5.78}$$

We now come to the results of ω_n and Ω_n in six and ten dimensions. Due to the different chirality structure of the charge conjugation matrix ω_n in six dimensions, is the direct relative of Ω_n in four dimensions, where in addition one power of (5.76) has to be taken into account:

$$\omega_{n,D=6} = \frac{\left[\Theta_{\vec{b}}^{\vec{a}}\!\left(\tfrac{1}{2}\int_{z_B}^{z_A}\vec{\omega}\right)\right]^{3-n}}{\sqrt{2}\,\left[\Theta_{\vec{b}}^{\vec{a}}(\vec{0})\right]^3 E_{AB}^{1/4}\, \prod_{i=1}^{2n-1}(E_{iA}\,E_{iB})^{1/2}}\, \sum_{l=0}^{n-1}\left(\frac{E_{AB}}{2\,\Theta_{\vec{b}}^{\vec{a}}\!\left(\tfrac{1}{2}\int_{z_B}^{z_A}\vec{\omega}\right)}\right)^l$$

5.3 Results of RNS Loop Correlators

$$\times \sum_{\rho \in S_{2n-1}/\mathcal{P}_{n,l}} \text{sgn}(\rho) \left(\gamma^{\mu_{\rho(1)}} \bar{\gamma}^{\mu_{\rho(2)}} \ldots \bar{\gamma}^{\mu_{\rho(2l)}} \gamma^{\mu_{\rho(2l+1)}} C\right)_{\alpha\beta} \prod_{k=1}^{2l+1} \Theta_{\vec{b}}^{\vec{a}}\left(\frac{1}{2}\int_{z_{\rho(k)}}^{z_A} \vec{\omega} + \frac{1}{2}\int_{z_{\rho(k)}}^{z_B} \vec{\omega}\right)$$

$$\times \prod_{j=1}^{n-l-1} \frac{\eta^{\mu_{\rho(2l+2j)}\mu_{\rho(2l+2j+1)}}}{E_{\rho(2l+2j),\rho(2l+2j+1)}} E_{\rho(2l+2j),A} E_{\rho(2l+2j+1),B} \Theta_{\vec{b}}^{\vec{a}}\left(\int_{z_{\rho(2l+2j+1)}}^{z_{\rho(2l+2j)}} \vec{\omega} + \frac{1}{2}\int_{z_B}^{z_A} \vec{\omega}\right). \quad (5.79)$$

Similarly, $\Omega_{n,D=6}$ differs from $\omega_{n,D=4}$ by one power of (5.76):

$$\Omega_{n,D=6} = \frac{\left[\Theta_{\vec{b}}^{\vec{a}}\left(\frac{1}{2}\int_{z_B}^{z_A} \vec{\omega}\right)\right]^{4-n}}{\left[\left[\Theta_{\vec{b}}^{\vec{a}}(\vec{0})\right]^3 E_{AB}^{3/4} \prod_{i=1}^{2n-2}(E_{iA} E_{iB})^{1/2}\right]} \sum_{l=0}^{n-1} \left(\frac{E_{AB}}{2\Theta_{\vec{b}}^{\vec{a}}\left(\frac{1}{2}\int_{z_B}^{z_A} \vec{\omega}\right)}\right)^l$$

$$\times \sum_{\rho \in S_{2n-2}/\mathcal{Q}_{n,l}} \text{sgn}(\rho) \left(\gamma^{\mu_{\rho(1)}} \bar{\gamma}^{\mu_{\rho(2)}} \ldots \gamma^{\mu_{\rho(2l-1)}} \bar{\gamma}^{\mu_{\rho(2l)}} C\right)_\alpha{}^{\dot\beta} \prod_{k=1}^{2l} \Theta_{\vec{b}}^{\vec{a}}\left(\frac{1}{2}\int_{z_{\rho(k)}}^{z_A} \vec{\omega} + \frac{1}{2}\int_{z_{\rho(k)}}^{z_B} \vec{\omega}\right)$$

$$\times \prod_{j=1}^{n-l-1} \frac{\eta^{\mu_{\rho(2l+2j-1)}\mu_{\rho(2l+2j)}}}{E_{\rho(2l+2j-1),\rho(2l+2j)}} E_{\rho(2l+2j-1),A} E_{\rho(2l+2j),B} \Theta_{\vec{b}}^{\vec{a}}\left(\int_{z_{\rho(2l+2j)}}^{z_{\rho(2l+2j-1)}} \vec{\omega} + \frac{1}{2}\int_{z_B}^{z_A} \vec{\omega}\right). \quad (5.80)$$

From the six-dimensional calculations it is easy to obtain the results in $D = 10$ dimensions. They simply differ by two powers of (5.76):

$$\omega_{n,D=10} = \frac{\left[\Theta_{\vec{b}}^{\vec{a}}\left(\frac{1}{2}\int_{z_B}^{z_A} \vec{\omega}\right)\right]^{5-n}}{\sqrt{2}\left[\Theta_{\vec{b}}^{\vec{a}}(\vec{0})\right]^5 E_{AB}^{3/4} \prod_{i=1}^{2n-1}(E_{iA} E_{iB})^{1/2}} \sum_{l=0}^{n-1} \left(\frac{E_{AB}}{2\Theta_{\vec{b}}^{\vec{a}}\left(\frac{1}{2}\int_{z_B}^{z_A} \vec{\omega}\right)}\right)^l$$

$$\times \sum_{\rho \in S_{2n-1}/\mathcal{P}_{n,l}} \text{sgn}(\rho) \left(\gamma^{\mu_{\rho(1)}} \bar{\gamma}^{\mu_{\rho(2)}} \ldots \bar{\gamma}^{\mu_{\rho(2l)}} \gamma^{\mu_{\rho(2l+1)}} C\right)_{\alpha\beta} \prod_{k=1}^{2l+1} \Theta_{\vec{b}}^{\vec{a}}\left(\frac{1}{2}\int_{z_{\rho(k)}}^{z_A} \vec{\omega} + \frac{1}{2}\int_{z_{\rho(k)}}^{z_B} \vec{\omega}\right)$$

$$\times \prod_{j=1}^{n-l-1} \frac{\eta^{\mu_{\rho(2l+2j)}\mu_{\rho(2l+2j+1)}}}{E_{\rho(2l+2j),\rho(2l+2j+1)}} E_{\rho(2l+2j),A} E_{\rho(2l+2j+1),B} \Theta_{\vec{b}}^{\vec{a}}\left(\int_{z_{\rho(2l+2j+1)}}^{z_{\rho(2l+2j)}} \vec{\omega} + \frac{1}{2}\int_{z_B}^{z_A} \vec{\omega}\right), \quad (5.81)$$

$$\Omega_{n,D=10} = \frac{\left[\Theta_{\vec{b}}^{\vec{a}}\left(\frac{1}{2}\int_{z_B}^{z_A} \vec{\omega}\right)\right]^{6-n}}{\left[\Theta_{\vec{b}}^{\vec{a}}(\vec{0})\right]^5 E_{AB}^{5/4} \prod_{i=1}^{2n-2}(E_{iA} E_{iB})^{1/2}} \sum_{l=0}^{n-1} \left(\frac{E_{AB}}{2\Theta_{\vec{b}}^{\vec{a}}\left(\frac{1}{2}\int_{z_B}^{z_A} \vec{\omega}\right)}\right)^l$$

$$\times \sum_{\rho \in S_{2n-2}/\mathcal{Q}_{n,l}} \text{sgn}(\rho) \left(\gamma^{\mu_{\rho(1)}} \bar{\gamma}^{\mu_{\rho(2)}} \ldots \gamma^{\mu_{\rho(2l-1)}} \bar{\gamma}^{\mu_{\rho(2l)}} C\right)_\alpha{}^{\dot\beta} \prod_{k=1}^{2l} \Theta_{\vec{b}}^{\vec{a}}\left(\frac{1}{2}\int_{z_{\rho(k)}}^{z_A} \vec{\omega} + \frac{1}{2}\int_{z_{\rho(k)}}^{z_B} \vec{\omega}\right)$$

$$\times \prod_{j=1}^{n-l-1} \frac{\eta^{\mu_{\rho(2l+2j-1)}\mu_{\rho(2l+2j)}}}{E_{\rho(2l+2j-1),\rho(2l+2j)}} E_{\rho(2l+2j-1),A} E_{\rho(2l+2j),B} \Theta_{\vec{b}}^{\vec{a}}\left(\int_{z_{\rho(2l+2j)}}^{z_{\rho(2l+2j-1)}} \vec{\omega} + \frac{1}{2}\int_{z_B}^{z_A} \vec{\omega}\right). \quad (5.82)$$

The proof of these results for Ω_n and ω_n in six, eight and ten dimensions can be carried over almost literally from the four-dimensional case. The only explicit dependence on the number of dimensions D lies in the pre-factors

$$\frac{\left[\Theta^{\vec{a}}_{\vec{b}}\left(\frac{1}{2}\int_{z_B}^{z_A}\vec{\omega}\right)\right]^{D/2-n}}{\left[\Theta^{\vec{a}}_{\vec{b}}(\vec{0})\right]^{D/2} E_{AB}^{D/8-1/2}} \quad \text{and} \quad \frac{\left[\Theta^{\vec{a}}_{\vec{b}}\left(\frac{1}{2}\int_{z_B}^{z_A}\vec{\omega}\right)\right]^{D/2+1-n}}{\left[\Theta^{\vec{a}}_{\vec{b}}(\vec{0})\right]^{D/2} E_{AB}^{D/8}}. \tag{5.83}$$

These are constructed such to match the leading behavior in the OPEs of the two spin fields for $z_A \to z_B$ and can be explained from (5.76).

CHAPTER 6

A Full Amplitude

The previous considerations of RNS correlation functions in various dimensions enable us to evaluate the tree-level amplitude of two gauge fields and four gauginos. The motivation for considering this six-point amplitude lies in the fact that it can be related to the amplitude involving two gauge fields and two RR moduli. We quickly recap the idea of mapping an amplitude involving only open strings onto an open-closed amplitude [83] and present the calculation of the open string amplitude. Mathematical details are deferred to Appendix D. The calculations in the following constitute work in progress and are based on [4].

6.1 Open vs. Open-Closed Amplitudes

A generalization of the KLT relations [82] between open and open-closed amplitudes on the disk has been achieved in [83]. In contrast to their open string counterparts the vertex operators of closed strings are not inserted on the boundary of the disk $\mathbb{D}$, but somewhere in the complex plane $\mathbb{C}$. This makes the integration over the closed vertex operator positions more involved and consequently the calculation of amplitudes involving closed strings more difficult. These new relations between open and open-closed amplitudes provide an interesting approach to investigate the latter because pure open string amplitudes are well-studied [113, 115, 156, 157].

A general amplitude involving N_o open and N_c closed strings assumes the form[1]

$$\mathcal{A} = \sum_{\rho \in S_{N_o-1}} \int_{\mathbb{R}} \prod_{i=1}^{N_o} \mathrm{d}x_i \int_{\mathbb{H}} \prod_{j=1}^{N_c} \mathrm{d}^2 z_j \, \mathcal{V}_{\mathrm{CKG}}^{-1} \left\langle \prod_{i=1}^{N_o} V_o(x_i) \prod_{j=1}^{N_o} V_c(x_j, \bar{x}_j) \right\rangle. \tag{6.1}$$

The vertex operators of the open strings are inserted on the boundary of the disk and are

---
[1]The upper half complex plane $\mathbb{H}$ can be obtained from $\mathbb{C}$ via a $\mathbb{Z}_2$ identification $z = \bar{z}$.

6. A Full Amplitude

integrated along the real axis $\mathbb{R}$. The closed strings, however, are inserted at points inside the upper half plane $\mathbb{H}$. As described in detail by Stieberger the integrals over the holomorphic and antiholomorphic coordinates z and $\bar{z}$ in (6.1) can be deformed for a certain kinematic $\mathcal{K}^i$ in such a way that they also run along the real axis. The resulting amplitude can be interpreted as a pure open string amplitude involving $N_o + 2N_c$ open strings. Polynomials in the amplitude of the form $(x_i - \xi_j)^a$, where ξ_j are new variables for the z_j, create branch points and one has to include phase factors from the contour deformations.

In particular, the amplitude $\mathcal{A}$ of two gauge fields and two massless NS NS string modes, as well as the amplitude A involving six gauge fields, has been considered in [83]. The vertex operator of the closed string mode reads in the $(-1, -1)$ ghost picture

$$V_G(z, \bar{z}, \varepsilon, q) = g_c \epsilon_{\mu\nu} \psi^\mu(z) \widetilde{\psi}^\nu(\bar{z}) e^{-\phi(z)} e^{-\tilde{\phi}(\bar{z})} e^{iqX(z,\bar{z})}. \tag{6.2}$$

The polarization tensor $\epsilon_{\mu\nu}$ fulfills the on-shell constraint $\epsilon_{\mu\nu} q^\mu = 0$. Comparing this with (2.57) shows that the NS NS vertex operator has twice the field content than that of a gauge field. Assigning the polarization vectors ξ and momenta k_i of two gauge field vertices in the following way to q and $\epsilon_{\mu\nu}$ in (6.2),

$$k_1 = \frac{1}{2} q, \qquad k_2 = \frac{1}{2} q, \qquad \xi_\mu \otimes \xi_\nu = \epsilon_{\mu\nu}, \tag{6.3}$$

one finds that the open-closed and open amplitude satisfy the relation

$$\mathcal{A}(1,2;3,4) = \sin\left(\frac{\pi s}{2}\right) \sin(\pi s) A(1,6,3,5,4,2) - \sin\left(\frac{\pi s}{2}\right) \sin(\pi t) A(1,3,5,4,2,6), \tag{6.4}$$

where $s \equiv 2\alpha' k_1 k_2$, $t \equiv 2\alpha' k_1 k_3$ are the Mandelstam variables. The focus of [4] is the extension of this formula to the amplitude involving two gauge fields and two RR moduli. The vertex operators of the RR field strength F_{n+1} is given by [43]

$$V_{F_{n+1}}^{(-1/2,-1/2)}(z, \bar{z}, f, q) = g_c f_{\mu_0 \ldots \mu_n} \left(P_- \Gamma^{[\mu_0} \ldots \Gamma^{\mu_n]}\right)^{\alpha\beta} \times S_\alpha(z) \widetilde{S}_\beta(\bar{z}) e^{-\phi(z)/2} e^{-\tilde{\phi}(\bar{z})/2} e^{iqX(z,\bar{z})}. \tag{6.5}$$

In this formula the spin fields S, the ten-dimensional gamma matrices Γ and the chiral projection operator $P_- = (1 - \Gamma_{11})/2$ appear. The polarization tensor $f_{\mu_0 \ldots \mu_n}$ is the Fourier transform of the RR n-form potential c_n and satisfies the on-shell condition $c_{\mu\mu_2 \ldots \mu_n} q^\mu = 0$. On the open string side we must consider the amplitude involving two gauge fields and four gauginos as can be seen by comparing the field content in (6.5) with (2.48). The calculation of the pure open amplitude is presented in the following.

6.2 Prerequisites

In the following we consider compactifications of type I or type II string theory which yield $\mathcal{N} = 4$ SYM in four space-time dimensions. The vector multiplet in this case consists of the following massless fields [158]: one gauge field A^μ, four gauginos λ^I and three complex scalars ϕ^J. The vertex operators of these fields have been introduced in Chapter 2.2. For gauginos $\lambda^a, \bar{\lambda}^a$ of negative and positive helicity these have the form

$$V^{(-1/2)}_{\lambda^{a,I}}(z, u, k) = g_\lambda\, T^a\, u^\alpha\, S_\alpha(z)\, \Sigma^I(z)\, e^{-\phi(z)/2}\, e^{ikX(z)},$$
$$V^{(-1/2)}_{\bar{\lambda}^{a,I}}(z, u, k) = g_\lambda\, T^a\, \bar{u}_{\dot{\beta}}\, S^{\dot{\beta}}(z)\, \bar{\Sigma}^I(z)\, e^{-\phi(z)/2}\, e^{ikX(z)}. \tag{6.6}$$

In the following calculation of the tree-level amplitude involving two gluons and four gauginos, we insert the latter in the $-1/2$ ghost picture. This is already enough to cancel the superghost background charge of $+2$ on the disk. The gluons must therefore be inserted in the 0 ghost picture. This vertex operator can be derived from the canonical vertex via the BRST operator as shown previously. One finds:

$$V^{(0)}_{A^a}(z, \xi, k) = \frac{g_A}{(2\alpha')^{1/2}}\, T^a\, \xi_\mu \left[i\, \partial X^\mu(z) + 2\alpha'\, (k\, \psi(z))\, \psi^\mu(z) \right] e^{ikX(z)}. \tag{6.7}$$

In the expressions above ψ and S are the four-dimensional RNS fields, X is the bosonic space-time coordinate and ϕ is a scalar bosonizing the superghost system. Furthermore, T^a are the Chan–Paton factors accounting for the gauge degrees of freedom of the two open string ends. The internal spin fields Σ have an explicit realization as pure exponentials [159]:

$$\Sigma^1 = e^{\frac{i}{2}(H_1 + H_2 + H_3)}, \qquad \Sigma^2 = e^{\frac{i}{2}(H_1 - H_2 - H_3)},$$
$$\Sigma^3 = e^{\frac{i}{2}(-H_1 + H_2 - H_3)}, \qquad \Sigma^4 = e^{\frac{i}{2}(-H_1 - H_2 + H_3)}. \tag{6.8}$$

The vertex operators (6.6) and (6.7) are BRST closed and the polarization vectors and spinors are therefore subject to the on-shell constraints:

$$\text{tansversality:} \qquad \xi_\mu k^\mu = 0,$$
$$\text{Dirac equation:} \qquad \slashed{k} u = \slashed{k} \bar{u} = 0. \tag{6.9}$$

As we are dealing with massless particles, $k^2 = 0$. The open string vertex couplings are related by

$$g_A = (2\alpha')^{1/2}\, g_{\text{YM}}, \qquad g_\lambda = (2\alpha')^{1/2}\, \alpha'^{1/4}\, g_{\text{YM}}. \tag{6.10}$$

The four-dimensional gauge coupling g_{YM} can be expressed in terms of the ten-dimensional coupling g_{10} and the dilaton field ϕ_{10} through the relation $g_{\text{YM}} = g_{10}\, e^{\phi_{10}/2}$ [36]. As derived in [35], we also have to include the factor

$$C_{D_2} = \frac{1}{g_{\text{YM}}^2\, \alpha'^2}\,. \tag{6.11}$$

It has been described in Chapter 2 that the world-sheet of the tree-level scattering process is equivalent to the upper half plane. The vertex operators of the respective fields are inserted on the real axis. The color-stripped partial amplitude then takes the form

$$\mathcal{A}(A^{a_1}, A^{a_2}, \lambda^{I_3}, \bar\lambda^{I_4}, \lambda^{I_5}, \bar\lambda^{I_6}) = C_{D_2}^{-1} \int_{z_1 < \ldots < z_6} \prod_{k=1}^{6} dz_k\, \mathcal{V}_{CKG}^{-1}$$
$$\times \left\langle V_{A^{a_1}}(z_1)\, V_{A^{a_2}}(z_2)\, V_{\lambda^{a_3, I_3}}(z_3)\, V_{\bar\lambda^{a_4, I_4}}(z_4)\, V_{\lambda^{a_5, I_5}}(z_5)\, V_{\bar\lambda^{a_6, I_6}}(z_6) \right\rangle. \tag{6.12}$$

The positions of the vertex operators are integrated over the real axis, where the cyclic ordering $z_1 < \ldots < z_6$ is retained. The CFTs of the different fields appearing in the vertex operators have been greatly discussed in Chapter 2. Only the four-dimensional RNS fields interact with each other, all others decouple. Hence, (6.12) splits into separate correlators:

$$\mathcal{A} = 4\, \alpha'\, g_{YM}^4 \int_{z_1 < \cdots < z_6} \prod_{k=1}^{6} dz_k\, \mathcal{V}_{CKG}^{-1}\, \xi_{\mu_1}\, \xi_{\mu_2}\, u_{I_3}^{\alpha}\, \bar u_{I_4\, \dot\beta}\, u_{I_5}^{\alpha}\, \bar u_{I_6\, \dot\delta}$$
$$\times \left\langle e^{-\phi(z_3)/2}\, e^{-\phi(z_4)/2}\, e^{-\phi(z_5)/2}\, e^{-\phi(z_6)/2} \right\rangle \left\langle \Sigma^{I_3}(z_3)\, \bar\Sigma^{I_4}(z_4)\, \Sigma^{I_5}(z_5)\, \bar\Sigma^{I_6}(z_6) \right\rangle$$
$$\times \left\langle \prod_{i=1}^{6} e^{i k_i \rho_i X^{\rho_i}(z_i)} \left[i\partial X^{\mu_1}(z_1) + 2\alpha'\,(k_1\psi_1)\, \psi^{\mu_1}(z_1) \right] \left[i\partial X^{\mu_2}(z_2) + 2\alpha'\,(k_2\psi_2)\, \psi^{\mu_2}(z_2) \right] \right.$$
$$\left. S_\alpha(z_3)\, S^{\dot\beta}(z_4)\, S_\gamma(z_5)\, S^{\dot\delta}(z_6) \right\rangle. \tag{6.13}$$

6.3 The Separate Correlators

Let us consider each correlation function in (6.13) separately. The ghost correlator which only consists of exponentials is easy to calculate,

$$\left\langle \prod_{i=1}^{N} e^{q_i \phi(z_i)} \right\rangle = \prod_{\substack{i,j=1 \\ i<j}}^{n} z_{ij}^{-q_i q_j}, \tag{6.14}$$

where we make again use of the notation $z_{ij} \equiv z_i - z_j$.

6.3 The Separate Correlators

The correlator consisting of the internal spin fields can be calculated from the pure spin field RNS correlator in six dimensions (3.88):

$$\left\langle \Sigma^{I_3}(z_3)\,\bar{\Sigma}^{I_4}(z_4)\,\Sigma^{I_5}(z_5)\,\bar{\Sigma}^{I_6}(z_6) \right\rangle = \left(\frac{z_{34}\,z_{36}\,z_{45}\,z_{56}}{z_{35}\,z_{46}} \right)^{1/4} \times \left[\frac{\delta^{I_3 I_4}\,\delta^{I_5 I_6}}{z_{34}\,z_{56}} + \frac{\delta^{I_3 I_6}\,\delta^{I_4 I_5}}{z_{36}\,z_{45}} \right]. \quad (6.15)$$

Here we have switched to Euclidean signature and replaced the charge conjugation matrices $C_\alpha{}^{\dot\beta}$ by Kronecker deltas. The same result can be obtained using bosonization. For this purpose (6.8) is inserted into (2.16) and the Kroneckers are tediously put in afterwards by hand. This shows the elegance of our covariant approach to the calculation of RNS correlators.

The bosonic string coordinate fields X^μ do not couple to the RNS fields ψ^μ and $S_\alpha, S^{\dot\beta}$. The last correlator in (6.13) therefore factorizes into an RNS part and a string coordinate part. The relevant correlators in the latter sector are nicely summarized, e.g., in Appendix A of [67]:

$$\left\langle \prod_{i=1}^n e^{ik_i X(z_i)} \right\rangle = \prod_{\substack{i,j=1 \\ i<j}}^n |z_{ij}|^{2\alpha' k_i k_j}, \quad (6.16a)$$

$$\left\langle \partial X^\mu(z_a) \prod_{i=1}^n e^{ik_i X(z_i)} \right\rangle = \left(-2i\alpha' \sum_{r=1}^n \frac{k_r^\mu}{z_{ar}} \right) \left\langle \prod_{i=1}^n e^{ik_i X(z_i)} \right\rangle, \quad (6.16b)$$

$$\left\langle \partial X^\mu(z_1)\,\partial X^\nu(z_2) \prod_{i=1}^n e^{ik_i X(z_i)} \right\rangle = \left(-4\alpha'^2 \sum_{r,s=1}^n \frac{k_r^\mu\,k_s^\nu}{z_{1r}\,z_{2s}} - 2\alpha'\,\frac{\eta^{\mu\nu}}{z_{12}^2} \right) \left\langle \prod_{i=1}^n e^{ik_i X(z_i)} \right\rangle. \quad (6.16c)$$

The powers of the z-coefficients in (6.16a) are given by the Mandelstam variables $s_{ij} \equiv \alpha'(k_i + k_j)^2 = 2\alpha'\,k_i\,k_j$. The number of independent kinematic quantities for the scattering of n massless particles is given by $1/2\,n\,(n-3)$ according to [160]. In accordance with [114] we choose to work with the following nine terms:

$$\begin{aligned}
s_1 &\equiv s_{12}, & s_4 &\equiv s_{45}, & t_1 &\equiv \alpha'(k_1+k_2+k_3)^2, \\
s_2 &\equiv s_{23}, & s_5 &\equiv s_{56}, & t_2 &\equiv \alpha'(k_2+k_3+k_4)^2, \\
s_3 &\equiv s_{34}, & s_6 &\equiv s_{61}, & t_3 &\equiv \alpha'(k_3+k_4+k_5)^2.
\end{aligned} \quad (6.17)$$

The expressions for other kinematic quantities s_{ij} in terms of these nine are found in Table 6.1. The correlator (6.16a) then reads:

$$\left\langle \prod_{i=1}^n e^{ik_i X(z_i)} \right\rangle = \left| \frac{z_{12}\,z_{36}}{z_{13}\,z_{26}} \right|^{s_1} \left| \frac{z_{14}\,z_{23}}{z_{13}\,z_{24}} \right|^{s_2} \left| \frac{z_{25}\,z_{34}}{z_{24}\,z_{35}} \right|^{s_3} \left| \frac{z_{36}\,z_{45}}{z_{35}\,z_{46}} \right|^{s_4} \left| \frac{z_{14}\,z_{56}}{z_{15}\,z_{46}} \right|^{s_5} \\
\left| \frac{z_{16}\,z_{25}}{z_{15}\,z_{26}} \right|^{s_6} \left| \frac{z_{13}\,z_{46}}{z_{14}\,z_{36}} \right|^{t_1} \left| \frac{z_{15}\,z_{24}}{z_{14}\,z_{25}} \right|^{t_2} \left| \frac{z_{26}\,z_{35}}{z_{25}\,z_{36}} \right|^{t_3}. \quad (6.18)$$

i,j	2	3	4	5	6
1	s_1	$-s_1-s_2+t_1$	$s_2+s_5-t_1-t_2$	$-s_5-s_6+t_2$	s_6
2		s_2	$-s_2-s_3+t_2$	$s_3+s_6-t_2-t_3$	$-s_1-s_6+t_3$
3			s_3	$-s_3-s_4+t_3$	$s_1+s_4-t_1-t_3$
4				s_4	$-s_4-s_5+t_1$
5					s_5

Table 6.1: Kinematic quantities s_{ij} expressed through s_k and t_k as defined in (6.17).

The sums over the momenta in (6.16b) and (6.16c) can be reduced using momentum conservation. By eliminating $k_6 = -\sum_{i=1}^{5} k_i$ we obtain

$$\sum_{r=1}^{6} \frac{k_r^\mu}{z_{ar}} = \sum_{r=1}^{5} k_r^\mu \frac{z_{r6}}{z_{ar} z_{a6}} ,$$

$$\sum_{r,s}^{6} \frac{k_r^\mu k_s^\nu}{z_{1r} z_{2s}} = \sum_{r,s}^{5} k_r^\mu k_s^\nu \frac{z_{r6} z_{s6}}{z_{1r} z_{16} z_{2s} z_{26}} , \qquad (6.19)$$

where once again the z-crossing relation

$$z_{ij} z_{kl} = z_{ik} z_{jl} + z_{il} z_{kj} \qquad (6.20)$$

has been used.

The last and most difficult correlators to calculate are the four-dimensional RNS correlation functions. Out of (6.13) three distinct types arise, namely the pure spin field correlator

$$\langle S_\alpha(z_3) S_{\dot\beta}(z_4) S_\gamma(z_5) S_{\dot\delta}(z_6) \rangle , \qquad (6.21)$$

the two six-point functions consisting of two fermions and four spin fields,

$$\langle \psi^{\rho_1}(z_1) \psi^{\mu_1}(z_1) S_\alpha(z_3) S_{\dot\beta}(z_4) S_\gamma(z_5) S_{\dot\delta}(z_6) \rangle , \qquad (6.22a)$$

$$\langle \psi^{\rho_2}(z_2) \psi^{\mu_2}(z_2) S_\alpha(z_3) S_{\dot\beta}(z_4) S_\gamma(z_5) S_{\dot\delta}(z_6) \rangle , \qquad (6.22b)$$

as well as the eight-point function

$$\langle \psi^{\rho_1}(z_1) \psi^{\mu_1}(z_1) \psi^{\rho_2}(z_2) \psi^{\mu_2}(z_2) S_\alpha(z_3) S_{\dot\beta}(z_4) S_\gamma(z_5) S_{\dot\delta}(z_6) \rangle . \qquad (6.23)$$

This last correlator is the most difficult to obtain and therefore poses the bottle neck of the calculation of the complete amplitude. Nevertheless results for all these correlation functions

6.3 The Separate Correlators

have been derived in Chapter 4. Using (4.5) one finds:

$$\langle S_\alpha(z_3)\, S_{\dot\beta}(z_4)\, S_\gamma(z_5)\, S_{\dot\delta}(z_6) \rangle \;=\; -\frac{1}{(z_{35}\, z_{46})^{1/2}}\, \varepsilon_{\alpha_3\alpha_5}\, \varepsilon_{\dot\beta\dot\delta}\,. \tag{6.24}$$

The fermions ψ^{ρ_i} and ψ^{μ_i} in (6.22) and (6.23) stem from the same gauge boson vertex and therefore come with the same world-sheet position z_i. This leads to simplifications as certain coefficients in the result will vanish due to $z_{ii} = 0$. Further cancellations arise if $\eta^{\rho_i\mu_i}$ appears in the index terms. On account of (6.13) this gives rise to the Lorentz product $k_i\, \xi_i$ which vanishes as a consequence of transversality.

Keeping this in mind it is possible to evaluate the correlators (6.22a) and (6.22b). At first, the total expression seems to become singular when adopting the coefficients in (4.12), i.e. $z_2 \to z_1$ for (6.22a) and $z_1 \to z_2$ for (6.22b). However, using the non-minimal form (4.13) all terms remain finite while only the last term involving $\eta^{\mu_i\rho_i}$ becomes singular. Transversality causes this term to vanish and one obtains for the relevant non-singular parts of the correlation functions:

$$\langle \psi^{\rho_1}(z_1)\, \psi^{\mu_1}(z_1)\, S_\alpha(z_3)\, S_{\dot\beta}(z_4)\, S_\gamma(z_5)\, S_{\dot\delta}(z_6) \rangle = \frac{-1}{2\, (z_{13}\, z_{14}\, z_{15}\, z_{16})\, (z_{35}\, z_{46})^{1/2}}$$
$$\times \left(\sigma^{\rho_1}_{\alpha\dot\beta}\, \sigma^{\mu_1}_{\gamma\dot\delta}\, z_{14}\, z_{15}\, z_{36} - \sigma^{\rho_1}_{\alpha\dot\delta}\, \sigma^{\mu_1}_{\gamma\dot\beta}\, z_{15}\, z_{16}\, z_{34} - \sigma^{\rho_1}_{\gamma\dot\delta}\, \sigma^{\mu_1}_{\alpha\dot\beta}\, z_{13}\, z_{16}\, z_{45} - \sigma^{\rho_1}_{\gamma\dot\beta}\, \sigma^{\mu_1}_{\alpha\dot\delta}\, z_{13}\, z_{14}\, z_{56} \right), \tag{6.25a}$$

$$\langle \psi^{\rho_2}(z_2)\, \psi^{\mu_2}(z_2)\, S_\alpha(z_3)\, S_{\dot\beta}(z_4)\, S_\gamma(z_5)\, S_{\dot\delta}(z_6) \rangle = \frac{-1}{2\, (z_{23}\, z_{24}\, z_{25}\, z_{26})\, (z_{35}\, z_{46})^{1/2}}$$
$$\times \left(\sigma^{\rho_2}_{\alpha\dot\beta}\, \sigma^{\mu_2}_{\gamma\dot\delta}\, z_{24}\, z_{25}\, z_{36} - \sigma^{\rho_2}_{\alpha\dot\delta}\, \sigma^{\mu_2}_{\gamma\dot\beta}\, z_{25}\, z_{26}\, z_{34} - \sigma^{\rho_2}_{\gamma\dot\delta}\, \sigma^{\mu_2}_{\alpha\dot\beta}\, z_{23}\, z_{26}\, z_{45} - \sigma^{\rho_2}_{\gamma\dot\beta}\, \sigma^{\mu_2}_{\alpha\dot\delta}\, z_{23}\, z_{24}\, z_{56} \right). \tag{6.25b}$$

Cancellations of the same type occur for the correlator (6.23), however the result remains rather lengthy:

$$\langle \psi^{\rho_1}(z_1)\, \psi^{\mu_1}(z_1)\, \psi^{\rho_2}(z_2)\, \psi^{\mu_2}(z_2)\, S_\alpha(z_3)\, S_{\dot\beta}(z_4)\, S_\gamma(z_5)\, S_{\dot\delta}(z_6) \rangle$$
$$= \frac{-1}{(z_{13}\, z_{14}\, z_{15}\, z_{16}\, z_{23}\, z_{24}\, z_{25}\, z_{26})(z_{35}\, z_{46})^{1/2}} \times \left[\left(\eta^{\rho_1\mu_2}\, \eta^{\mu_1\rho_2} - \eta^{\rho_1\rho_2}\, \eta^{\mu_1\mu_2} \right) \left(\frac{z_{13}\, z_{14}\, z_{25}\, z_{26}}{z_{12}} \right)^2 \right.$$
$$- \frac{1}{2} \left[\eta^{\rho_1\rho_2}\, (\varepsilon\, \bar\sigma^{\mu_1}\, \sigma^{\mu_2})_{\dot\beta\dot\delta} + \eta^{\mu_1\mu_2}\, (\varepsilon\, \bar\sigma^{\rho_1}\, \sigma^{\rho_2})_{\dot\beta\dot\delta} - \eta^{\rho_1\mu_2}\, (\varepsilon\, \bar\sigma^{\mu_1}\, \sigma^{\rho_2})_{\dot\beta\dot\delta} - \eta^{\mu_1\rho_2}\, (\varepsilon\, \bar\sigma^{\rho_1}\, \sigma^{\mu_2})_{\dot\beta\dot\delta} \right] \varepsilon_{\alpha\gamma}$$
$$\times \frac{z_{13}^2\, z_{25}^2\, z_{14}\, z_{26}\, z_{46}}{z_{12}}$$
$$- \frac{1}{2} \left[\eta^{\rho_1\rho_2}\, (\sigma^{\mu_1}\, \bar\sigma^{\mu_2}\, \varepsilon)_{\alpha\gamma} + \eta^{\mu_1\mu_2}\, (\sigma^{\rho_1}\, \bar\sigma^{\rho_2}\, \varepsilon)_{\alpha\gamma} - \eta^{\rho_1\mu_2}\, (\sigma^{\mu_1}\, \bar\sigma^{\rho_2}\, \varepsilon)_{\alpha\gamma} - \eta^{\mu_1\rho_2}\, (\sigma^{\rho_1}\, \bar\sigma^{\mu_2}\, \varepsilon)_{\alpha\gamma} \right] \varepsilon_{\dot\beta\dot\delta}$$
$$\times \frac{z_{14}^2\, z_{26}^2\, z_{13}\, z_{25}\, z_{35}}{z_{12}}$$

$$+ \frac{1}{2} \left[\eta^{\mu_1\mu_2} \left(\varepsilon \, \bar{\sigma}^{\rho_1} \, \sigma^{\rho_2} \right)_{\dot{\beta}\dot{\delta}} \varepsilon_{\alpha\gamma} - \eta^{\mu_1\rho_2} \left(\varepsilon \, \bar{\sigma}^{\rho_1} \, \sigma^{\mu_2} \right)_{\dot{\beta}\dot{\delta}} \varepsilon_{\alpha\gamma} + \eta^{\rho_1\rho_2} \, \sigma^{\mu_1}_{\alpha\dot{\beta}} \, \sigma^{\mu_2}_{\gamma\dot{\delta}} - \eta^{\rho_1\mu_2} \, \sigma^{\mu_1}_{\alpha\dot{\beta}} \, \sigma^{\rho_2}_{\gamma\dot{\delta}} \right]$$
$$\times z_{12} \, z_{13} \, z_{26} \, z_{35} \, z_{45} \, z_{46}$$
$$+ \frac{1}{2} \left[\eta^{\rho_1\rho_2} \, \sigma^{\mu_1}_{\alpha\dot{\beta}} \, \sigma^{\mu_2}_{\gamma\dot{\delta}} + \eta^{\mu_1\mu_2} \, \sigma^{\rho_1}_{\alpha\dot{\beta}} \, \sigma^{\rho_2}_{\gamma\dot{\delta}} - \eta^{\rho_1\mu_2} \, \sigma^{\mu_1}_{\alpha\dot{\beta}} \, \sigma^{\rho_2}_{\gamma\dot{\delta}} - \eta^{\mu_1\rho_2} \, \sigma^{\rho_1}_{\alpha\dot{\beta}} \, \sigma^{\mu_2}_{\gamma\dot{\delta}} \right] z_{13} \, z_{15} \, z_{24} \, z_{26} \, z_{35} \, z_{46}$$
$$+ \frac{1}{4} \left[\sigma^{\rho_1}_{\alpha\dot{\beta}} \left(\sigma^{\mu_1} \, \bar{\sigma}^{\rho_2} \, \sigma^{\mu_2} \right)_{\gamma\dot{\delta}} - \sigma^{\mu_2}_{\gamma\dot{\delta}} \left(\sigma^{\rho_1} \, \bar{\sigma}^{\mu_1} \, \sigma^{\rho_2} \right)_{\alpha\dot{\delta}} \right] z_{13} \, z_{14} \, z_{25} \, z_{26} \, z_{35} \, z_{46}$$
$$+ \frac{1}{4} \left(\varepsilon \, \bar{\sigma}^{\rho_1} \, \sigma^{\mu_1} \, \bar{\sigma}^{\rho_2} \, \sigma^{\mu_2} \right)_{\dot{\beta}\dot{\delta}} \varepsilon_{\alpha\gamma} \, z_{46}^2 \, z_{13} \, z_{15} \, z_{23} \, z_{25} + \frac{1}{4} \left(\sigma^{\rho_1} \, \bar{\sigma}^{\mu_1} \, \sigma^{\rho_2} \, \bar{\sigma}^{\mu_2} \, \varepsilon \right)_{\alpha\gamma} \varepsilon_{\dot{\beta}\dot{\delta}} \, z_{35}^2 \, z_{14} \, z_{16} \, z_{24} \, z_{26}$$
$$+ \frac{1}{4} \sigma^{\mu_2}_{\gamma\dot{\delta}} \left(\sigma^{\rho_1} \, \bar{\sigma}^{\mu_1} \, \sigma^{\rho_2} \right)_{\alpha\dot{\beta}} z_{14} \, z_{16} \, z_{23} \, z_{25} \, z_{35} \, z_{46} - \frac{1}{4} \sigma^{\mu_1}_{\gamma\dot{\delta}} \left(\sigma^{\rho_2} \, \bar{\sigma}^{\mu_2} \, \sigma^{\rho_1} \right)_{\alpha\dot{\beta}} z_{12} \, z_{13} \, z_{26} \, z_{35} \, z_{45} \, z_{46}$$
$$- \frac{1}{4} \sigma^{\rho_1}_{\gamma\dot{\beta}} \left(\sigma^{\mu_1} \, \bar{\sigma}^{\rho_2} \, \sigma^{\mu_2} \right)_{\alpha\dot{\delta}} z_{13} \, z_{15} \, z_{24} \, z_{26} \, z_{35} \, z_{46} + \frac{1}{4} \sigma^{\rho_2}_{\gamma\dot{\beta}} \left(\sigma^{\rho_1} \, \bar{\sigma}^{\mu_1} \, \sigma^{\mu_2} \right)_{\alpha\dot{\delta}} z_{12} \, z_{14} \, z_{25} \, z_{35} \, z_{36} \, z_{46} \right] . \quad (6.26)$$

All z coefficients in the RNS correlators (6.24)-(6.26) appear with integer powers apart from the factor $(z_{35} \, z_{46})^{-1/2}$. Combining this term with the ghost correlator (6.14) and the expression for the internal spin fields (6.15),

$$\mathcal{F} \equiv \frac{1}{z_{35} \, z_{46}} \left(\frac{\delta^{I_3 I_4} \, \delta^{I_5 I_6}}{z_{34} \, z_{56}} + \frac{\delta^{I_3 I_6} \, \delta^{I_4 I_5}}{z_{36} \, z_{45}} \right), \quad (6.27)$$

shows that in total only integer powers of z's enter the amplitude. By careful bookkeeping we find that in total every z_i appears with two negative powers. This coincides with the requirement that the amplitude must behave as $z_i^{-2h} = z_i^{-2}$ for $z_i \to \pm\infty$, where $h = 1$ is the conformal weight of the vertex operators.

This fact is important for gauging the residual $SL(2, \mathbb{R})/\mathbb{Z}_2$ symmetry of the amplitude. The volume factor of the conformal Killing group $\mathcal{V}_{CKG}^{-1}$ is canceled by fixing three positions z_i, z_j, z_k and inserting the c-ghost correlator

$$\langle c(z_i) \, c(z_j) \, c(z_k) \rangle = z_{ij} \, z_{ik} \, z_{jk} . \quad (6.28)$$

For the moment we only fix z_1 at minus infinity, two further finite points z_A, z_B will be specified later. Equation (6.28) adds two positive powers in z_1 to the amplitude. This contains already two negative powers which we denote as $(z_{1m} \, z_{1n})^{-1}$. The powers simply cancel in total:

$$\lim_{z_1 \to -\infty} \frac{z_{1A} \, z_{1B}}{z_{1m} \, z_{1n}} = \lim_{z_1 \to -\infty} \left(1 - \frac{z_{Am}}{z_{1m}} \right) \left(1 - \frac{z_{Bn}}{z_{1n}} \right) = 1 . \quad (6.29)$$

This proves that under $z_1 \to -\infty$ the amplitude remains mathematically well defined and that no further cancellations appear[2]. One can then strike out all appearing z_1 terms in the

---
[2] All cancellations are thus due to transversality and momentum conservation.

6.4 A First Result

numerators and denominators of (6.16b), (6.16c), (6.18), (6.25) and (6.26), but has to be alert to potential minus signs. We are now ready to construct the full amplitude out of the separate correlators.

6.4 A First Result

Combining carefully the various correlators calculated in Chapter 6.3 it is then possible to obtain a first result for the partial amplitude (6.12):

$$\mathcal{A} = 4\,\alpha'^3\,g_{YM}^4 \int_{z_2 < \cdots < z_6} \prod_{k=2}^{6} dz_k\,\delta(z_i - z_A)\,\delta(z_j - z_B)\,z_{AB}\,\mathcal{F}$$

$$\times \left|\frac{z_{36}}{z_{26}}\right|^{s_1} \left|\frac{z_{23}}{z_{24}}\right|^{s_2} \left|\frac{z_{25}\,z_{34}}{z_{24}\,z_{35}}\right|^{s_3} \left|\frac{z_{36}\,z_{45}}{z_{35}\,z_{46}}\right|^{s_4} \left|\frac{z_{56}}{z_{46}}\right|^{s_5} \left|\frac{z_{25}}{z_{26}}\right|^{s_6} \left|\frac{z_{46}}{z_{36}}\right|^{t_1} \left|\frac{z_{24}}{z_{25}}\right|^{t_2} \left|\frac{z_{26}\,z_{35}}{z_{25}\,z_{36}}\right|^{t_3}$$

$$\times \left[\mathcal{K}^1 + \frac{z_{36}}{z_{23}}\mathcal{K}^2 + \frac{z_{46}}{z_{24}}\mathcal{K}^3 + \frac{z_{56}}{z_{25}}\mathcal{K}^4 + \frac{z_{36}}{z_{26}}\mathcal{K}^5 + \frac{z_{46}}{z_{26}}\mathcal{K}^6 + \frac{z_{56}}{z_{26}}\mathcal{K}^7 + \frac{z_{25}\,z_{26}}{z_{23}\,z_{24}}\mathcal{K}^8 \right.$$

$$+ \frac{z_{25}\,z_{46}}{z_{23}\,z_{24}}\mathcal{K}^9 + \frac{z_{26}\,z_{35}}{z_{23}\,z_{24}}\mathcal{K}^{10} + \frac{z_{35}\,z_{46}}{z_{23}\,z_{25}}\mathcal{K}^{11} + \frac{z_{36}}{z_{23}\,z_{26}}\left(z_{46}\,\mathcal{K}^{12} + z_{56}\,\mathcal{K}^{13}\right)$$

$$+ \frac{z_{46}}{z_{24}\,z_{26}}\left(z_{36}\,\mathcal{K}^{14} + z_{56}\,\mathcal{K}^{15}\right) + \frac{z_{56}}{z_{25}\,z_{26}}\left(z_{36}\,\mathcal{K}^{16} + z_{46}\,\mathcal{K}^{17}\right)$$

$$+ \frac{z_{34}}{z_{23}\,z_{24}}\left(z_{36}\,\mathcal{K}^{18} + z_{46}\,\mathcal{K}^{19} + z_{56}\,\mathcal{K}^{20}\right) + \frac{z_{35}^2}{z_{23}\,z_{25}}\mathcal{K}^{21}$$

$$\left. + \frac{z_{45}}{z_{24}\,z_{25}}\left(z_{36}\,\mathcal{K}^{22} + z_{46}\,\mathcal{K}^{23} + z_{56}\,\mathcal{K}^{24}\right) \right]. \tag{6.30}$$

In this result 24 kinematical terms $\mathcal{K}^i$ appear. These consist of index terms of the four-dimensional RNS correlators which are multiplied with momenta, polarization vectors and spinors, like

$$\mathcal{K}^{18} = 2\,(k_3\,\xi_1)\,k_{2\mu}\,\xi_{2\nu}\,(u_3\,\sigma^\mu\,\bar{u}_6)\,(u_5\,\sigma^\nu\,\bar{u}_4)\,. \tag{6.31}$$

Their explicit forms are collected in Appendix D.1. The delta function and the term z_{AB} come into play from the two vertex positions that can still be fixed. It is useful to introduce the following short-hand notation:

$$\mathcal{A} \equiv 4\,\alpha'^3\,g_{YM}^4 \sum_{i=1}^{24} H_i\,\mathcal{K}^i\,. \tag{6.32}$$

The terms H_i include the integrals over the vertex operator positions and the corresponding z coefficient for each kinematic factor.

6.5 Gauge Invariance

Two gauge bosons enter the amplitude. Therefore $\mathcal{A}$ has to be invariant under gauge transformations acting on these gluon fields, $A_\mu \to A_\mu + \partial_\mu \chi$. Consequently the amplitude must vanish for pure gauge configurations of the gluons, $A_\mu \sim \partial_\mu \chi$, as this cannot yield any non-zero physical observables.

An on-shell gauge transformation is implemented in the form that the amplitude must not change under shifting the polarization $\xi^\mu \to \xi^\mu + k^\mu$. A pure gauge configuration is then obviously given for $\xi^\mu = k^\mu$ and hence the amplitude must vanish. Many of the terms entering the kinematics $\mathcal{K}^i$ in this case coincide. If an independent set is chosen the vanishing of the amplitude must then be due to relations between the integrals. This turns out to be a powerful tool to check the consistency of the expression (6.30) and furthermore yields relations between the integral terms which can be used to cast the amplitude into shorter form.

The detailed calculation of inspecting gauge invariance for the first gluon field is given in Appendix D.2. One sets $\xi_1 = k_1$, evaluates the kinematical terms $\mathcal{K}^i$ for this choice and deduces relations for the integral terms H_i. The following identities are found:

$$\begin{aligned}
0 =\ & (1 - s_1)\, H_1 + (s_1 + s_2 - t_1)\, H_5 - (s_2 + s_5 - t_1 - t_2)\, H_6 + (s_5 + s_6 - t_2)\, H_7\,, \\
0 =\ & -(s_2 - t_1)\, H_2 + (s_1 + s_2 - t_1)\, H_5 + (s_2 + s_5 - t_1 - t_2)\, H_{12} - (s_5 + s_6 - t_2)\, H_{13}\,, \\
0 =\ & (s_1 + s_2 + s_5 - t_1 - t_2)\, H_3 - (s_2 + s_5 - t_1 - t_2)\, H_6 - (s_1 + s_2 - t_1)\, H_{14} \\
    & - (s_5 + s_6 - t_2)\, H_{15}\,, \\
0 =\ & (s_1 - s_5 - s_6 + t_2)\, H_4 + (s_5 + s_6 - t_2)\, H_7 - (s_1 + s_2 - t_1)\, H_{16} \\
    & + (s_2 + s_5 - t_1 - t_2)\, H_{17}\,, \\
0 =\ & s_1(H_2 - H_3) - (s_1 + s_2 - t_1)\, H_{18} + (s_2 + s_5 - t_1 - t_2)\, H_{19} - (s_5 + s_6 - t_2)\, H_{20}\,, \\
0 =\ & s_1(H_3 - H_4) - (s_1 + s_2 - t_1)\, H_{22} + (s_2 + s_5 - t_1 - t_2)\, H_{23} - (s_5 + s_6 - t_2)\, H_{24}\,. \quad (6.33)
\end{aligned}$$

A similar calculation for the second gluon yields the following identities:

$$\begin{aligned}
0 =\ & (1 - s_1)\, H_1 + s_2\, H_2 - (s_2 + s_3 - t_2)\, H_3 + (s_3 + s_6 - t_2 - t_3)\, H_4\,, \\
0 =\ & s_2\, H_2 - (s_1 + s_2)\, H_5 - (s_2 + s_3 - t_2)\, H_{14} + (s_3 + s_6 - t_2 - t_3)\, H_{16}\,, \\
0 =\ & -(s_2 + s_3 - t_2)\, H_3 - (s_1 - s_2 - s_3 + t_2)\, H_6 + s_2\, H_{12} + (s_3 + s_6 - t_2 - t_3)\, H_{17}\,, \\
0 =\ & (s_3 + s_6 - t_2 - t_3)\, H_4 - (s_1 + s_3 + s_6 - t_2 - t_3)\, H_7 + s_2\, H_{13} - (s_2 + s_3 - t_2)\, H_{15}\,. \quad (6.34)
\end{aligned}$$

These equations provide important consistency checks for the calculation. Their correctness can be checked directly by evaluating the integrals appearing in H_i and expressing the results in a minimal basis of hypergeometric functions [118]. Apart from these, we obtain relations of a different type,

$$\begin{aligned}
0 &= H_2 - H_5 - H_{14} - H_{18}\,, & 0 &= H_{14} - H_{16} - H_{22}\,, \\
0 &= H_4 - H_7 - H_{15} + H_{24}\,, & 0 &= H_{13} - H_{15} - H_{20}\,, \\
0 &= H_{11} - H_{19} - H_{23}\,, & 0 &= H_{11} - H_{12} + H_{17}\,,
\end{aligned} \qquad (6.35)$$

where no Mandelstam variables appear. One can check easily that these are correct. We only have to insert the correct expressions for H_i from (6.30) and merge the partial fractions. The last equation in (6.35) for example holds on account of

$$z_{46}\left(\frac{z_{35}}{z_{23}\,z_{25}} - \frac{z_{36}}{z_{23}\,z_{26}} + \frac{z_{56}}{z_{25}\,z_{26}}\right) = 0\,. \qquad (6.36)$$

The equations in (6.35) simply state that we have not fully reduced the z coefficients in (6.30), but another six can be eliminated. Nevertheless, they are strong consistency checks and together with (6.33) and (6.34) can be used to cast $\mathcal{A}$ into shorter form.

6.6 Spinor Products

It is well-known from field theory that compact expressions for (partial) scattering amplitudes can be obtained when the spinor helicity formalism is used. Especially the MHV amplitude of n gluons in field theory, also known as Parke–Taylor amplitude [119, 161], assumes a particular simple form in this notation. The spinor helicity formalism has first been introduced in [162] to describe multiple bremsstrahlung processes in massless gauge theories. Further details on the topic are found in [163, 164] and in Appendix E.

In the spinor helicity formalism one considers interactions where the massless external particles are definite left- or right-handed polarization states, i.e. helicity $-$ or $+$ respectively. The key quantities are two-component chiral spinors which we simply denote by $k_{i\alpha}$ and $\bar{k}_{i\dot\alpha}$ for the $i = 1,\ldots,n$ particles. Products of these spinors like

$$\langle i\,j\rangle \equiv k_i{}^\alpha\,k_{j\alpha}\,, \qquad [i\,j] \equiv \bar{k}_{i\dot\alpha}\,\bar{k}_j{}^{\dot\alpha} \qquad (6.37)$$

appear as the new kinematic variables in the amplitude. In our conventions the spinors are commuting and due to the antisymmetry of the ε tensor

$$\langle i\,j\rangle = -\langle j\,i\rangle\,, \qquad [i\,j] = -[j\,i]\,, \qquad \langle i\,i\rangle = [j\,j] = 0\,. \qquad (6.38)$$

Let us now discuss the origin of the spinors. Any four-vector k_μ can be contracted with the matrices $\sigma^\mu = (-1, -\sigma^i)$, $\bar\sigma^\mu = (-1, \sigma^i)$,

$$k_{\alpha\dot\alpha} \equiv k_\mu \sigma^\mu_{\alpha\dot\alpha}, \qquad k^{\dot\alpha\alpha} \equiv k_\mu \bar\sigma^{\mu\,\dot\alpha\alpha}, \tag{6.39}$$

where σ^i are the standard Pauli matrices. If k_μ is the on-shell momentum of a massless particle and therefore satisfies $k^2 = 0$, it is possible to show that the expressions in (6.39) factorize as

$$k_{\alpha\dot\alpha} = k_\alpha\, \bar k_{\dot\alpha}, \qquad \bar k^{\dot\alpha\alpha} = \bar k^{\dot\alpha}\, k^\alpha. \tag{6.40}$$

The spinors k_α, $\bar k_{\dot\alpha}$ are therefore called momentum spinors. If the particle with momentum k_μ is a fermion the momentum spinors additionally satisfy the Dirac equation and we can identify

$$u_\alpha(k) = k_\alpha, \qquad \bar u_{\dot\alpha}(k) = \bar k_{\dot\alpha}. \tag{6.41}$$

In the spinor helicity formalism the polarization vectors of left- and right-handed gluons with momentum k are given by

$$\xi^{\mu-}(k,r) = -\frac{1}{\sqrt{2}} \frac{\bar\sigma^{\mu\,\dot\alpha\alpha} \bar r_{\dot\alpha} k_\alpha}{[k\,r]}, \qquad \xi^{\mu+}(k,r) = -\frac{1}{\sqrt{2}} \frac{\bar\sigma^{\mu\,\dot\alpha\alpha} \bar k_{\dot\alpha} r_\alpha}{\langle r\,k\rangle}, \tag{6.42}$$

where r is an arbitrary reference momentum. One can derive that ξ^μ is shifted by an amount proportional to k^μ if the reference momentum is changed. This corresponds to an on-shell gauge transformation and therefore the choice of r is in fact a gauge choice.

In the following we apply the spinor helicity formalism to the expression (6.30). We choose the two gluons in (6.12) to have different helicity. Then the amplitude can be related via SUSY Ward identities to the six gluon NMHV amplitude [165, 166]. It is convenient to choose negative helicity for the first gluon and set its reference momentum to k_4. The second gluon then has positive helicity and we take k_5 as reference momentum. Many of the kinematics $\mathcal{K}^i$ then vanish due to $\langle ii\rangle = [jj] = 0$ as shown in Appendix D.3. The remaining ones are

$$\mathcal{K}^1 = 2\,(1 - 1/s_1)\, \frac{\langle 12\rangle [12] \langle 15\rangle \langle 35\rangle [24][46]}{[14]\langle 52\rangle},$$

$$\mathcal{K}^2 = -2\, \frac{\langle 35\rangle [23][46]}{[14]\langle 52\rangle} \Big(\langle 15\rangle \langle 23\rangle [24] + \langle 13\rangle \langle 56\rangle [46]\Big),$$

$$\mathcal{K}^3 = -2\, \frac{\langle 12\rangle [35][24]^2}{[14]\langle 52\rangle} \Big(\langle 52\rangle [26] + \langle 54\rangle [46]\Big),$$

$$\mathcal{K}^7 = 2\, \frac{\langle 15\rangle^2 [12][46]}{[14]\langle 52\rangle} \Big(\langle 31\rangle [14] + \langle 35\rangle [54]\Big),$$

$$\mathcal{K}^{13} = 2\, \frac{\langle 15\rangle \langle 35\rangle [23][46]}{[14]\langle 52\rangle} \Big(\langle 31\rangle [14] + \langle 35\rangle [54]\Big),$$

$$\mathcal{K}^5 = 2\frac{\langle 13\rangle[46]}{[14]\langle 52\rangle}\Big(\langle 51\rangle[14] + \langle 53\rangle[34]\Big)\Big(\langle 51\rangle[12] + \langle 53\rangle[32]\Big),$$

$$\mathcal{K}^{14} = 2\frac{\langle 13\rangle[24]}{[14]\langle 52\rangle}\Big(\langle 51\rangle[14] + \langle 53\rangle[34]\Big)\Big(\langle 52\rangle[26] + \langle 54\rangle[46]\Big),$$

$$\mathcal{K}^{15} = -2\frac{\langle 15\rangle[24]}{[14]\langle 52\rangle}\Big(\langle 31\rangle[14] + \langle 35\rangle[54]\Big)\Big(\langle 52\rangle[26] + \langle 54\rangle[46]\Big). \tag{6.43}$$

Together with the integral terms H_i this constitutes the result of the partial amplitude for a specific helicity choice.

6.7 The Field Theory Limit

In order to discuss the integral terms in more detail it is necessary to fix the two remaining vertex operator positions in (6.30). For $z_2 = 0$ and $z_3 = 1$ the integration region of the remaining positions is

$$\mathcal{I} = \{z_4, z_5, z_6 \in \mathbb{R} : 1 < z_4 < z_5 < z_6 < \infty\}. \tag{6.44}$$

Performing the change of variables $z_4 \to 1/x$, $z_5 \to 1/xy$, $z_6 \to 1/xyz$, where x, y, z run from 0 to 1, the integral terms H_i assume the generic form [118]:

$$F\begin{bmatrix} a,b,d,e,g \\ c,f,h,j \end{bmatrix} \equiv \int_0^1 dx \int_0^1 dy \int_0^1 dz \, x^a \, y^b \, z^c \, (1-x)^d \, (1-y)^e \, (1-z)^f$$
$$(1-xy)^g \, (1-yz)^h \, (1-xyz)^j. \tag{6.45}$$

The function F can be expressed through triple hypergeometric functions [118, 167–169], where the powers $a, \ldots, j$ are some combinations of the nine Mandelstam variables (6.17). Expanding F with respect to these powers yields the momentum expansion of the integral which is given by a Euler-Zagier series [170]. The lowest-order terms correspond to the field theory limit of the amplitude, whereas higher-order terms, coming with powers of α'^k and (multi-valued) zeta factors of order k, are stringy corrections to the interaction. For simplicity we state in the following only the field theory limit of the integral terms H_i that we have calculated for the choice $I_3 = \bar{I}_4 = I_5 = \bar{I}_6$ of the internal spin fields. The term $\mathcal{F}$ in (6.30) then becomes:

$$\mathcal{F} = \frac{1}{z_{34}\, z_{36}\, z_{45}\, z_{56}}. \tag{6.46}$$

Performing the momentum expansion of the integrals corresponding to the kinematics (6.43) yields at lowest order[3]:

$$\begin{aligned}
H_1 &= \frac{1}{s_1 s_3 s_5} + \frac{1}{s_1 s_4 t_1} + \frac{1}{s_1 s_5 t_1} + \frac{1}{s_1 s_3 t_3} + \frac{1}{s_1 s_4 t_3} - \frac{1}{s_1 s_1 s_3} - \frac{1}{s_1 s_1 s_4} \\
&\quad - \frac{1}{s_1 s_1 s_5} + \frac{s_2/s_1}{s_1 s_4 t_1} + \frac{s_2/s_1}{s_1 s_5 t_1} + \frac{s_6/s_1}{s_1 s_3 t_3} + \frac{s_6/s_1}{s_1 s_4 t_3} + \frac{t_2/s_1}{s_1 s_3 s_5}, \\
H_2 &= \frac{1}{s_1 s_3 s_5} + \frac{1}{s_2 s_4 s_6} + \frac{1}{s_1 s_4 t_1} + \frac{1}{s_2 s_4 t_1} + \frac{1}{s_1 s_5 t_1} + \frac{1}{s_2 s_5 t_1} + \frac{1}{s_2 s_5 t_2} \\
&\quad + \frac{1}{s_3 s_5 t_2} + \frac{1}{s_2 s_6 t_2} + \frac{1}{s_3 s_6 t_2} + \frac{1}{s_1 s_3 t_3} + \frac{1}{s_1 s_4 t_3} + \frac{1}{s_3 s_6 t_3} + \frac{1}{s_4 s_6 t_3}, \\
H_3 &= \frac{1}{s_1 s_3 s_5} + \frac{1}{s_3 s_5 t_2} + \frac{1}{s_3 s_6 t_2} + \frac{1}{s_1 s_3 t_3} + \frac{1}{s_1 s_4 t_3} + \frac{1}{s_3 s_6 t_3} + \frac{1}{s_4 s_6 t_3}, \\
H_5 &= \frac{1}{s_1 s_3 s_5} + \frac{1}{s_1 s_4 t_1} + \frac{1}{s_1 s_5 t_1} + \frac{1}{s_1 s_3 t_3} + \frac{1}{s_1 s_4 t_3}, \\
H_7 &= \frac{1}{s_1 s_3 t_3} + \frac{1}{s_1 s_4 t_3}, \\
H_{13} &= \frac{1}{s_2 s_4 s_6} + \frac{1}{s_2 s_6 t_2} + \frac{1}{s_3 s_6 t_2} + \frac{1}{s_3 s_6 t_3} + \frac{1}{s_4 s_6 t_3}, \\
H_{14} &= \frac{1}{s_3 s_5 t_2} + \frac{1}{s_3 s_6 t_2} + \frac{1}{s_3 s_6 t_3} + \frac{1}{s_4 s_6 t_3}, \\
H_{15} &= \frac{1}{s_3 s_6 t_2} + \frac{1}{s_3 s_6 t_3} + \frac{1}{s_4 s_6 t_3}.
\end{aligned}$$

(6.47)

The kinematical quantities s_i and t_j are defined with a power of α' in (6.17). The prefactor $4\,\alpha'^3 g_{\text{YM}}^4$ in (6.32) cancels all powers of α' in the denominators as required in the field theory limit.

The results presented in this Chapter are the starting points for further investigations [4]. Dual-Ward identities [163], Kleiss–Kuijf [171] and BCJ identities [84] between different partial amplitudes can be checked as well as their generalizations in string theory. Most important, the exact mapping between (6.32) and the amplitude of two gauge fields and four RR bulk fields can be studied as proposed by [83].

---

[3]Here we have absorbed the term $(s_1 - 1)/s_1$ appearing in $\mathcal{K}^1$ into H_1.

CHAPTER 7

Conclusion

Scattering amplitudes in string theory enjoy a great range of applications. Compactification scenarios of string theory with a string scale in the TeV range are natural realizations of the ADD proposal to address the hierarchy problem of the SM. String amplitudes in such a model with perturbative string coupling predict signals which can be detected at LHC and future hadron colliders. From a conceptual point of view scattering amplitudes can also provide new insights into the "gravity side" of string theory via the KLT relations and the recently discovered duality between open and open-closed amplitudes [83].

7.1 Summary

There are four main results obtained in this thesis. In the calculation of scattering amplitudes in the RNS formalism correlation functions of NS fermions and R spin fields must be evaluated, but their interacting nature forms an obstacle. In Chapters 3 and 4 we focused on the calculation of such correlators at tree-level. We showed how these n-point functions in four, six, eight and ten space-time dimensions can be evaluated in general by analyzing their Lorentz structure and singular properties dictated by the underlying CFT. Another method, which we presented, rests on replacing NS fermions by spin fields and then evaluating the pure spin field correlator.

In the separate dimensions we could profit from special properties. In four dimensions one is able to completely solve the RNS CFT at tree-level. Fermions can be replaced by spin fields and the resulting spin field correlator factorizes into a left- and right-handed part, for which a general formula could be derived by induction. In this sense (3.77) and (3.82) are important results of our work. Explicit expressions for all non-vanishing correlators in four dimensions up to eight-point level were presented as well. Although the factorization feature does not carry over to higher dimensions, we were able to calculate large classes of tree-level correlators in six dimensions and find relations between distinct correlation functions in eight dimensions via $SO(8)$ triality.

Another focus of this work was the evaluation of RNS correlators for arbitrary loop order. The necessary methods, in particular re-expressing the RNS fields in terms of $SO(2)$ spin systems and calculating their loop correlators, were presented in Chapter 5. We calculated all n-point functions up to at least $n = 6$. Of particular interest are the findings for the correlators Ω and ω as given in (5.74)-(5.82). Such correlators in four dimensions are the essential ingredients for string scattering amplitudes involving arbitrary many gluons and at most two chiral fermions or gauginos. For string compactifications with low string scale and weak string coupling these amplitudes give corrections to SM processes which can be measured at hadron colliders. Most important, these amplitudes are completely independent of the compactification details.

The index terms consisting of gamma and charge conjugation matrices play a key role in the calculation of RNS correlators. Apart from carrying all space-time indices and thus determining the Lorentz structure of the string interaction they can be regarded as Clebsch–Gordan coefficients associated with a particular scalar representation in the tensor product of the correlator. From this perspective the number of independent index terms is simply given by the number of scalar representations. Determining this number via group theory and then deriving relations between different terms was one of the most important steps in the calculation of the RNS correlation functions. All identities which we derived for the evaluations are collected in Appendix B.

While ten-dimensional RNS correlators enter the calculation of scattering amplitudes in the non-compactified theory, the cases $D = 4$ and $D = 6$ apply to phenomenological purposes, where they describe the interaction of the external and internal RNS fields. Equipped with the previous result we were able to calculate the partial string amplitude involving two gauge fields and four gauginos. The result (6.30) is gauge invariant which yields various relations for the hypergeometric functions after integration over the vertex operator positions. The kinematical terms in the partial amplitude were expressed through spinor products for certain helicity choices and the field theory limit was presented.

7.2 Outlook

The findings in the last Chapter of this work are the starting point for further investigations, especially matching the integrals to those stemming from the amplitude of two gauge fields and two RR moduli. It will then be possible to explore the relations between open and open-closed amplitudes as proposed in [83] in more detail and furthermore express the brane-bulk couplings of the latter amplitude in terms of pure open couplings. This is the main topic of [4].

A further issue worth exploring are RNS correlators that arise from the scattering of chiral scalars and fermions. In type II these states stem from strings stretching between different brane stacks near their intersection. The vertex operators involve internal spin fields that have

7.2 Outlook

an explicit representation in terms of bosonic and fermionic twist operators. Their conformal weight depends on the intersection angle, which makes their correlation functions more difficult to evaluate. In particular it would be interesting whether one can also calculate general classes of correlators as it was possible for the standard fermions and spin fields in six dimensions.

APPENDIX A

Gamma Matrices in D Dimensions

Gamma matrices Γ^μ play a key role in the interplay of vector- and spinor representations of the Lorentz group $SO(1, D-1)$ in D dimensions. First of all, they can be viewed as operators acting on the space of spinors whose anti-commutation relations are given by the Clifford algebra. On the other hand, antisymmetric products $\Gamma^{\mu_1...\mu_p}$ of p gamma matrices, multiplied by the charge conjugation matrix $\mathcal{C}$, appear as Clebsch–Gordan coefficients in the decomposition of bi-spinors to p forms. The tensor structure of an RNS correlation function is expressed in terms of these products ($\Gamma^{\mu_1...\mu_p}\mathcal{C}$). Many of their properties, which we present in the following, are needed for the calculation of RNS correlators. For further information on Clifford algebras and spinors in higher dimensions the reader might refer to [36, 136–138].

A.1 Notation and conventions

First of all let us fix our notation and conventions. In contrast to [172] we use the sign convention of Wess & Bagger [21] for the Clifford algebra

$$\{\Gamma^\mu, \Gamma^\nu\} = -2\,\eta^{\mu\nu}. \tag{A.1}$$

The Minkowski metric contains "mostly plus" entries, $\eta^{\mu\nu} = \text{diag}(-1, +1, \ldots, +1)$. A concrete representation of the Clifford algebra is not necessary for the following general discussion. However, we state the gamma matrices which were used in the calculations of the correlators in Chapters 4 and 5 at the end of this Chapter.

Dirac spinors furnish a representation of the Clifford algebra (A.1) and form a complex vector space of dimensions $2^{D/2}$. In the case of even space-time dimensions $D = 2m$ this representation is reducible and can be decomposed into two irreducible representations of dimension $2^{D/2-1}$

each. These are referred to as left- and right-handed, their elements are called Weyl spinors of positive and negative chirality. Generic Dirac spinors Ξ live in the direct sum of both irreducible subspaces and are written in component notation as

$$\Xi_A = \begin{pmatrix} \psi_\alpha \\ \chi^{\dot\alpha} \end{pmatrix}, \qquad \psi_\alpha \equiv \text{left-handed}, \qquad \chi^{\dot\alpha} \equiv \text{right-handed}. \tag{A.2}$$

The gamma matrices transform left-handed spinors into right-handed ones and vice versa. Therefore one can write them as

$$(\Gamma^\mu)_A{}^B = \begin{pmatrix} 0 & \gamma^\mu_{\alpha\dot\beta} \\ \bar\gamma^{\mu\,\dot\alpha\beta} & 0 \end{pmatrix}, \tag{A.3}$$

where the off-diagonal matrices are known as 'generalized' Pauli matrices. The Clifford algebra (A.1) translates then into

$$\gamma^\mu_{\alpha\dot\beta}\bar\gamma^{\nu\,\dot\beta\gamma} + \gamma^\nu_{\alpha\dot\beta}\bar\gamma^{\mu\,\dot\beta\gamma} = -2\,\delta^\gamma_\alpha\,\eta^{\mu\nu}, \qquad \bar\gamma^{\mu\,\dot\alpha\beta}\gamma^\nu_{\beta\dot\gamma} + \bar\gamma^{\nu\,\dot\alpha\beta}\gamma^\mu_{\beta\dot\gamma} = -2\,\delta^{\dot\alpha}_{\dot\gamma}\,\eta^{\mu\nu}. \tag{A.4}$$

The action of Γ on a Dirac spinor Ξ in index notation reads

$$(\Gamma^\mu)_A{}^B\,\Xi_B = \begin{pmatrix} 0 & \gamma^\mu_{\alpha\dot\beta} \\ \bar\gamma^{\mu\,\dot\alpha\beta} & 0 \end{pmatrix}\begin{pmatrix} \psi_\beta \\ \chi^{\dot\beta} \end{pmatrix} = \begin{pmatrix} \gamma^\mu_{\alpha\dot\beta}\chi^{\dot\beta} \\ \bar\gamma^{\mu\,\dot\alpha\beta}\psi_\beta \end{pmatrix}. \tag{A.5}$$

Products of an even (odd) number of Γ matrices carry alternating products of γ and $\bar\gamma$ matrices in their diagonal (off-diagonal) blocks:

$$(\Gamma^{\mu_1}\Gamma^{\mu_2}\dots\Gamma^{\mu_p})_A{}^B = \begin{cases} \begin{pmatrix} (\gamma^{\mu_1}\bar\gamma^{\mu_2}\dots\bar\gamma^{\mu_p})_\alpha{}^\beta & 0 \\ 0 & (\bar\gamma^{\mu_1}\gamma^{\mu_2}\dots\gamma^{\mu_p})^{\dot\alpha}{}_{\dot\beta} \end{pmatrix} & : p \text{ even}, \\[2ex] \begin{pmatrix} 0 & (\gamma^{\mu_1}\bar\gamma^{\mu_2}\dots\gamma^{\mu_p})_{\alpha\dot\beta} \\ (\bar\gamma^{\mu_1}\gamma^{\mu_2}\dots\bar\gamma^{\mu_p})^{\dot\alpha\beta} & 0 \end{pmatrix} & : p \text{ odd}. \end{cases} \tag{A.6}$$

Totally antisymmetric products of Γ matrices are defined in this work as

$$\Gamma^{\mu_1\dots\mu_p} \equiv \frac{1}{p!}\sum_{\rho\in S_p}\text{sgn}(\rho)\,\Gamma^{\mu_{\rho(1)}}\dots\Gamma^{\mu_{\rho(p)}}. \tag{A.7}$$

The Γ matrices alone obviously have the wrong index structure to serve as Clebsch–Gordan coefficients for bi-spinors, especially for such constructed from spinors of the same chirality. Some kind of metric on spinor space is needed. This metric is known as the charge conjugation

matrix $\mathcal{C}$:
$$(\Gamma^\mu)_A{}^B \mathcal{C}_{BD} \equiv (\Gamma^\mu \mathcal{C})_{AD}. \tag{A.8}$$

The chirality structure of $\mathcal{C}$ depends on the number of dimensions D due to the representation theory of the associated Lorentz group $SO(1, D-1)$. In dimensions $D = 0$ mod 4, only spinor representations of alike chiralities contain a scalar in their tensor product whereas for $D = 2$ mod 4 dimensions opposite chiralities are required to form a singlet. Therefore the charge conjugation matrix written with spinor indices becomes

$$D = 0 \text{ mod } 4: \quad \mathcal{C}_{AB} = \begin{pmatrix} C_{\alpha\beta} & 0 \\ 0 & C^{\dot\alpha\dot\beta} \end{pmatrix}, \quad (\Gamma^\mu \mathcal{C})_{AB} = \begin{pmatrix} 0 & (\gamma^\mu C)_\alpha{}^{\dot\beta} \\ (\bar\gamma^\mu C)^{\dot\alpha}{}_\beta & 0 \end{pmatrix},$$

$$D = 2 \text{ mod } 4: \quad \mathcal{C}_{AB} = \begin{pmatrix} 0 & C_\alpha{}^{\dot\beta} \\ C^{\dot\alpha}{}_\beta & 0 \end{pmatrix}, \quad (\Gamma^\mu \mathcal{C})_{AB} = \begin{pmatrix} (\gamma^\mu C)_{\alpha\beta} & 0 \\ 0 & (\bar\gamma^\mu C)^{\dot\alpha\dot\beta} \end{pmatrix}. \tag{A.9}$$

The inverse of the charge conjugation matrix $\mathcal{C}^{-1}$ is then denoted by

$$D = 0 \text{ mod } 4: \quad (\mathcal{C}^{-1})^{AB} = \begin{pmatrix} (C^{-1})^{\alpha\beta} & 0 \\ 0 & (C^{-1})_{\dot\alpha\dot\beta} \end{pmatrix},$$

$$(\mathcal{C}^{-1}\Gamma^\mu)^{AB} = \begin{pmatrix} 0 & (C^{-1}\gamma^\mu)^\alpha{}_{\dot\beta} \\ (C^{-1}\bar\gamma^\mu)_{\dot\alpha}{}^\beta & 0 \end{pmatrix},$$

$$D = 2 \text{ mod } 4: \quad (\mathcal{C}^{-1})^{AB} = \begin{pmatrix} 0 & (C^{-1})^\alpha{}_{\dot\beta} \\ (C^{-1})_{\dot\alpha}{}^\beta & 0 \end{pmatrix},$$

$$(\mathcal{C}^{-1}\Gamma^\mu)^{AB} = \begin{pmatrix} (C^{-1}\bar\gamma^\mu)^{\alpha\beta} & 0 \\ 0 & (C^{-1}\gamma^\mu)_{\dot\alpha\dot\beta} \end{pmatrix}. \tag{A.10}$$

A.2 Symmetry properties

The transposed gamma matrices Γ^t also satisfy the Clifford algebra (A.1). Schur's lemma guarantees that Γ and Γ^t must be related by a similarity transformation. This transformation is given by the charge conjugation matrix and its inverse[1]

$$\mathcal{C}^{-1}\Gamma^\mu \mathcal{C} = -(\Gamma^\mu)^t. \tag{A.11}$$

---

[1]In even dimensions $D = 2m$, the signs in (A.11) and (A.13) are a matter of convention due to the freedom to redefine $\mathcal{C} \to \Gamma^D \mathcal{C}$, where Γ^D is the chirality matrix. The absence of a chirality matrix in $D = 2m-1$ dimensions, leads to a unique choice.

On the level of chiral blocks this leads to two different scenarios. For $D = 0 \mod 4$ (A.11) intertwines the two classes of matrices γ^μ, $\bar\gamma^\mu$, while in $D = 2 \mod 4$ dimensions one obtains a consistency condition:

$$D = 0 \mod 4: \quad \gamma^\mu_{\beta\dot\alpha} = -(C^{-1})_{\dot\alpha\dot\gamma}\, \bar\gamma^{\mu\,\dot\gamma\gamma}\, C_{\gamma\beta}\,, \qquad \bar\gamma^{\mu\,\dot\beta\alpha} = -(C^{-1})^{\alpha\gamma}\, \gamma^\mu_{\gamma\dot\gamma}\, C^{\dot\gamma\dot\beta}\,,$$

$$D = 2 \mod 4: \quad \gamma^\mu_{\beta\dot\alpha} = -(C^{-1})_{\dot\alpha}{}^\gamma\, \gamma^\mu_{\gamma\dot\gamma}\, C^{\dot\gamma}{}_\beta\,, \qquad \bar\gamma^{\mu\,\dot\beta\alpha} = -(C^{-1})^\alpha{}_{\dot\gamma}\, \bar\gamma^{\mu\,\dot\gamma\gamma}\, C_\gamma{}^{\dot\beta}\,. \quad (A.12)$$

We now give a unified way of understanding these conditions: The symmetry property of $(\Gamma^\mu C)$ is opposite to that of the charge conjugation matrix,

$$C^t = \wp_D\, C \quad \Rightarrow \quad (\Gamma^\mu C)^t = -\wp_D\, (\Gamma^\mu C)\,, \qquad (A.13)$$

where $\wp_D$ is a dimension-dependent phase which we determine later. Applying this argument iteratively we can determine the symmetry properties of general gamma products. As these fulfill $(\Gamma^{\mu_1}\ldots\Gamma^{\mu_p} C)^t = \wp_D(-1)^p\, (\Gamma^{\mu_p}\ldots\Gamma^{\mu_1} C)$ antisymmetric chains of Γ matrices satisfy

$$(\Gamma^{\mu_1\ldots\mu_p} C)^t = \wp_D\, (-1)^{\frac{p}{2}(p+1)}\, (\Gamma^{\mu_1\ldots\mu_p} C) = \begin{cases} +\wp_D\, (\Gamma^{\mu_1\ldots\mu_p} C) & : p = 0, 3 \mod 4\,, \\ -\wp_D\, (\Gamma^{\mu_1\ldots\mu_p} C) & : p = 1, 2 \mod 4\,. \end{cases} \quad (A.14)$$

In order to fix the phase $\wp_D$ in (A.13) one has to make use of the fact that in $D = 2m$ dimensions the set $\{(\Gamma^{\mu_1\ldots\mu_p} C) : 0 \le p \le D\}$ forms a basis of the $2^m \times 2^m$ matrices. In particular, there must be $2^m(2^m - 1)/2$ antisymmetric matrices, and this fixes

$$\wp_D = (-1)^{\frac{m}{2}(m+1)} \quad \Rightarrow \quad C^t = \begin{cases} +C & : D = 0, 6 \mod 8\,, \\ -C & : D = 2, 4 \mod 8\,. \end{cases} \quad (A.15)$$

The equations (A.11) and (A.15) can be used to explicitly construct the charge conjugation matrix for a given representation of the Clifford algebra.

To avoid over-counting of the independent symmetric matrices, one should be aware of the self-dualities of $D/2$-fold products:

$$(\gamma^{\mu_1\ldots\mu_{D/2}} C)_{\alpha\beta} = \frac{e^{i\phi_D}}{(D/2)!}\, \varepsilon^{\mu_1\ldots\mu_{D/2}\nu_1\ldots\nu_{D/2}}\, (\gamma_{\nu_1\ldots\nu_{D/2}} C)_{\alpha\beta}\,,$$

$$(\bar\gamma^{\mu_1\ldots\mu_{D/2}} C)^{\dot\alpha\dot\beta} = -\frac{e^{i\phi_D}}{(D/2)!}\, \varepsilon^{\mu_1\ldots\mu_{D/2}\nu_1\ldots\nu_{D/2}}\, (\bar\gamma_{\nu_1\ldots\nu_{D/2}} C)^{\dot\alpha\dot\beta}\,. \quad (A.16)$$

Here ϕ_D denotes a phase that depends on the number of dimensions.

A.3 Fierz identities

Antisymmetrized gamma products $\Gamma^{\mu_1\cdots\mu_p}$ with $0 \leq p \leq D$ form a complete set of all $2^{D/2} \times 2^{D/2}$ matrices. Therefore, it is possible to expand any bi-spinor in terms of forms. The expansion prescriptions are referred to as Fierz identities. Within the chiral blocks γ^μ, $\bar{\gamma}^\mu$, it is sufficient to consider forms up to degree $D/2$ since any p-fold product $\gamma^{\mu_1\cdots\mu_p}$, $p \leq D$, is related to $(D-p)$-fold products via Hodge duality. Weyl bi-spinors can therefore be expanded in the following way [136, 146]:

- $D = 0 \bmod 4$:

$$\psi_\alpha \chi_\beta = 2^{-D/2} \sum_{\substack{p=0 \\ p \text{ even}}}^{D/2-2} \frac{1}{p!} (\gamma^{\mu_1\cdots\mu_p} C)_{\beta\alpha} (\psi C^{-1} \gamma_{\mu_p\cdots\mu_1} \chi)$$
$$+ \frac{2^{-D/2}}{2(D/2)!} (\gamma^{\mu_1\cdots\mu_{D/2}} C)_{\beta\alpha} (\psi C^{-1} \gamma_{\mu_{D/2}\cdots\mu_1} \chi), \tag{A.17}$$

$$\psi_\alpha \bar{\chi}^{\dot{\beta}} = -2^{-D/2} \sum_{\substack{p=1 \\ p \text{ odd}}}^{D/2-1} \frac{1}{p!} (\bar{\gamma}^{\mu_1\cdots\mu_p} C)^{\dot{\beta}}{}_\alpha (\psi C^{-1} \gamma_{\mu_p\cdots\mu_1} \bar{\chi}), \tag{A.18}$$

- $D = 2 \bmod 4$:

$$\psi_\alpha \chi_\beta = -2^{-D/2} \sum_{\substack{p=1 \\ p \text{ odd}}}^{D/2-2} \frac{1}{p!} (\gamma^{\mu_1\cdots\mu_p} C)_{\beta\alpha} (\psi C^{-1} \bar{\gamma}_{\mu_p\cdots\mu_1} \chi)$$
$$- \frac{2^{-D/2}}{2(D/2)!} (\gamma^{\mu_1\cdots\mu_{D/2}} C)_{\beta\alpha} (\psi C^{-1} \bar{\gamma}_{\mu_{D/2}\cdots\mu_1} \chi), \tag{A.19}$$

$$\psi_\alpha \bar{\chi}^{\dot{\beta}} = 2^{-D/2} \sum_{\substack{p=0 \\ p \text{ even}}}^{D/2-1} \frac{1}{p!} (\bar{\gamma}^{\mu_1\cdots\mu_p} C)^{\dot{\beta}}{}_\alpha (\psi C^{-1} \bar{\gamma}_{\mu_p\cdots\mu_1} \bar{\chi}). \tag{A.20}$$

The proof of this identities relies on the fact that the matrices $\gamma^{\mu_1\cdots\mu_p}$ are orthonormal with respect to the trace as inner product, up to the subtlety that in some traces the symbol $\varepsilon^{\mu_1\cdots\mu_{D/2}}{}_{\nu_1\cdots\nu_{D/2}}$ appears. Let us display the Fierz identities in $D = 4, 6, 8, 10$ dimensions explicitly:

- $D = 4$:

$$\psi_\alpha \chi_\beta = \frac{1}{2} C_{\beta\alpha} (\psi C^{-1} \chi) + \frac{1}{8} (\gamma^{\mu\nu} C)_{\beta\alpha} (\psi C^{-1} \gamma_{\nu\mu} \chi), \tag{A.21}$$

$$\psi_\alpha \bar{\chi}^{\dot{\beta}} = -\frac{1}{2} (\bar{\gamma}^\mu C)^{\dot{\beta}}{}_\alpha (\psi C^{-1} \gamma_\mu \bar{\chi}), \tag{A.22}$$

- $D = 6$:

$$\psi_\alpha \chi_\beta = -\frac{1}{4} (\gamma^\mu C)_{\beta\alpha} (\psi C^{-1} \bar{\gamma}_\mu \chi) - \frac{1}{48} (\gamma^{\mu\nu\lambda} C)_{\beta\alpha} (\psi C^{-1} \bar{\gamma}_{\lambda\nu\mu} \chi) , \qquad (A.23)$$

$$\psi_\alpha \bar{\chi}^{\dot{\beta}} = \frac{1}{4} C^{\dot{\beta}}{}_\alpha (\psi C^{-1} \bar{\chi}) + \frac{1}{8} (\bar{\gamma}^{\mu\nu} C)^{\dot{\beta}}{}_\alpha (\psi C^{-1} \bar{\gamma}_{\nu\mu} \bar{\chi}) , \qquad (A.24)$$

- $D = 8$:

$$\psi_\alpha \chi_\beta = \frac{1}{8} C_{\beta\alpha} (\psi C^{-1} \chi) + \frac{1}{16} (\gamma^{\mu\nu} C)_{\beta\alpha} (\psi C^{-1} \gamma_{\nu\mu} \chi)$$
$$+ \frac{1}{384} (\gamma^{\mu\nu\lambda\rho} C)_{\beta\alpha} (\psi C^{-1} \gamma_{\rho\lambda\nu\mu} \chi) , \qquad (A.25)$$

$$\psi_\alpha \bar{\chi}^{\dot{\beta}} = -\frac{1}{8} (\bar{\gamma}^\mu C)^{\dot{\beta}}{}_\alpha (\psi C^{-1} \gamma_\mu \bar{\chi}) - \frac{1}{48} (\bar{\gamma}^{\mu\nu\lambda} C)^{\dot{\beta}}{}_\alpha (\psi C^{-1} \gamma_{\lambda\nu\mu} \bar{\chi}) , \qquad (A.26)$$

- $D = 10$:

$$\psi_\alpha \chi_\beta = -\frac{1}{16} (\gamma^\mu C)_{\beta\alpha} (\psi C^{-1} \bar{\gamma}_\mu \chi) - \frac{1}{96} (\gamma^{\mu\nu\lambda} C)_{\beta\alpha} (\psi C^{-1} \bar{\gamma}_{\lambda\nu\mu} \chi)$$
$$- \frac{1}{3840} (\gamma^{\mu\nu\lambda\rho\tau} C)_{\beta\alpha} (\psi C^{-1} \bar{\gamma}_{\tau\rho\lambda\nu\mu} \chi) , \qquad (A.27)$$

$$\psi_\alpha \bar{\chi}^{\dot{\beta}} = \frac{1}{16} C^{\dot{\beta}}{}_\alpha (\psi C^{-1} \bar{\chi}) + \frac{1}{32} (\bar{\gamma}^{\mu\nu} C)^{\dot{\beta}}{}_\alpha (\psi C^{-1} \bar{\gamma}_{\nu\mu} \bar{\chi})$$
$$+ \frac{1}{384} (\bar{\gamma}^{\mu\nu\lambda\rho} C)^{\dot{\beta}}{}_\alpha (\psi C^{-1} \bar{\gamma}_{\rho\lambda\nu\mu} \bar{\chi}) . \qquad (A.28)$$

Fierz identities allow to derive relations between different $SO(1, D - 1)$ Clebsch–Gordan coefficients by making appropriate choices for ψ_α, χ_β and $\bar{\chi}^{\dot{\beta}}$. Two examples of this method have been shown in Chapter 3.3. This techniques is further used in Appendix B where we collect all necessary relations for the correlators calculated in this work.

A.4 A Concrete Representation

The results for the RNS correlation functions stated in Chapters 4 and 5 are written down in Lorentz covariant form and are therefore valid for all representations of the Clifford algebra. However for some issues in the calculations, e.g. checking relations between different Clebsch–Gordan coefficients and especially for obtaining loop results, an explicit representation of the gamma matrices is helpful.

It is well-known that the gamma matrices in $D = 2m$ dimensions, Γ_D^μ, and the Pauli matrices σ^i can be used to construct Γ_{D+2}^μ, the Dirac matrices in $D + 2$ dimensions. It is easily verified

A.4 A Concrete Representation

that for $k = 0, \ldots, D-1$

$$\Gamma^k_{D+2} \equiv \sigma^1 \otimes \Gamma^k_D, \qquad \Gamma^D_{D+2} \equiv i\sigma^1 \otimes \Gamma^D_D, \qquad \Gamma^{D+1}_{D+2} \equiv i\sigma^2 \otimes 1_{2^m} \tag{A.29}$$

satisfy the Clifford algebra (A.1) in $D+2$ dimensions, if Γ^μ_D satisfy the Clifford algebra in D space-time dimensions[2]. The matrix Γ^D_D is the chirality matrix defined as

$$\Gamma^D_D \equiv (-1)^{\frac{1}{2}(m+1)} \prod_{i=1}^{D-1} \Gamma^i_D. \tag{A.31}$$

We start in $D = 2$ dimensions with

$$\Gamma^0_2 \equiv \sigma^1 = \begin{pmatrix} 0 & 1 \\ 1 & 0 \end{pmatrix}, \qquad \Gamma^1_2 \equiv i\sigma^2 = \begin{pmatrix} 0 & 1 \\ -1 & 0 \end{pmatrix}. \tag{A.32}$$

The Dirac matrices for $D = 4, 6, 8, 10$ can then be constructed by recursively using (A.29). Our construction is chosen such that we obtain the Γ^μ_D's in Weyl basis, i.e. Γ^D_D always has the form

$$\Gamma^D_D = \begin{pmatrix} 1_{2^{m-1}} & 0 \\ 0 & -1_{2^{m-1}} \end{pmatrix}. \tag{A.33}$$

The charge conjugation matrix $\mathcal{C}_D$ is determined from the equations (A.11) and (A.15). This yields

$$\mathcal{C}_2 = i\sigma^2, \qquad \mathcal{C}_4 = 1_2 \otimes \mathcal{C}_2, \qquad \mathcal{C}_6 = -i\sigma^2 \otimes \mathcal{C}_4,$$
$$\mathcal{C}_8 = 1_2 \otimes \mathcal{C}_6, \qquad \mathcal{C}_{10} = -i\sigma^2 \otimes \mathcal{C}_8. \tag{A.34}$$

As 1 is diagonal and σ^2 off-diagonal, this results exactly in the diagonal and off-diagonal structure stated in (A.9).

---

[2] The product $\otimes$ in (A.29) is the Kronecker product which takes an $m \times n$ matrix A and a $p \times q$ matrix B to the $mp \times nq$ matrix

$$A \otimes B \equiv \begin{pmatrix} a_{11}B & \cdots & a_{1n}B \\ \vdots & \ddots & \vdots \\ a_{m1}B & \cdots & a_{mn}B \end{pmatrix}. \tag{A.30}$$

APPENDIX B

Relations between Index Terms

In this Appendix we collect relations between different index terms, that enter the correlators calculated in Chapters 4 and 5. The numbers of independent Clebsch–Gordan coefficients for each correlator are summarized in Table 3.3. As we have explained before these relations can be derived from Fierz identities given in Appendix A, multiplying known equations with further γ and C matrices, as well as demanding that a tensor in D dimensions, which is anti-symmetric in $2^{D/2-1}+1$ Weyl indices of the same type, has to vanish. We have checked the validity of the following relations by (anti-)symmetry arguments and furthermore by explicit verification using the representation of the Clifford algebra given in Appendix A.4.

Before we start listing the relations we recap our conventions. The Clifford algebra establishes that

$$\gamma^{\mu}_{\alpha\dot{\beta}}\,\bar{\gamma}^{\nu\dot{\beta}\gamma} + \gamma^{\nu}_{\alpha\dot{\beta}}\,\bar{\gamma}^{\mu\dot{\beta}\gamma} = -2\,\delta^{\gamma}_{\alpha}\,\eta^{\mu\nu}\,, \qquad \bar{\gamma}^{\mu\dot{\alpha}\beta}\,\gamma^{\nu}_{\beta\dot{\gamma}} + \bar{\gamma}^{\nu\dot{\alpha}\beta}\,\gamma^{\mu}_{\beta\dot{\gamma}} = -2\,\delta^{\dot{\alpha}}_{\dot{\gamma}}\,\eta^{\mu\nu}\,. \qquad (B.1)$$

These can be used to interchange Lorentz indices in expressions where chains of γ matrices appear. Antisymmetric products of γ matrices are defined with a normalization factor $1/p!$ in contrast to [21],

$$\gamma^{\mu_1\ldots\mu_p} \equiv \frac{1}{p!}\sum_{\rho\in S_p}\mathrm{sgn}(\rho)\begin{cases}\gamma^{\mu_{\rho(1)}}\,\bar{\gamma}^{\mu_{\rho(2)}}\ldots\gamma^{\mu_{\rho(p)}} & : p \text{ odd}\,,\\ \gamma^{\mu_{\rho(1)}}\,\bar{\gamma}^{\mu_{\rho(2)}}\ldots\bar{\gamma}^{\mu_{\rho(p)}} & : p \text{ even}\,.\end{cases} \qquad (B.2)$$

On the right hand side of this equation we can interchange the Lorentz indices with the help of (B.1). Then the antisymmetric γ-product can be written as an ordered γ-product and further η terms. Let us illustrate this for the simplest case $p=2$:

$$\gamma^{\mu\nu} = \frac{1}{2}\left(\gamma^{\mu}\,\bar{\gamma}^{\nu} - \gamma^{\nu}\,\bar{\gamma}^{\mu}\right) = \eta^{\mu\nu} + \gamma^{\mu}\,\bar{\gamma}^{\nu}\,. \qquad (B.3)$$

In deriving the following relations the behavior of tensors under interchanging their spinor indices is very important. These can be derived by employing the identities from Table 3.4. Contractions of γ and C matrices in the following should always be understood as matrix multiplications. We refrain from explicitly denoting C^{-1} in index notation. This matrix differs from C in the positioning of its indices and hence can be distinguished in the respective contexts.

B.1 Relations for $D = 4$

In the following we state relations that allow to reduce the number of index terms for a given correlator in four space-time dimensions. We adopt the widely used notation in four dimensions and denote the blocks of the Γ matrices and the charge conjugation matrix $\mathcal{C}$ by $\sigma^\mu_{\alpha\dot\alpha}$ and $\varepsilon_{\alpha\beta}$. The Fierz identity

$$\varepsilon_{\alpha\gamma}\,\varepsilon_{\beta\delta} = \varepsilon_{\alpha\beta}\,\varepsilon_{\gamma\delta} + \varepsilon_{\alpha\delta}\,\varepsilon_{\beta\gamma} \tag{B.4}$$

is a good starting point for deriving further relations.

Correlator $\langle \psi^\mu \psi^\nu S_\alpha S_{\dot\beta} S_\gamma S_{\dot\delta} \rangle$

The relevant index terms for this correlation function up to permutations in the spinor indices are $2\eta^{\mu\nu}\,\varepsilon_{\alpha\gamma}\,\varepsilon_{\dot\beta\dot\delta}$, $\sigma^\mu_{\alpha\dot\beta}\,\sigma^\nu_{\gamma\dot\delta}$, $(\sigma^\mu\bar\sigma^\nu\varepsilon)_{\alpha\gamma}\,\varepsilon_{\dot\beta\dot\delta}$ and $(\varepsilon\,\bar\sigma^\mu\sigma^\nu)_{\dot\beta\dot\delta}\,\varepsilon_{\alpha\gamma}$. The latter two terms can be eliminated by multiplying (B.4) with $\varepsilon_{\dot\beta\dot\alpha}\,\bar\sigma^{\mu\,\dot\alpha\beta}\,\varepsilon_{\dot\beta\dot\gamma}\,\bar\sigma^{\nu\,\dot\gamma\delta}$ and treating the anti-chiral version of (B.4) in the same manner:

$$(\sigma^\mu\bar\sigma^\nu\varepsilon)_{\alpha\gamma}\,\varepsilon_{\dot\beta\dot\delta} = \sigma^\mu_{\alpha\dot\beta}\,\sigma^\nu_{\gamma\dot\delta} - \sigma^\mu_{\alpha\dot\delta}\,\sigma^\nu_{\gamma\dot\beta}\,, \tag{B.5a}$$

$$(\varepsilon\,\bar\sigma^\mu\sigma^\nu)_{\dot\beta\dot\delta}\,\varepsilon_{\alpha\gamma} = \sigma^\mu_{\alpha\dot\beta}\,\sigma^\nu_{\gamma\dot\delta} - \sigma^\mu_{\gamma\dot\beta}\,\sigma^\nu_{\alpha\dot\delta}\,. \tag{B.5b}$$

By symmetrizing in the vector indices μ, ν we arrive at:

$$2\eta^{\mu\nu}\,\varepsilon_{\alpha\gamma}\,\varepsilon_{\dot\beta\dot\delta} = \sigma^\mu_{\alpha\dot\delta}\,\sigma^\nu_{\gamma\dot\beta} + \sigma^\mu_{\gamma\dot\beta}\,\sigma^\nu_{\alpha\dot\delta} - \sigma^\mu_{\alpha\dot\beta}\,\sigma^\nu_{\gamma\dot\delta} - \sigma^\mu_{\gamma\dot\delta}\,\sigma^\nu_{\alpha\dot\beta}\,. \tag{B.6}$$

Correlator $\langle \psi^\mu \psi^\nu S_\alpha S_\beta S_\gamma S_\delta \rangle$

This correlator can be expressed in terms of $\eta^{\mu\nu}\,\varepsilon_{\alpha\beta}\,\varepsilon_{\gamma\delta}$, $(\sigma^\mu\bar\sigma^\nu\varepsilon)_{\alpha\beta}\,\varepsilon_{\gamma\delta}$ and permutations in the spinor indices of these terms. Forming antisymmetric combinations of the latter, where one spinor index of the σ-chains is kept fixed, yields three independent relations:

$$(\sigma^\mu\bar\sigma^\nu\varepsilon)_{\alpha\delta}\,\varepsilon_{\beta\gamma} = -(\sigma^\mu\bar\sigma^\nu\varepsilon)_{\alpha\beta}\,\varepsilon_{\gamma\delta} + (\sigma^\mu\bar\sigma^\nu\varepsilon)_{\alpha\gamma}\,\varepsilon_{\beta\delta}\,, \tag{B.7a}$$

$$(\sigma^\mu\bar\sigma^\nu\varepsilon)_{\beta\delta}\,\varepsilon_{\alpha\gamma} = -(\sigma^\mu\bar\sigma^\nu\varepsilon)_{\alpha\beta}\,\varepsilon_{\gamma\delta} + (\sigma^\mu\bar\sigma^\nu\varepsilon)_{\beta\gamma}\,\varepsilon_{\alpha\delta} - 2\eta^{\mu\nu}\,\varepsilon_{\alpha\beta}\,\varepsilon_{\gamma\delta}\,, \tag{B.7b}$$

$$(\sigma^\mu\bar\sigma^\nu\varepsilon)_{\gamma\delta}\,\varepsilon_{\alpha\beta} = -(\sigma^\mu\bar\sigma^\nu\varepsilon)_{\alpha\gamma}\,\varepsilon_{\beta\delta} + (\sigma^\mu\bar\sigma^\nu\varepsilon)_{\beta\gamma}\,\varepsilon_{\alpha\delta} - 2\eta^{\mu\nu}\,\varepsilon_{\alpha\beta}\,\varepsilon_{\gamma\delta}\,. \tag{B.7c}$$

B.1 Relations for $D = 4$

The Fierz identity (B.4) can be used to eliminate $\eta^{\mu\nu}\varepsilon_{\alpha\gamma}\varepsilon_{\beta\delta}$ from the calculations and hence one arrives at five independent index terms. In order to write the correlator in terms of anti-symmetric σ-products the following equation is of use:

$$(\sigma^{\mu\nu}\varepsilon)_{\alpha\beta}\varepsilon_{\gamma\delta} - (\sigma^{\mu\nu}\varepsilon)_{\alpha\delta}\varepsilon_{\gamma\beta} + (\sigma^{\mu\nu}\varepsilon)_{\gamma\delta}\varepsilon_{\alpha\beta} - (\sigma^{\mu\nu}\varepsilon)_{\gamma\beta}\varepsilon_{\alpha\delta} = 0. \quad (B.8)$$

Correlator $\langle \psi^\mu \psi^\nu \psi^\lambda S_\alpha S_\beta S_\gamma S_{\dot\delta} \rangle$

The anti-symmetrization argument can be used also for this correlator to derive relations between different index terms,

$$(\sigma^\mu \bar\sigma^\nu \sigma^\lambda)_{\beta\dot\delta}\varepsilon_{\alpha\gamma} = (\sigma^\mu \bar\sigma^\nu \sigma^\lambda)_{\alpha\dot\delta}\varepsilon_{\beta\gamma} + (\sigma^\mu \bar\sigma^\nu \sigma^\lambda)_{\gamma\dot\delta}\varepsilon_{\alpha\beta}, \quad (B.9a)$$

$$\eta^{\nu\lambda}\sigma^\mu_{\beta\dot\delta}\varepsilon_{\alpha\gamma} = \eta^{\nu\lambda}\sigma^\mu_{\alpha\dot\delta}\varepsilon_{\beta\gamma} + \eta^{\nu\lambda}\sigma^\mu_{\gamma\dot\delta}\varepsilon_{\alpha\beta}. \quad (B.9b)$$

In the second equation it is possible to permute the Lorentz indices μ, ν, λ in order to derive two additional equations. Furthermore, applying $(\sigma^\mu \bar\sigma^\nu)_\alpha{}^\kappa \varepsilon_{[\kappa\beta}\sigma^\lambda_{\gamma]\dot\delta} = 0$ yields identities that mix different σ configurations:

$$\varepsilon_{\beta\gamma}(\sigma^\mu \bar\sigma^\nu \sigma^\lambda)_{\alpha\dot\delta} = (\sigma^\mu \bar\sigma^\nu \varepsilon)_{\alpha\gamma}\sigma^\lambda_{\beta\dot\delta} - (\sigma^\mu \bar\sigma^\nu \varepsilon)_{\alpha\beta}\sigma^\lambda_{\gamma\dot\delta}, \quad (B.10a)$$

$$\varepsilon_{\alpha\beta}(\sigma^\mu \bar\sigma^\nu \sigma^\lambda)_{\gamma\dot\delta} = (\sigma^\mu \bar\sigma^\nu \varepsilon)_{\beta\gamma}\sigma^\lambda_{\alpha\dot\delta} - (\sigma^\mu \bar\sigma^\nu \varepsilon)_{\alpha\gamma}\sigma^\lambda_{\beta\dot\delta} - 2\eta^{\mu\nu}\varepsilon_{\alpha\beta}\sigma^\lambda_{\gamma\dot\delta}. \quad (B.10b)$$

Hence, the triple products $(\sigma^\mu\bar\sigma^\nu\sigma^\lambda)$ can be completely eliminated. By permuting the Lorentz indices in (B.10) one obtains four linearly independent relations:

$$2\eta^{\mu\lambda}\sigma^\nu_{\alpha\dot\delta}\varepsilon_{\beta\gamma} - 2\eta^{\mu\nu}\sigma^\lambda_{\alpha\dot\delta}\varepsilon_{\beta\gamma} = (\sigma^\nu\bar\sigma^\lambda\varepsilon)_{\alpha\beta}\sigma^\mu_{\gamma\dot\delta} - (\sigma^\nu\bar\sigma^\lambda\varepsilon)_{\alpha\gamma}\sigma^\mu_{\beta\dot\delta}$$
$$-(\sigma^\mu\bar\sigma^\nu\varepsilon)_{\alpha\beta}\sigma^\lambda_{\gamma\dot\delta} + (\sigma^\mu\bar\sigma^\nu\varepsilon)_{\alpha\gamma}\sigma^\lambda_{\beta\dot\delta}, \quad (B.11a)$$

$$2\eta^{\mu\lambda}\sigma^\nu_{\gamma\dot\delta}\varepsilon_{\alpha\beta} - 2\eta^{\nu\lambda}\sigma^\mu_{\gamma\dot\delta}\varepsilon_{\alpha\beta} = (\sigma^\nu\bar\sigma^\lambda\varepsilon)_{\alpha\gamma}\sigma^\mu_{\beta\dot\delta} - (\sigma^\nu\bar\sigma^\lambda\varepsilon)_{\beta\gamma}\sigma^\mu_{\alpha\dot\delta}$$
$$+(\sigma^\mu\bar\sigma^\nu\varepsilon)_{\beta\gamma}\sigma^\lambda_{\alpha\dot\delta} - (\sigma^\mu\bar\sigma^\nu\varepsilon)_{\alpha\gamma}\sigma^\lambda_{\beta\dot\delta}, \quad (B.11b)$$

$$2\eta^{\nu\lambda}\sigma^\mu_{\alpha\dot\delta}\varepsilon_{\beta\gamma} = (\sigma^\mu\bar\sigma^\nu\varepsilon)_{\alpha\beta}\sigma^\lambda_{\gamma\dot\delta} - (\sigma^\mu\bar\sigma^\nu\varepsilon)_{\alpha\gamma}\sigma^\lambda_{\beta\dot\delta}$$
$$+(\sigma^\mu\bar\sigma^\lambda\varepsilon)_{\alpha\beta}\sigma^\nu_{\gamma\dot\delta} - (\sigma^\mu\bar\sigma^\lambda\varepsilon)_{\alpha\gamma}\sigma^\nu_{\beta\dot\delta}, \quad (B.11c)$$

$$2\eta^{\mu\nu}\sigma^\lambda_{\gamma\dot\delta}\varepsilon_{\alpha\beta} = (\sigma^\mu\bar\sigma^\lambda\varepsilon)_{\beta\gamma}\sigma^\nu_{\alpha\dot\delta} - (\sigma^\mu\bar\sigma^\lambda\varepsilon)_{\alpha\gamma}\sigma^\nu_{\beta\dot\delta}$$
$$+(\sigma^\nu\bar\sigma^\lambda\varepsilon)_{\beta\gamma}\sigma^\mu_{\alpha\dot\delta} - (\sigma^\nu\bar\sigma^\lambda\varepsilon)_{\alpha\gamma}\sigma^\mu_{\beta\dot\delta}. \quad (B.11d)$$

Another relation is necessary in order to reduce the number of Clebsch–Gordan coefficients to ten. This is achieved by multiplying (B.5a) with $\bar\sigma^{\lambda\dot\beta\delta}\varepsilon_{\delta\beta}$. After a further permutation in the spinor indices one finds:

$$(\sigma^\mu \bar\sigma^\nu \varepsilon)_{\alpha\beta} \sigma^\lambda_{\gamma\dot\delta} - (\sigma^\mu \bar\sigma^\lambda \varepsilon)_{\alpha\gamma} \sigma^\nu_{\beta\dot\delta} + (\sigma^\nu \bar\sigma^\lambda \varepsilon)_{\beta\gamma} \sigma^\mu_{\alpha\dot\delta} = 0 \,. \tag{B.12}$$

Correlator $\langle \psi^\mu \, S_\alpha \, S_\beta \, S_\gamma \, S_\delta \, S_\epsilon \, S_\zeta \rangle$

The starting point for relating different index term for this correlator is

$$\varepsilon_{\alpha\beta}\,\varepsilon_{\gamma\delta}\,\varepsilon_{\epsilon\zeta} - \varepsilon_{\alpha\beta}\,\varepsilon_{\gamma\zeta}\,\varepsilon_{\epsilon\delta} + \varepsilon_{\alpha\delta}\,\varepsilon_{\gamma\zeta}\,\varepsilon_{\epsilon\beta} - \varepsilon_{\alpha\delta}\,\varepsilon_{\gamma\beta}\,\varepsilon_{\epsilon\zeta} + \varepsilon_{\alpha\zeta}\,\varepsilon_{\gamma\beta}\,\varepsilon_{\epsilon\delta} - \varepsilon_{\alpha\zeta}\,\varepsilon_{\gamma\delta}\,\varepsilon_{\epsilon\beta} = 0 \,, \tag{B.13}$$

which stems from $\delta^\alpha_{[\beta} \delta^\gamma_\delta \delta^\epsilon_{\zeta]} = 0$. Contracting this equation with $\varepsilon^{\zeta\kappa} \sigma^\mu_{\kappa\dot\zeta}$ yields the only needed relation for putting the correlator into minimal form:

$$\sigma^\mu_{\epsilon\dot\zeta}\,\varepsilon_{\alpha\beta}\,\varepsilon_{\gamma\delta} + \sigma^\mu_{\epsilon\dot\zeta}\,\varepsilon_{\alpha\delta}\,\varepsilon_{\beta\gamma} + \sigma^\mu_{\alpha\dot\zeta}\,\varepsilon_{\beta\epsilon}\,\varepsilon_{\gamma\delta} + \sigma^\mu_{\alpha\dot\zeta}\,\varepsilon_{\beta\gamma}\,\varepsilon_{\delta\epsilon} + \sigma^\mu_{\gamma\dot\zeta}\,\varepsilon_{\alpha\beta}\,\varepsilon_{\delta\epsilon} - \sigma^\mu_{\gamma\dot\zeta}\,\varepsilon_{\alpha\delta}\,\varepsilon_{\beta\epsilon} = 0 \,. \tag{B.14}$$

Correlator $\langle \psi^\mu \, \psi^\nu \, \psi^\lambda \, \psi^\rho \, S_\alpha \, S_{\dot\beta} \, S_\gamma \, S_{\dot\delta} \rangle$

This correlation function can be expressed in terms of permutations in the spinor indices of $\eta^{\mu\nu}\,\sigma^\lambda_{\alpha\dot\beta}\,\sigma^\rho_{\gamma\dot\delta}$, $(\sigma^\mu \bar\sigma^\nu \varepsilon)_{\alpha\gamma}\,(\varepsilon\,\bar\sigma^\lambda\,\sigma^\rho)_{\dot\beta\dot\delta}$, $(\sigma^\mu \bar\sigma^\nu \sigma^\lambda)_{\alpha\dot\beta}\,\sigma^\rho_{\gamma\dot\delta}$ and $(\sigma^\mu \bar\sigma^\nu \sigma^\lambda \bar\sigma^\rho \varepsilon)_{\alpha\gamma}\,\varepsilon_{\dot\beta\dot\delta}$. The two expressions containing four σ matrices can be eliminated with the help of (B.5):

$$(\sigma^\mu \bar\sigma^\nu \sigma^\lambda \bar\sigma^\rho \varepsilon)_{\alpha\gamma}\,\varepsilon_{\dot\beta\dot\delta} = (\sigma^\mu \bar\sigma^\nu \sigma^\lambda)_{\alpha\dot\beta}\,\sigma^\rho_{\gamma\dot\delta} - (\sigma^\mu \bar\sigma^\nu \sigma^\lambda)_{\alpha\dot\delta}\,\sigma^\rho_{\gamma\dot\beta} \,, \tag{B.15a}$$

$$(\varepsilon\,\bar\sigma^\mu\,\sigma^\nu\,\bar\sigma^\lambda\,\sigma^\rho)_{\dot\beta\dot\delta}\,\varepsilon_{\alpha\gamma} = (\sigma^\nu\,\bar\sigma^\lambda\,\sigma^\rho)_{\gamma\dot\delta}\,\sigma^\mu_{\alpha\dot\beta} - (\sigma^\nu\,\bar\sigma^\lambda\,\sigma^\rho)_{\alpha\dot\delta}\,\sigma^\mu_{\gamma\dot\beta} \,. \tag{B.15b}$$

Relations between the Clebsch–Gordan coefficients consisting of three σ matrices are found by decomposing terms of the type $(\sigma\,\bar\sigma\,\varepsilon)\,(\varepsilon\,\bar\sigma\,\sigma)$ using (B.5). Applying these relations either to the first or the second σ-chain results in

$$(\sigma^\mu \bar\sigma^\nu \varepsilon)_{\alpha\gamma}\,(\varepsilon\,\bar\sigma^\lambda\,\sigma^\rho)_{\dot\beta\dot\delta} = \begin{cases} (\sigma^\mu \bar\sigma^\nu \sigma^\lambda)_{\alpha\dot\beta}\,\sigma^\rho_{\gamma\dot\delta} - (\sigma^\mu \bar\sigma^\nu \sigma^\rho)_{\alpha\dot\delta}\,\sigma^\lambda_{\gamma\dot\beta} \,, \\ (\sigma^\nu\,\bar\sigma^\lambda\,\sigma^\rho)_{\gamma\dot\delta}\,\sigma^\mu_{\alpha\dot\beta} - (\sigma^\mu\,\bar\sigma^\lambda\,\sigma^\rho)_{\alpha\dot\delta}\,\sigma^\nu_{\gamma\dot\beta} \,. \end{cases} \tag{B.16}$$

Now we can write down relations between the terms on the right hand side of (B.16):

$$(\sigma^\mu \bar\sigma^\nu \sigma^\lambda)_{\alpha\dot\beta}\,\sigma^\rho_{\gamma\dot\delta} = (\sigma^\nu\,\bar\sigma^\lambda\,\sigma^\rho)_{\gamma\dot\delta}\,\sigma^\mu_{\alpha\dot\beta} - (\sigma^\mu\,\bar\sigma^\lambda\,\sigma^\rho)_{\alpha\dot\delta}\,\sigma^\nu_{\gamma\dot\beta} + (\sigma^\mu\,\bar\sigma^\nu\,\sigma^\rho)_{\alpha\dot\delta}\,\sigma^\lambda_{\gamma\dot\beta} \,, \tag{B.17a}$$

$$(\sigma^\mu \bar\sigma^\nu \sigma^\lambda)_{\alpha\dot\delta}\,\sigma^\rho_{\gamma\dot\beta} = (\sigma^\nu\,\bar\sigma^\lambda\,\sigma^\rho)_{\gamma\dot\beta}\,\sigma^\mu_{\alpha\dot\delta} - (\sigma^\mu\,\bar\sigma^\lambda\,\sigma^\rho)_{\alpha\dot\beta}\,\sigma^\nu_{\gamma\dot\delta} + (\sigma^\mu\,\bar\sigma^\nu\,\sigma^\rho)_{\alpha\dot\beta}\,\sigma^\lambda_{\gamma\dot\delta} \,, \tag{B.17b}$$

$$(\sigma^\mu \bar\sigma^\nu \sigma^\lambda)_{\gamma\dot\delta}\,\sigma^\rho_{\alpha\dot\beta} = (\sigma^\nu\,\bar\sigma^\lambda\,\sigma^\rho)_{\alpha\dot\beta}\,\sigma^\mu_{\gamma\dot\delta} - (\sigma^\mu\,\bar\sigma^\lambda\,\sigma^\rho)_{\gamma\dot\beta}\,\sigma^\nu_{\alpha\dot\delta} + (\sigma^\mu\,\bar\sigma^\nu\,\sigma^\rho)_{\gamma\dot\beta}\,\sigma^\lambda_{\alpha\dot\delta} \,, \tag{B.17c}$$

$$(\sigma^\mu \bar\sigma^\nu \sigma^\lambda)_{\gamma\dot\beta}\,\sigma^\rho_{\alpha\dot\delta} = (\sigma^\nu\,\bar\sigma^\lambda\,\sigma^\rho)_{\alpha\dot\delta}\,\sigma^\mu_{\gamma\dot\beta} - (\sigma^\mu\,\bar\sigma^\lambda\,\sigma^\rho)_{\gamma\dot\delta}\,\sigma^\nu_{\alpha\dot\beta} + (\sigma^\mu\,\bar\sigma^\nu\,\sigma^\rho)_{\gamma\dot\delta}\,\sigma^\lambda_{\alpha\dot\beta} \,. \tag{B.17d}$$

B.1 Relations for $D = 4$

The last three equations were found by permuting the spinor indices in (B.16). However, one can also perform permutations in the Lorentz indices. This yields

$$2\eta^{\mu\nu}(\sigma^\lambda_{\alpha\dot\beta}\sigma^\rho_{\gamma\dot\delta} - \sigma^\lambda_{\gamma\dot\beta}\sigma^\rho_{\alpha\dot\delta}) = (\sigma^\nu\bar\sigma^\lambda\sigma^\rho)_{\alpha\dot\delta}\sigma^\mu_{\gamma\dot\beta} - (\sigma^\nu\bar\sigma^\lambda\sigma^\rho)_{\gamma\dot\delta}\sigma^\mu_{\alpha\dot\beta}$$
$$+(\sigma^\mu\bar\sigma^\lambda\sigma^\rho)_{\alpha\dot\delta}\sigma^\nu_{\gamma\dot\beta} - (\sigma^\mu\bar\sigma^\lambda\sigma^\rho)_{\gamma\dot\delta}\sigma^\nu_{\alpha\dot\beta}, \quad \text{(B.18a)}$$

$$2\eta^{\mu\nu}(\sigma^\lambda_{\alpha\dot\delta}\sigma^\rho_{\gamma\dot\beta} - \sigma^\lambda_{\gamma\dot\delta}\sigma^\rho_{\alpha\dot\beta}) = (\sigma^\nu\bar\sigma^\lambda\sigma^\rho)_{\alpha\dot\beta}\sigma^\mu_{\gamma\dot\delta} - (\sigma^\nu\bar\sigma^\lambda\sigma^\rho)_{\gamma\dot\beta}\sigma^\mu_{\alpha\dot\delta}$$
$$+(\sigma^\mu\bar\sigma^\lambda\sigma^\rho)_{\alpha\dot\beta}\sigma^\nu_{\gamma\dot\delta} - (\sigma^\mu\bar\sigma^\lambda\sigma^\rho)_{\gamma\dot\beta}\sigma^\nu_{\alpha\dot\delta}, \quad \text{(B.18b)}$$

$$-2\eta^{\lambda\rho}(\sigma^\mu_{\alpha\dot\beta}\sigma^\nu_{\gamma\dot\delta} - \sigma^\mu_{\alpha\dot\delta}\sigma^\nu_{\gamma\dot\beta}) = (\sigma^\nu\bar\sigma^\lambda\sigma^\rho)_{\gamma\dot\delta}\sigma^\mu_{\alpha\dot\beta} - (\sigma^\nu\bar\sigma^\lambda\sigma^\rho)_{\gamma\dot\beta}\sigma^\mu_{\alpha\dot\delta}$$
$$+(\sigma^\mu\bar\sigma^\lambda\sigma^\rho)_{\alpha\dot\beta}\sigma^\nu_{\gamma\dot\delta} - (\sigma^\mu\bar\sigma^\lambda\sigma^\rho)_{\alpha\dot\delta}\sigma^\nu_{\gamma\dot\beta}, \quad \text{(B.18c)}$$

$$-2\eta^{\lambda\rho}(\sigma^\mu_{\gamma\dot\delta}\sigma^\nu_{\alpha\dot\beta} - \sigma^\mu_{\gamma\dot\beta}\sigma^\nu_{\alpha\dot\delta}) = (\sigma^\nu\bar\sigma^\lambda\sigma^\rho)_{\alpha\dot\beta}\sigma^\mu_{\gamma\dot\delta} - (\sigma^\nu\bar\sigma^\lambda\sigma^\rho)_{\alpha\dot\delta}\sigma^\mu_{\gamma\dot\beta}$$
$$+(\sigma^\mu\bar\sigma^\lambda\sigma^\rho)_{\gamma\dot\delta}\sigma^\nu_{\alpha\dot\beta} - (\sigma^\mu\bar\sigma^\lambda\sigma^\rho)_{\gamma\dot\beta}\sigma^\nu_{\alpha\dot\delta}, \quad \text{(B.18d)}$$

and

$$2(\eta^{\mu\lambda}\sigma^\nu_{\alpha\dot\beta}\sigma^\rho_{\gamma\dot\delta} - \eta^{\mu\lambda}\sigma^\nu_{\gamma\dot\beta}\sigma^\rho_{\alpha\dot\delta} - \eta^{\nu\lambda}\sigma^\mu_{\alpha\dot\beta}\sigma^\rho_{\gamma\dot\delta} + \eta^{\nu\lambda}\sigma^\mu_{\gamma\dot\beta}\sigma^\rho_{\alpha\dot\delta})$$
$$= -(\sigma^\nu\bar\sigma^\lambda\sigma^\rho)_{\alpha\dot\delta}\sigma^\mu_{\gamma\dot\beta} + (\sigma^\nu\bar\sigma^\lambda\sigma^\rho)_{\gamma\dot\delta}\sigma^\mu_{\alpha\dot\beta} + (\sigma^\mu\bar\sigma^\nu\sigma^\rho)_{\alpha\dot\delta}\sigma^\lambda_{\gamma\dot\beta} - (\sigma^\mu\bar\sigma^\nu\sigma^\rho)_{\gamma\dot\delta}\sigma^\lambda_{\alpha\dot\beta}, \quad \text{(B.19a)}$$

$$2(\eta^{\mu\lambda}\sigma^\nu_{\alpha\dot\delta}\sigma^\rho_{\gamma\dot\beta} - \eta^{\mu\lambda}\sigma^\nu_{\gamma\dot\delta}\sigma^\rho_{\alpha\dot\beta} - \eta^{\nu\lambda}\sigma^\mu_{\alpha\dot\delta}\sigma^\rho_{\gamma\dot\beta} + \eta^{\nu\lambda}\sigma^\mu_{\gamma\dot\delta}\sigma^\rho_{\alpha\dot\beta})$$
$$= -(\sigma^\nu\bar\sigma^\lambda\sigma^\rho)_{\alpha\dot\beta}\sigma^\mu_{\gamma\dot\delta} + (\sigma^\nu\bar\sigma^\lambda\sigma^\rho)_{\gamma\dot\beta}\sigma^\mu_{\alpha\dot\delta} + (\sigma^\mu\bar\sigma^\nu\sigma^\rho)_{\alpha\dot\beta}\sigma^\lambda_{\gamma\dot\delta} - (\sigma^\mu\bar\sigma^\nu\sigma^\rho)_{\gamma\dot\beta}\sigma^\lambda_{\alpha\dot\delta}, \quad \text{(B.19b)}$$

$$2(\eta^{\nu\lambda}\sigma^\mu_{\alpha\dot\beta}\sigma^\rho_{\gamma\dot\delta} - \eta^{\nu\lambda}\sigma^\mu_{\alpha\dot\delta}\sigma^\rho_{\gamma\dot\beta}) - \eta^{\nu\rho}\sigma^\mu_{\alpha\dot\beta}\sigma^\lambda_{\gamma\dot\delta} + \eta^{\nu\rho}\sigma^\mu_{\alpha\dot\delta}\sigma^\lambda_{\gamma\dot\beta})$$
$$= -(\sigma^\nu\bar\sigma^\lambda\sigma^\rho)_{\gamma\dot\delta}\sigma^\mu_{\alpha\dot\beta} + (\sigma^\nu\bar\sigma^\lambda\sigma^\rho)_{\gamma\dot\beta}\sigma^\mu_{\alpha\dot\delta} + (\sigma^\mu\bar\sigma^\nu\sigma^\rho)_{\alpha\dot\beta}\sigma^\lambda_{\gamma\dot\delta} - (\sigma^\mu\bar\sigma^\nu\sigma^\rho)_{\alpha\dot\delta}\sigma^\lambda_{\gamma\dot\beta}, \quad \text{(B.19c)}$$

$$2(\eta^{\nu\lambda}\sigma^\mu_{\gamma\dot\delta}\sigma^\rho_{\alpha\dot\beta} - \eta^{\nu\lambda}\sigma^\mu_{\gamma\dot\beta}\sigma^\rho_{\alpha\dot\delta} - \eta^{\nu\rho}\sigma^\mu_{\gamma\dot\delta}\sigma^\lambda_{\alpha\dot\beta} + \eta^{\nu\rho}\sigma^\mu_{\gamma\dot\beta}\sigma^\lambda_{\alpha\dot\delta})$$
$$= -(\sigma^\nu\bar\sigma^\lambda\sigma^\rho)_{\alpha\dot\beta}\sigma^\mu_{\gamma\dot\delta} + (\sigma^\nu\bar\sigma^\lambda\sigma^\rho)_{\alpha\dot\delta}\sigma^\mu_{\gamma\dot\beta} + (\sigma^\mu\bar\sigma^\nu\sigma^\rho)_{\gamma\dot\delta}\sigma^\lambda_{\alpha\dot\beta} - (\sigma^\mu\bar\sigma^\nu\sigma^\rho)_{\gamma\dot\beta}\sigma^\lambda_{\alpha\dot\delta}. \quad \text{(B.19d)}$$

In addition the following relations hold,

$$2(\eta^{\mu\rho}\sigma^\nu_{\alpha\dot\beta}\sigma^\lambda_{\gamma\dot\delta} - \eta^{\mu\rho}\sigma^\nu_{\gamma\dot\beta}\sigma^\lambda_{\alpha\dot\delta} - \eta^{\nu\rho}\sigma^\mu_{\alpha\dot\beta}\sigma^\lambda_{\gamma\dot\delta} + \eta^{\nu\rho}\sigma^\mu_{\gamma\dot\beta}\sigma^\lambda_{\alpha\dot\delta} - \eta^{\lambda\rho}\sigma^\mu_{\alpha\dot\beta}\sigma^\nu_{\gamma\dot\delta} + \eta^{\lambda\rho}\sigma^\mu_{\gamma\dot\beta}\sigma^\nu_{\alpha\dot\delta})$$
$$= -(\sigma^\nu\bar\sigma^\lambda\sigma^\rho)_{\alpha\dot\delta}\sigma^\mu_{\gamma\dot\beta} + (\sigma^\nu\bar\sigma^\lambda\sigma^\rho)_{\alpha\dot\delta}\sigma^\mu_{\gamma\dot\beta} - (\sigma^\nu\bar\sigma^\lambda\sigma^\rho)_{\gamma\dot\delta}\sigma^\mu_{\alpha\dot\beta} + (\sigma^\nu\bar\sigma^\lambda\sigma^\rho)_{\gamma\dot\beta}\sigma^\mu_{\alpha\dot\delta}$$
$$-(\sigma^\mu\bar\sigma^\lambda\sigma^\rho)_{\alpha\dot\delta}\sigma^\nu_{\gamma\dot\beta} + (\sigma^\mu\bar\sigma^\lambda\sigma^\rho)_{\gamma\dot\beta}\sigma^\nu_{\alpha\dot\delta} + (\sigma^\mu\bar\sigma^\nu\sigma^\rho)_{\alpha\dot\beta}\sigma^\lambda_{\gamma\dot\delta} - (\sigma^\mu\bar\sigma^\nu\sigma^\rho)_{\gamma\dot\beta}\sigma^\lambda_{\alpha\dot\delta}, \quad \text{(B.20a)}$$

$$2(\eta^{\mu\rho}\sigma^\nu_{\alpha\dot\delta}\sigma^\lambda_{\gamma\dot\beta} - \eta^{\mu\rho}\sigma^\nu_{\gamma\dot\delta}\sigma^\lambda_{\alpha\dot\beta} - \eta^{\nu\rho}\sigma^\mu_{\alpha\dot\delta}\sigma^\lambda_{\gamma\dot\beta} + \eta^{\nu\rho}\sigma^\mu_{\gamma\dot\delta}\sigma^\lambda_{\alpha\dot\beta} + \eta^{\lambda\rho}\sigma^\mu_{\alpha\dot\delta}\sigma^\nu_{\gamma\dot\beta} - \eta^{\lambda\rho}\sigma^\mu_{\gamma\dot\delta}\sigma^\nu_{\alpha\dot\beta})$$
$$= (\sigma^\nu\bar\sigma^\lambda\sigma^\rho)_{\alpha\dot\beta}\sigma^\mu_{\gamma\dot\delta} - (\sigma^\nu\bar\sigma^\lambda\sigma^\rho)_{\alpha\dot\delta}\sigma^\mu_{\gamma\dot\beta} + (\sigma^\nu\bar\sigma^\lambda\sigma^\rho)_{\gamma\dot\delta}\sigma^\mu_{\alpha\dot\beta} - (\sigma^\nu\bar\sigma^\lambda\sigma^\rho)_{\gamma\dot\beta}\sigma^\mu_{\alpha\dot\delta}$$
$$-(\sigma^\mu\bar\sigma^\lambda\sigma^\rho)_{\alpha\dot\delta}\sigma^\nu_{\gamma\dot\beta} + (\sigma^\mu\bar\sigma^\lambda\sigma^\rho)_{\gamma\dot\delta}\sigma^\nu_{\alpha\dot\beta} + (\sigma^\mu\bar\sigma^\nu\sigma^\rho)_{\alpha\dot\delta}\sigma^\lambda_{\gamma\dot\beta} - (\sigma^\mu\bar\sigma^\nu\sigma^\rho)_{\gamma\dot\delta}\sigma^\lambda_{\alpha\dot\beta}, \quad \text{(B.20b)}$$

as well as

$$2 \left(\eta^{\mu\nu} \sigma^{\lambda}_{\alpha\dot{\beta}} \sigma^{\rho}_{\gamma\dot{\delta}} - \eta^{\mu\nu} \sigma^{\lambda}_{\gamma\dot{\delta}} \sigma^{\rho}_{\alpha\dot{\beta}} - \eta^{\mu\lambda} \sigma^{\nu}_{\alpha\dot{\beta}} \sigma^{\rho}_{\gamma\dot{\delta}} + \eta^{\mu\lambda} \sigma^{\nu}_{\alpha\dot{\delta}} \sigma^{\rho}_{\gamma\dot{\beta}} + \eta^{\mu\rho} \sigma^{\nu}_{\alpha\dot{\beta}} \sigma^{\lambda}_{\gamma\dot{\delta}} - \eta^{\mu\rho} \sigma^{\nu}_{\alpha\dot{\delta}} \sigma^{\lambda}_{\gamma\dot{\beta}} \right)$$
$$= (\sigma^{\nu} \bar{\sigma}^{\lambda} \sigma^{\rho})_{\alpha\dot{\delta}} \sigma^{\mu}_{\gamma\dot{\beta}} - (\sigma^{\mu} \bar{\sigma}^{\lambda} \sigma^{\rho})_{\gamma\dot{\beta}} \sigma^{\nu}_{\alpha\dot{\delta}} + (\sigma^{\mu} \bar{\sigma}^{\nu} \sigma^{\rho})_{\alpha\dot{\beta}} \sigma^{\lambda}_{\gamma\dot{\delta}} - (\sigma^{\mu} \bar{\sigma}^{\nu} \sigma^{\lambda})_{\alpha\dot{\beta}} \sigma^{\rho}_{\gamma\dot{\delta}}. \qquad (B.21)$$

Using the identities stated above it is possible to arrive at a set of 25 independent index terms for this correlator.

Correlator $\langle \psi^{\mu} \psi^{\nu} \psi^{\lambda} \psi^{\rho} S_{\alpha} S_{\beta} S_{\gamma} S_{\delta} \rangle$

The index terms $\eta^{\mu\nu} \eta^{\lambda\rho} \varepsilon_{\alpha\beta} \varepsilon_{\gamma\delta}$, $\eta^{\mu\nu} (\sigma^{\lambda} \bar{\sigma}^{\rho} \varepsilon)_{\alpha\beta} \varepsilon_{\gamma\delta}$, $(\sigma^{\mu} \bar{\sigma}^{\nu} \varepsilon)_{\alpha\beta} (\sigma^{\lambda} \bar{\sigma}^{\rho} \varepsilon)_{\gamma\delta}$ and the four σ-chains $(\sigma^{\mu} \bar{\sigma}^{\nu} \sigma^{\lambda} \bar{\sigma}^{\rho} \varepsilon)_{\alpha\beta} \varepsilon_{\gamma\delta}$ arise for this correlator. Additionally, permutations of these terms in the vector and spinor indices appear. Again, the Fierz identity (B.4) can be used to eliminate terms of the type $\eta\eta \varepsilon_{\alpha\gamma} \varepsilon_{\beta\delta}$. Further eliminations can be accomplished by applying (B.7). In particular (B.7b) gives rise to new relations between the terms $\eta (\sigma \bar{\sigma} \varepsilon) \varepsilon$:

$$-2 \eta^{\mu\nu} \eta^{\lambda\rho} \varepsilon_{\alpha\beta} \varepsilon_{\gamma\delta} = \begin{cases} \eta^{\mu\nu} (\sigma^{\lambda} \bar{\sigma}^{\rho} \varepsilon)_{\alpha\beta} \varepsilon_{\gamma\delta} - \eta^{\mu\nu} (\sigma^{\lambda} \bar{\sigma}^{\rho} \varepsilon)_{\beta\gamma} \varepsilon_{\alpha\delta} + \eta^{\mu\nu} (\sigma^{\lambda} \bar{\sigma}^{\rho} \varepsilon)_{\beta\delta} \varepsilon_{\alpha\gamma}, \\ \eta^{\lambda\rho} (\sigma^{\mu} \bar{\sigma}^{\nu} \varepsilon)_{\alpha\beta} \varepsilon_{\gamma\delta} - \eta^{\lambda\rho} (\sigma^{\mu} \bar{\sigma}^{\nu} \varepsilon)_{\beta\gamma} \varepsilon_{\alpha\delta} + \eta^{\lambda\rho} (\sigma^{\mu} \bar{\sigma}^{\nu} \varepsilon)_{\beta\delta} \varepsilon_{\alpha\gamma}. \end{cases} \qquad (B.22)$$

Tensors consisting of two separate σ-chains satisfy

$$(\sigma^{\mu} \bar{\sigma}^{\nu} \varepsilon)_{\gamma\beta} (\sigma^{\lambda} \bar{\sigma}^{\rho} \varepsilon)_{\alpha\delta} - (\sigma^{\mu} \bar{\sigma}^{\nu} \varepsilon)_{\alpha\beta} (\sigma^{\lambda} \bar{\sigma}^{\rho} \varepsilon)_{\gamma\delta}$$
$$= \varepsilon_{\alpha\gamma} \left[(\sigma^{\mu} \bar{\sigma}^{\nu} \sigma^{\lambda} \bar{\sigma}^{\rho} \varepsilon)_{\beta\delta} + 2 \eta^{\mu\nu} (\sigma^{\lambda} \bar{\sigma}^{\rho} \varepsilon)_{\beta\delta} \right], \qquad (B.23a)$$

$$(\sigma^{\mu} \bar{\sigma}^{\nu} \varepsilon)_{\alpha\delta} (\sigma^{\lambda} \bar{\sigma}^{\rho} \varepsilon)_{\gamma\beta} - (\sigma^{\mu} \bar{\sigma}^{\nu} \varepsilon)_{\gamma\delta} (\sigma^{\lambda} \bar{\sigma}^{\rho} \varepsilon)_{\alpha\beta}$$
$$= \varepsilon_{\alpha\gamma} \left[(\sigma^{\mu} \bar{\sigma}^{\nu} \sigma^{\lambda} \bar{\sigma}^{\rho} \varepsilon)_{\beta\delta} - 4 \eta^{\mu\rho} \eta^{\nu\rho} \varepsilon_{\beta\delta} + 4 \eta^{\mu\rho} \eta^{\nu\lambda} \varepsilon_{\beta\delta} \right.$$
$$- 2 \eta^{\mu\lambda} (\sigma^{\nu} \bar{\sigma}^{\rho} \varepsilon)_{\beta\delta} + 2 \eta^{\mu\rho} (\sigma^{\nu} \bar{\sigma}^{\lambda} \varepsilon)_{\beta\delta} + 2 \eta^{\lambda\rho} (\sigma^{\mu} \bar{\sigma}^{\nu} \varepsilon)_{\beta\delta}$$
$$\left. - 2 \eta^{\nu\rho} (\sigma^{\mu} \bar{\sigma}^{\lambda} \varepsilon)_{\beta\delta} + 2 \eta^{\nu\lambda} (\sigma^{\mu} \bar{\sigma}^{\rho} \varepsilon)_{\beta\delta} \right], \qquad (B.23b)$$

which stem from $(\sigma^{\mu} \bar{\sigma}^{\nu})_{\alpha}{}^{\kappa} \varepsilon_{[\kappa\beta} (\sigma^{\lambda} \bar{\sigma}^{\rho} \varepsilon)_{\gamma]\delta} = 0$. Applying the same technique to the four-σ terms, namely $(\sigma^{\mu} \bar{\sigma}^{\nu} \sigma^{\lambda} \bar{\sigma}^{\rho} \varepsilon)_{\alpha[\beta} \varepsilon_{\gamma\delta]} = 0$, yields the equations

$$(\sigma^{\mu} \bar{\sigma}^{\nu} \sigma^{\lambda} \bar{\sigma}^{\rho} \varepsilon)_{\alpha\gamma} \varepsilon_{\beta\delta} = (\sigma^{\mu} \bar{\sigma}^{\nu} \sigma^{\lambda} \bar{\sigma}^{\rho} \varepsilon)_{\alpha\beta} \varepsilon_{\gamma\delta} - (\sigma^{\mu} \bar{\sigma}^{\nu} \sigma^{\lambda} \bar{\sigma}^{\rho} \varepsilon)_{\alpha\delta} \varepsilon_{\gamma\beta}, \qquad (B.24a)$$

$$(\sigma^{\mu} \bar{\sigma}^{\nu} \sigma^{\lambda} \bar{\sigma}^{\rho} \varepsilon)_{\beta\delta} \varepsilon_{\alpha\gamma} = (\sigma^{\mu} \bar{\sigma}^{\nu} \sigma^{\lambda} \bar{\sigma}^{\rho} \varepsilon)_{\gamma\delta} \varepsilon_{\alpha\beta} - (\sigma^{\mu} \bar{\sigma}^{\nu} \sigma^{\lambda} \bar{\sigma}^{\rho} \varepsilon)_{\gamma\beta} \varepsilon_{\alpha\delta}. \qquad (B.24b)$$

From permutations of the spinor indices in the relations above we find

$$(\sigma^{\mu} \bar{\sigma}^{\nu} \sigma^{\lambda} \bar{\sigma}^{\rho} \varepsilon)_{\alpha\beta} \varepsilon_{\gamma\delta} - (\sigma^{\mu} \bar{\sigma}^{\nu} \sigma^{\lambda} \bar{\sigma}^{\rho} \varepsilon)_{\alpha\delta} \varepsilon_{\gamma\beta} + (\sigma^{\mu} \bar{\sigma}^{\nu} \sigma^{\lambda} \bar{\sigma}^{\rho} \varepsilon)_{\gamma\beta} \varepsilon_{\alpha\delta} - (\sigma^{\mu} \bar{\sigma}^{\nu} \sigma^{\lambda} \bar{\sigma}^{\rho} \varepsilon)_{\gamma\delta} \varepsilon_{\alpha\beta}$$

B.1 Relations for $D = 4$

$$= -\varepsilon_{\beta\delta} \Big[2\,\eta^{\mu\nu} \,(\sigma^\lambda\,\bar\sigma^\rho\,\varepsilon)_{\alpha\gamma} - 2\,\eta^{\mu\lambda} \,(\sigma^\nu\,\bar\sigma^\rho\,\varepsilon)_{\alpha\gamma} + 2\,\eta^{\mu\rho} \,(\sigma^\nu\,\bar\sigma^\lambda\,\varepsilon)_{\alpha\gamma}$$
$$+ 2\,\eta^{\lambda\rho} \,(\sigma^\mu\,\bar\sigma^\nu\,\varepsilon)_{\alpha\gamma} - 2\,\eta^{\nu\rho} \,(\sigma^\mu\,\bar\sigma^\lambda\,\varepsilon)_{\alpha\gamma} + 2\,\eta^{\nu\lambda} \,(\sigma^\mu\,\bar\sigma^\rho\,\varepsilon)_{\alpha\gamma}$$
$$+ 4\,\eta^{\mu\nu}\,\eta^{\lambda\rho}\,\varepsilon_{\alpha\gamma} - 4\,\eta^{\mu\lambda}\,\eta^{\nu\rho}\,\varepsilon_{\alpha\gamma} + 4\,\eta^{\mu\rho}\,\eta^{\nu\lambda}\,\varepsilon_{\alpha\gamma} \Big] \,. \tag{B.25}$$

Note that the spinor indices attached to the four-σ term can be interchanged by

$$(\sigma^\mu\,\bar\sigma^\nu\,\sigma^\lambda\,\bar\sigma^\rho\,\varepsilon)_{\beta\alpha} = -(\sigma^\rho\,\bar\sigma^\lambda\,\sigma^\nu\,\bar\sigma^\mu\,\varepsilon)_{\alpha\beta}$$
$$= -(\sigma^\mu\,\bar\sigma^\nu\,\sigma^\lambda\,\bar\sigma^\rho\,\varepsilon)_{\alpha\beta} - 2\,\eta^{\mu\nu}\,(\sigma^\lambda\,\bar\sigma^\rho\,\varepsilon)_{\alpha\beta} + 2\,\eta^{\mu\lambda}\,(\sigma^\nu\,\bar\sigma^\rho\,\varepsilon)_{\alpha\beta} - 2\,\eta^{\mu\rho}\,(\sigma^\nu\,\bar\sigma^\lambda\,\varepsilon)_{\alpha\beta}$$
$$- 2\,\eta^{\lambda\rho}\,(\sigma^\mu\,\bar\sigma^\nu\,\varepsilon)_{\alpha\beta} + 2\,\eta^{\nu\rho}\,(\sigma^\mu\,\bar\sigma^\lambda\,\varepsilon)_{\alpha\beta} - 2\,\eta^{\nu\lambda}\,(\sigma^\mu\,\bar\sigma^\rho\,\varepsilon)_{\alpha\beta}$$
$$- 4\,\eta^{\mu\nu}\,\eta^{\lambda\rho}\,\varepsilon_{\alpha\beta} + 4\,\eta^{\mu\lambda}\,\eta^{\nu\rho}\,\varepsilon_{\alpha\beta} - 4\,\eta^{\mu\rho}\,\eta^{\nu\lambda}\,\varepsilon_{\alpha\beta} \,, \tag{B.26}$$

where we have successively made use of the Clifford algebra (B.1). Finally, poles in $z_{13}\,z_{24}$ in the result of this eight-point function can be removed by the following identities:

$$-2\,\eta^{\nu\rho}\,(\sigma^\mu\,\bar\sigma^\lambda\,\varepsilon)_{\alpha\gamma}\,\varepsilon_{\beta\delta} = (\sigma^\mu\,\bar\sigma^\nu\,\varepsilon)_{\alpha\beta}\,(\sigma^\lambda\,\bar\sigma^\rho\,\varepsilon)_{\gamma\delta} - (\sigma^\mu\,\bar\sigma^\nu\,\varepsilon)_{\alpha\delta}\,(\sigma^\lambda\,\bar\sigma^\rho\,\varepsilon)_{\gamma\beta}$$
$$+ (\sigma^\mu\,\bar\sigma^\rho\,\varepsilon)_{\alpha\beta}\,(\sigma^\lambda\,\bar\sigma^\nu\,\varepsilon)_{\gamma\delta} - (\sigma^\mu\,\bar\sigma^\rho\,\varepsilon)_{\alpha\delta}\,(\sigma^\lambda\,\bar\sigma^\nu\,\varepsilon)_{\gamma\beta} \,, \tag{B.27a}$$

$$-2\,\eta^{\mu\lambda}\,(\sigma^\nu\,\bar\sigma^\rho\,\varepsilon)_{\beta\delta}\,\varepsilon_{\alpha\gamma} = (\sigma^\mu\,\bar\sigma^\nu\,\varepsilon)_{\alpha\beta}\,(\sigma^\lambda\,\bar\sigma^\rho\,\varepsilon)_{\gamma\delta} - (\sigma^\mu\,\bar\sigma^\nu\,\varepsilon)_{\gamma\beta}\,(\sigma^\lambda\,\bar\sigma^\rho\,\varepsilon)_{\alpha\delta}$$
$$+ (\sigma^\lambda\,\bar\sigma^\nu\,\varepsilon)_{\alpha\beta}\,(\sigma^\mu\,\bar\sigma^\rho\,\varepsilon)_{\gamma\delta} - (\sigma^\lambda\,\bar\sigma^\nu\,\varepsilon)_{\gamma\beta}\,(\sigma^\mu\,\bar\sigma^\rho\,\varepsilon)_{\alpha\delta} \,, \tag{B.27b}$$

$$4\,\eta^{\mu\lambda}\,\eta^{\nu\rho}\,\varepsilon_{\alpha\gamma}\,\varepsilon_{\beta\delta} = (\sigma^\mu\,\bar\sigma^\nu\,\varepsilon)_{\alpha\beta}\,(\sigma^\lambda\,\bar\sigma^\rho\,\varepsilon)_{\gamma\delta} - (\sigma^\mu\,\bar\sigma^\nu\,\varepsilon)_{\gamma\beta}\,(\sigma^\lambda\,\bar\sigma^\rho\,\varepsilon)_{\alpha\delta}$$
$$+ (\sigma^\mu\,\bar\sigma^\nu\,\varepsilon)_{\gamma\delta}\,(\sigma^\lambda\,\bar\sigma^\rho\,\varepsilon)_{\alpha\beta} - (\sigma^\mu\,\bar\sigma^\nu\,\varepsilon)_{\alpha\delta}\,(\sigma^\lambda\,\bar\sigma^\rho\,\varepsilon)_{\beta\gamma}$$
$$+ (\sigma^\lambda\,\bar\sigma^\nu\,\varepsilon)_{\alpha\beta}\,(\sigma^\mu\,\bar\sigma^\rho\,\varepsilon)_{\gamma\delta} - (\sigma^\lambda\,\bar\sigma^\nu\,\varepsilon)_{\gamma\beta}\,(\sigma^\mu\,\bar\sigma^\rho\,\varepsilon)_{\alpha\delta}$$
$$+ (\sigma^\lambda\,\bar\sigma^\nu\,\varepsilon)_{\gamma\delta}\,(\sigma^\mu\,\bar\sigma^\rho\,\varepsilon)_{\alpha\beta} - (\sigma^\lambda\,\bar\sigma^\nu\,\varepsilon)_{\alpha\delta}\,(\sigma^\mu\,\bar\sigma^\rho\,\varepsilon)_{\gamma\beta} \,. \tag{B.27c}$$

Correlator $\langle \psi^\mu\,\psi^\nu\,S_\alpha\,S_\beta\,S_\gamma\,S_\delta\,S_\epsilon\,S_\zeta \rangle$

The index terms appearing in the calculation of this correlation function have either the form $\eta^{\mu\nu}\,\varepsilon_{\alpha\beta}\,\varepsilon_{\gamma\delta}\,\varepsilon_{\epsilon\zeta}$ or $(\sigma^\mu\,\bar\sigma^\nu\,\varepsilon)_{\alpha\beta}\,\varepsilon_{\gamma\delta}\,\varepsilon_{\delta\zeta}$. The former can be reduced to five terms by virtue of (B.13). For the latter terms we use (B.7a):

$$(\sigma^\mu\,\bar\sigma^\nu\,\varepsilon)_{\alpha\gamma}\,\varepsilon_{\beta\delta}\,\varepsilon_{\epsilon\zeta} = (\sigma^\mu\,\bar\sigma^\nu\,\varepsilon)_{\alpha\beta}\,\varepsilon_{\gamma\delta}\,\varepsilon_{\epsilon\zeta} - (\sigma^\mu\,\bar\sigma^\nu\,\varepsilon)_{\alpha\delta}\,\varepsilon_{\gamma\beta}\,\varepsilon_{\epsilon\zeta} \,. \tag{B.28}$$

Permuting the spinor indices in this identity provides, together with the Fierz identity (B.4), enough relations to reduce the index terms down to a minimal set of 14.

B.2 Relations for $D = 6$

In six dimensions, tensors with four spinor indices are severely constrained by Fierz identities presented in Appendix A. In the following we list the relevant relations between different index terms which are needed for our calculations of RNS correlators in six dimensions.

Correlator $\langle S_\alpha S_\beta S_\gamma S_\delta \rangle$

The most important relation between $(\gamma^\mu C)_{\alpha\beta} (\gamma_\mu C)_{\gamma\delta}$ and permutations in the spinor indices of this tensor arises from (A.23) with $\psi_\alpha = (\gamma^\mu C)_{\alpha\gamma}$ and $\chi_\beta = (\gamma_\mu C)_{\beta\delta}$. After some manipulations, one arrives at[1]

$$(\gamma^\mu C)_{\alpha\gamma} (\gamma_\mu C)_{\beta\delta} = (\gamma^\mu C)_{\beta\alpha} (\gamma_\mu C)_{\gamma\delta} \,. \tag{B.30}$$

Together with the antisymmetry of $(\gamma^\mu C)_{\alpha\beta} = (\gamma^\mu C)_{[\alpha\beta]}$, this implies that the contraction $(\gamma^\mu C)_{\alpha\beta} (\gamma_\mu C)_{\gamma\delta}$ is totally antisymmetric in the four spinor indices and therefore proportional to the ε tensor in the four-dimensional chiral spinor representations. Normalizing $\varepsilon_{1234} = 1$, we get

$$(\gamma^\mu C)_{\alpha\beta} (\gamma_\mu C)_{\gamma\delta} = -2\, \varepsilon_{\alpha\beta\gamma\delta} \,. \tag{B.31}$$

Correlator $\langle S_\alpha S_\beta S^{\dot\gamma} S^{\dot\delta} \rangle$

The choices $\psi_\alpha = C_\alpha{}^{\dot\gamma}$, $\chi_\beta = C_\beta{}^{\dot\delta}$ in (A.23) and $\psi_\alpha = (\gamma^\mu C)_{\alpha\gamma}$, $\bar\chi^{\dot\beta} = (\bar\gamma_\mu C)^{\dot\beta\dot\delta}$ in (A.24) yield the set of equations:

$$(\gamma^\mu C)_{\alpha\beta} (\bar\gamma_\mu C)^{\dot\gamma\dot\delta} = 2 \left(C_\alpha{}^{\dot\gamma} C_\beta{}^{\dot\delta} - C_\alpha{}^{\dot\delta} C_\beta{}^{\dot\gamma} \right) , \tag{B.32a}$$

$$(\gamma^{\mu\nu} C)_\alpha{}^{\dot\gamma} (\gamma_{\mu\nu} C)_\beta{}^{\dot\delta} = 2\, C_\alpha{}^{\dot\gamma} C_\beta{}^{\dot\delta} - 8\, C_\alpha{}^{\dot\delta} C_\beta{}^{\dot\gamma} , \tag{B.32b}$$

$$(\gamma^{\mu\nu\lambda} C)_{\alpha\beta} (\bar\gamma_{\mu\nu\lambda} C)^{\dot\gamma\dot\delta} = 24 \left(C_\alpha{}^{\dot\delta} C_\beta{}^{\dot\gamma} + C_\alpha{}^{\dot\gamma} C_\beta{}^{\dot\delta} \right) . \tag{B.32c}$$

Hence, only $C_\alpha{}^{\dot\delta} C_\beta{}^{\dot\gamma}$ and $C_\alpha{}^{\dot\gamma} C_\beta{}^{\dot\delta}$ remain as independent Clebsch–Gordan coefficients.

Correlator $\langle \psi^\mu S_\alpha S_\beta S_\gamma S^{\dot\delta} \rangle$

For this correlation function equations of the type

$$(\gamma^\mu \bar\gamma^\nu C)_\gamma{}^{\dot\delta} (\gamma_\nu C)_{\alpha\beta} = 2\, (\gamma^\mu C)_{\beta\gamma} C_\alpha{}^{\dot\delta} - 2\, (\gamma^\mu C)_{\alpha\gamma} C_\beta{}^{\dot\delta} \tag{B.33}$$

---
[1] Useful relations in the derivation are

$$\Gamma^\mu \, \Gamma^{\nu_1 \ldots \nu_m} \, \Gamma_\mu = (-1)^{m-1} (D - 2m) \, \Gamma^{\nu_1 \ldots \nu_m} \,, \tag{B.29a}$$

$$\Gamma^{\mu\nu} \, \Gamma^{\lambda_1 \ldots \lambda_m} \, \Gamma_{\mu\nu} = \left(D - (D - 2m)^2 \right) \Gamma^{\lambda_1 \ldots \lambda_m} \,. \tag{B.29b}$$

B.2 Relations for $D = 6$

prove to be useful in order to eliminate index terms containing three γ matrices. These relations can be derived by multiplying (B.32a) with $\gamma^\mu_{\gamma\dot\gamma}$.

Correlator $\langle S_\alpha S_\beta S_\gamma S_\delta S_\epsilon S^i \rangle$

As Weyl spinors in six dimensions only have $2^{6/2-1} = 4$ independent components, the relation

$$(\gamma^\mu C)_{[\alpha\beta} (\gamma_\mu C)_{\gamma\delta} C_\epsilon\right]^i = 0 \tag{B.34}$$

holds for this correlator.

Correlator $\langle \psi^\mu \psi^\nu S_\alpha S_\beta S_\gamma S_\delta \rangle$

This correlator can be expressed in terms of $\eta^{\mu\nu} (\gamma^\lambda C)_{\alpha\beta} (\gamma_\lambda C)_{\gamma\delta}$, $(\gamma^\mu C)_{\alpha\beta} (\gamma^\nu C)_{\gamma\delta}$ and permutations in $\alpha, \beta, \gamma, \delta$ thereof. These are related by the equation

$$\eta^{\mu\nu} (\gamma^\lambda C)_{\alpha\beta} (\gamma_\lambda C)_{\gamma\delta} = -(\gamma^\mu C)_{[\alpha\beta} (\gamma^\nu C)_{\gamma\delta]}. \tag{B.35}$$

Correlator $\langle \psi^\mu \psi^\nu S_\alpha S_\beta S^{\dot\gamma} S^{\dot\delta} \rangle$

In this case ten index terms appear. They are of the form $\eta^{\mu\nu} C_\alpha{}^{\dot\gamma} C_\beta{}^{\dot\delta}$, $(\gamma^\mu C)_{\alpha\beta} (\bar\gamma^\nu C)^{\dot\gamma\dot\delta}$, $(\gamma^\nu C)_{\alpha\beta} (\bar\gamma^\mu C)^{\dot\gamma\dot\delta}$, $(\gamma^\mu \bar\gamma^\nu C)_\alpha{}^{\dot\gamma} C_\beta{}^{\dot\delta}$, $(\gamma^\mu \bar\gamma^\nu \gamma^\lambda C)_{\alpha\beta} (\bar\gamma_\lambda C)^{\dot\gamma\dot\delta}$ and $(\bar\gamma^\mu \gamma^\nu \bar\gamma^\lambda C)^{\dot\gamma\dot\delta} (\gamma_\lambda C)_{\alpha\beta}$ including permutations in $\alpha, \beta, \dot\gamma, \dot\delta$. Applying (B.32a) to $(\gamma^\lambda C)(\gamma_\lambda C)$ in the terms above yields

$$(\gamma^\mu \bar\gamma^\nu \gamma^\lambda C)_{\alpha\beta} (\bar\gamma_\lambda C)^{\dot\gamma\dot\delta} = 2 (\gamma^\mu \bar\gamma^\nu C)_\alpha{}^{\dot\gamma} C_\beta{}^{\dot\delta} - 2 (\gamma^\mu \bar\gamma^\nu C)_\alpha{}^{\dot\delta} C_\beta{}^{\dot\gamma}, \tag{B.36a}$$

$$(\bar\gamma^\mu \gamma^\nu \bar\gamma^\lambda C)^{\dot\gamma\dot\delta} (\gamma_\lambda C)_{\alpha\beta} = 2 (\gamma^\mu \bar\gamma^\nu C)_\beta{}^{\dot\gamma} C_\alpha{}^{\dot\delta} - 2 (\gamma^\mu \bar\gamma^\nu C)_\alpha{}^{\dot\gamma} C_\beta{}^{\dot\delta}$$
$$+ 4 \eta^{\mu\nu} C_\alpha{}^{\dot\delta} C_\beta{}^{\dot\gamma} - 4 \eta^{\mu\nu} C_\alpha{}^{\dot\gamma} C_\beta{}^{\dot\delta}. \tag{B.36b}$$

A further relation is obtained by making the previous result antisymmetric in the spinor indices α, β:

$$2 \eta^{\mu\nu} C_\alpha{}^{\dot\delta} C_\beta{}^{\dot\gamma} - 2 \eta^{\mu\nu} C_\alpha{}^{\dot\gamma} C_\beta{}^{\dot\delta} - (\gamma^\mu C)_{\alpha\beta} (\bar\gamma^\nu C)^{\dot\gamma\dot\delta} + (\gamma^\nu C)_{\alpha\beta} (\bar\gamma^\mu C)^{\dot\gamma\dot\delta} =$$
$$(\gamma^\mu \bar\gamma^\nu C)_\alpha{}^{\dot\gamma} C_\beta{}^{\dot\delta} - (\gamma^\mu \bar\gamma^\nu C)_\alpha{}^{\dot\delta} C_\beta{}^{\dot\gamma} - (\gamma^\mu \bar\gamma^\nu C)_\beta{}^{\dot\gamma} C_\alpha{}^{\dot\delta} + (\gamma^\mu \bar\gamma^\nu C)_\beta{}^{\dot\delta} C_\alpha{}^{\dot\gamma}. \tag{B.37}$$

Correlator $\langle \psi^\mu S_\alpha S_\beta S_\gamma S_\delta S^{\dot\epsilon} S^i \rangle$

The relevant index terms for this correlator are $(\bar\gamma^\mu C)^{\dot\epsilon i} \varepsilon_{\alpha\beta\gamma\delta}$, $(\gamma^\mu C)_{\alpha\beta} C_\gamma{}^{\dot\epsilon} C_\delta{}^i$ and permutations in $\alpha, \beta, \gamma, \delta$. By replacing $\dot\gamma, \dot\delta$ with $\dot\epsilon, i$ in (B.37) and multiplying with $(\gamma_\nu C)_{\gamma\delta}$ they turn out to be related:

$$2 (\bar\gamma^\mu C)^{\dot\epsilon i} \varepsilon_{\alpha\beta\gamma\delta} = -(\gamma^\mu C)_{[\alpha\beta} C_\gamma{}^{\dot\epsilon} C_{\delta]}{}^i. \tag{B.38}$$

B.3 Relations for $D = 8$

Certain tensor equations in eight dimensions can be related by $SO(8)$ triality. In addition, the Fierz identities from Appendix A.3 are very useful to derive new relations. In the following, we list the identities needed for the calculations of correlation functions with four or more spin fields.

Correlator $\langle S_\alpha S_\beta S^{\dot\gamma} S^{\dot\delta}\rangle$

Here, the choices $\psi_\alpha = (\gamma^\mu C)_\alpha{}^{\dot\gamma}$, $\chi_\beta = (\gamma_\mu C)_\beta{}^{\dot\delta}$ and $\tilde\psi_\alpha = C_{\alpha\gamma}$, $\tilde\chi^{\dot\beta} = C^{\dot\beta\dot\delta}$ in the eight-dimensional Fierz identities (A.25) and (A.26) yield

$$C_{\alpha\beta} C^{\dot\gamma\dot\delta} = \frac{1}{2}\left[(\gamma^\mu C)_\alpha{}^{\dot\gamma}(\gamma_\mu C)_\beta{}^{\dot\delta} + (\gamma^\mu C)_\alpha{}^{\dot\delta}(\gamma_\mu C)_\beta{}^{\dot\gamma}\right], \qquad (B.39a)$$

$$(\gamma^{\mu\nu} C)_{\alpha\beta}(\bar\gamma_{\mu\nu} C)^{\dot\gamma\dot\delta} = 2\left[(\gamma^\mu C)_\alpha{}^{\dot\gamma}(\gamma_\mu C)_\beta{}^{\dot\delta} - (\gamma^\mu C)_\alpha{}^{\dot\delta}(\gamma_\mu C)_\beta{}^{\dot\gamma}\right], \qquad (B.39b)$$

$$(\gamma^{\mu\nu\lambda} C)_\alpha{}^{\dot\gamma}(\gamma_{\mu\nu\lambda} C)_\beta{}^{\dot\delta} = 18\,(\gamma^\mu C)_\alpha{}^{\dot\gamma}(\gamma_\mu C)_\beta{}^{\dot\delta} + 24\,(\gamma^\mu C)_\alpha{}^{\dot\delta}(\gamma_\mu C)_\beta{}^{\dot\gamma}, \qquad (B.39c)$$

such that this correlator can be expressed through $(\gamma^\mu C)_\alpha{}^{\dot\gamma}(\gamma_\mu C)_\beta{}^{\dot\delta}$ and $(\gamma^\mu C)_\alpha{}^{\dot\delta}(\gamma_\mu C)_\beta{}^{\dot\gamma}$.

Correlator $\langle S_\alpha S_\beta S_\gamma S_\delta\rangle$

The Fierz identities for $\psi_\alpha = C_{\alpha\gamma}$ and $\chi_\beta = C_{\beta\delta}$ lead to

$$(\gamma^{\mu\nu} C)_{\alpha\beta}(\gamma_{\mu\nu} C)_{\gamma\delta} = 8\left(C_{\alpha\gamma}C_{\beta\delta} - C_{\alpha\delta}C_{\beta\gamma}\right), \qquad (B.40a)$$

$$(\gamma^{\mu\nu\rho} C)_{\alpha\beta}(\gamma_{\mu\nu\rho} C)_{\gamma\delta} = 192\left(C_{\alpha\gamma}C_{\beta\delta} + C_{\alpha\delta}C_{\beta\gamma}\right) - 48\,C_{\alpha\beta}C_{\gamma\delta}. \qquad (B.40b)$$

Correlator $\langle\psi^\mu S_\alpha S_\beta S_\gamma S^{\dot\delta}\rangle$

Multiplying (B.39a) with $\gamma^\mu_{\gamma\dot\gamma}$ gives

$$C_{\alpha\beta}(\gamma^\mu C)_\gamma{}^{\dot\delta} = -\frac{1}{2}\left[(\gamma^\lambda \bar\gamma^\mu C)_{\alpha\gamma}(\gamma_\lambda C)_\beta{}^{\dot\delta} + (\gamma^\lambda \bar\gamma^\mu C)_{\beta\gamma}(\gamma_\lambda C)_\alpha{}^{\dot\delta},\right]. \qquad (B.41)$$

Hence, it is possible to eliminate two out of the six tensors arising from permutations in the spinor indices of $C_{\alpha\beta}(\gamma^\mu C)_\gamma{}^{\dot\delta}$ and $(\gamma^\lambda \bar\gamma^\mu C)_{\alpha\gamma}(\gamma_\lambda C)_\beta{}^{\dot\delta}$.

Correlator $\langle\psi^\mu \psi^\nu S_\alpha S_\beta S_\gamma S_\delta\rangle$

Possible Clebsch–Gordan coefficients for this correlator are $(\gamma^\mu \bar\gamma^\lambda C)_{\alpha\beta}(\gamma_\lambda \bar\gamma^\nu C)_{\gamma\delta}$ and permutations in $\alpha, \beta, \gamma, \delta$ of this tensor. However, these are not independent as multiplication of (B.41)

B.3 Relations for $D=8$

with $\gamma^\nu_{\dot\delta\dot\delta}$ shows:

$$(\gamma^\mu \bar\gamma^\lambda C)_{\alpha\beta} (\gamma_\lambda \bar\gamma^\nu C)_{\gamma\delta} = -(\gamma^\mu \bar\gamma^\lambda C)_{\alpha\gamma} (\gamma_\lambda \bar\gamma^\nu C)_{\beta\delta} - 2 (\gamma^\mu \bar\gamma^\nu C)_{\alpha\delta} C_{\beta\gamma}. \quad (B.42)$$

Permuting the spinor indices in this identity and making antisymmetry in the vector indices manifest, one finds the following relation:

$$(\gamma^{\lambda[\mu} C)_{\alpha\beta} (\gamma^{\nu]}{}_\lambda C)_{\gamma\delta} = C_{\alpha\delta} (\gamma^{\mu\nu} C)_{\gamma\beta} - C_{\alpha\gamma} (\gamma^{\mu\nu} C)_{\delta\beta} - C_{\beta\delta} (\gamma^{\mu\nu} C)_{\gamma\alpha} + C_{\gamma\beta} (\gamma^{\mu\nu} C)_{\delta\alpha}. \quad (B.43)$$

Correlator $\langle \psi^\mu \psi^\nu S_\alpha S_\beta S^{\dot\gamma} S^{\dot\delta} \rangle$

This is a triality invariant index structure for which the tensor identities are particularly interesting. The result of this correlator is given in (5.56) in terms of nine index terms. There we only keep the antisymmetric part in μ,ν of $(\gamma^\mu{}_\lambda C)_{\alpha\beta} (\bar\gamma^{\nu\lambda} C)^{\dot\gamma\dot\delta}$ because the symmetric piece can be reduced to

$$(\gamma^{(\mu}{}_\lambda C)_{\alpha\beta} (\bar\gamma^{\nu)\lambda} C)^{\dot\gamma\dot\delta} = \eta^{\mu\nu} (\gamma_\lambda C)_{[\alpha}{}^{\dot\gamma} (\gamma^\lambda C)_{\beta]}{}^{\dot\delta} - (\gamma^\mu C)_\alpha{}^{\dot\gamma} (\gamma^\nu C)_\beta{}^{\dot\delta} + (\gamma^\mu C)_\alpha{}^{\dot\delta} (\gamma^\nu C)_\beta{}^{\dot\gamma}$$
$$- (\gamma^\mu C)_\beta{}^{\dot\delta} (\gamma^\nu C)_\alpha{}^{\dot\gamma} + (\gamma^\mu C)_\beta{}^{\dot\gamma} (\gamma^\nu C)_\alpha{}^{\dot\delta}. \quad (B.44)$$

This can be derived by multiplying (B.41) with $\bar\gamma^{\nu\dot\gamma\gamma}$. Triple products of γ matrices can be eliminated as follows:

$$(\gamma^\mu \bar\gamma^\nu \gamma^\lambda C)_\alpha{}^{\dot\gamma} (\gamma_\lambda C)_\beta{}^{\dot\delta} = -(\gamma^{[\mu}{}_\lambda C)_{\alpha\beta} (\bar\gamma^{\nu]\lambda} C)^{\dot\gamma\dot\delta} - \eta^{\mu\nu} (\gamma_\lambda C)_{[\alpha}{}^{\dot\gamma} (\gamma^\lambda C)_{\beta]}{}^{\dot\delta} - \eta^{\mu\nu} C_{\alpha\beta} C^{\dot\gamma\dot\delta}$$
$$- (\gamma^\mu C)_\alpha{}^{\dot\gamma} (\gamma^\nu C)_\beta{}^{\dot\delta} + (\gamma^\mu C)_\alpha{}^{\dot\delta} (\gamma^\nu C)_\beta{}^{\dot\gamma} + (\gamma^\mu C)_\beta{}^{\dot\delta} (\gamma^\nu C)_\alpha{}^{\dot\gamma}$$
$$- (\gamma^\mu C)_\beta{}^{\dot\gamma} (\gamma^\nu C)_\alpha{}^{\dot\delta} + (\gamma^{\mu\nu} C)_{\alpha\beta} C^{\dot\gamma\dot\delta} - C_{\alpha\beta} (\bar\gamma^{\mu\nu} C)^{\dot\gamma\dot\delta}. \quad (B.45)$$

Observe that upon adding the relations where α and β and/or $\dot\gamma$ and $\dot\delta$ are interchanged one obtains for the antisymmetric γ-products $(\gamma^\mu \bar\gamma^\nu \gamma^\lambda)$:

$$(\gamma^{\mu\nu\lambda} C)_\alpha{}^{\dot\gamma} (\gamma_\lambda C)_\beta{}^{\dot\delta} + (\gamma^{\mu\nu\lambda} C)_\beta{}^{\dot\gamma} (\gamma_\lambda C)_\alpha{}^{\dot\delta} + (\gamma^{\mu\nu\lambda} C)_\alpha{}^{\dot\delta} (\gamma_\lambda C)_\beta{}^{\dot\gamma} + (\gamma^{\mu\nu\lambda} C)_\beta{}^{\dot\delta} (\gamma_\lambda C)_\alpha{}^{\dot\gamma} = 0. \quad (B.46)$$

Correlator $\langle S_\alpha S_\beta S_\gamma S_\delta S^{\dot\epsilon} S^{\dot\zeta} \rangle$

The novel tensors here are of the type $(\gamma^\lambda C)_\alpha{}^{\dot\epsilon} (\gamma_{\lambda\rho} C)_{\beta\gamma} (\gamma^\rho C)_\delta{}^{\dot\zeta}$. The antisymmetric piece in the indices $\dot\epsilon, \dot\zeta$ can be expressed in terms of simpler combinations:

$$(\gamma^\lambda C)_\alpha{}^{[\dot\epsilon} (\gamma_{\lambda\rho} C)_{\beta\gamma} (\gamma^{|\rho|} C)_\delta{}^{\dot\zeta]} = C_{\alpha\gamma} (\gamma_\lambda C)_{[\beta}{}^{\dot\epsilon} (\gamma^\lambda C)_{\delta]}{}^{\dot\zeta} + C_{\delta\gamma} (\gamma_\lambda C)_{[\beta}{}^{\dot\epsilon} (\gamma^\lambda C)_{\alpha]}{}^{\dot\zeta}$$
$$- C_{\alpha\beta} (\gamma_\lambda C)_{[\gamma}{}^{\dot\epsilon} (\gamma^\lambda C)_{\delta]}{}^{\dot\zeta} - C_{\delta\beta} (\gamma_\lambda C)_{[\gamma}{}^{\dot\epsilon} (\gamma^\lambda C)_{\alpha]}{}^{\dot\zeta} - C_{\alpha\delta} (\gamma_\lambda C)_{[\beta}{}^{\dot\epsilon} (\gamma^\lambda C)_{\gamma]}{}^{\dot\zeta}. \quad (B.47)$$

B.4 Relations for $D = 10$

Many of the following tensor identities can be traced back to the fundamental relation

$$(\gamma^\mu C)_{\alpha\beta} (\gamma_\mu C)_{\gamma\delta} + (\gamma^\mu C)_{\beta\gamma} (\gamma_\mu C)_{\alpha\delta} + (\gamma^\mu C)_{\gamma\alpha} (\gamma_\mu C)_{\beta\delta} = 0, \tag{B.48}$$

due to the fact that $(S)^{\otimes_s 3} \otimes (S)$ does not contain any scalars. Here $(S)^{\otimes_s 3}$ denotes a totally symmetric threefold tensor product of the left-handed $SO(1,9)$ spinor representation (S). In general, correlators in $D = 10$ dimensions involve more independent Lorentz tensors which enter more difficult relations compared to their $D = 6$ relatives. Observe, for instance, that no direct analogs of the relations (B.31) and (B.32) hold.

Correlator $\langle S_\alpha S_\beta S_\gamma S_\delta \rangle$

The Fierz identity (A.27) with $\psi_\alpha = (\gamma^\mu C)_{\alpha\gamma}$ and $\chi_\beta = (\gamma_\mu C)_{\beta\delta}$ admits to eliminate

$$(\gamma^{\mu\nu\lambda} C)_{\alpha\beta} (\gamma_{\mu\nu\lambda} C)_{\gamma\delta} = 12 \left((\gamma^\mu C)_{\alpha\delta} (\gamma_\mu C)_{\beta\gamma} - (\gamma^\mu C)_{\alpha\gamma} (\gamma_\mu C)_{\beta\delta} \right), \tag{B.49}$$

and $(\gamma^\mu C)_{\alpha\gamma} (\gamma_\mu C)_{\beta\delta}$ is redundant on account of (B.48).

Correlator $\langle S_\alpha S_\beta S^{\dot\gamma} S^{\dot\delta} \rangle$

Setting $\psi_\alpha = C_\alpha{}^{\dot\gamma}$, $\chi_\beta = C_\beta{}^{\dot\delta}$ and $\psi_\alpha = (\gamma^{\mu\nu} C)_\alpha{}^{\dot\gamma}$, $\chi_\beta = (\gamma_{\mu\nu} C)_\beta{}^{\dot\delta}$ in (A.27) as well as $\psi_\alpha = C_\alpha{}^{\dot\gamma}$, $\bar\chi^{\dot\beta} = C_\delta{}^{\dot\beta}$ and $\psi_\alpha = (\gamma^\mu C)_{\alpha\gamma}$, $\bar\chi^{\dot\beta} = (\bar\gamma_\mu C)^{\dot\beta\dot\delta}$ in (A.28) gives rise to the following identities:

$$(\gamma^{\mu\nu} C)_\alpha{}^{\dot\gamma} (\gamma_{\mu\nu} C)_\beta{}^{\dot\delta} = -2\, C_\alpha{}^{\dot\gamma} C_\beta{}^{\dot\delta} - 8\, C_\alpha{}^{\dot\delta} C_\beta{}^{\dot\gamma} + 4\, (\gamma^\mu C)_{\alpha\beta} (\bar\gamma_\mu C)^{\dot\gamma\dot\delta}, \tag{B.50a}$$

$$(\gamma^{\mu\nu\lambda} C)_{\alpha\beta} (\bar\gamma_{\mu\nu\lambda} C)^{\dot\gamma\dot\delta} = 48 \left(C_\alpha{}^{\dot\gamma} C_\beta{}^{\dot\delta} - C_\alpha{}^{\dot\delta} C_\beta{}^{\dot\gamma} \right), \tag{B.50b}$$

$$(\gamma^{\mu\nu\lambda\rho} C)_\alpha{}^{\dot\gamma} (\gamma_{\mu\nu\lambda\rho} C)_\beta{}^{\dot\delta} = -48\, C_\alpha{}^{\dot\gamma} C_\beta{}^{\dot\delta} + 288\, C_\alpha{}^{\dot\delta} C_\beta{}^{\dot\gamma} + 48\, (\gamma^\mu C)_{\alpha\beta} (\bar\gamma_\mu C)^{\dot\gamma\dot\delta}, \tag{B.50c}$$

$$(\gamma^{\mu\nu\lambda\rho\tau} C)_{\alpha\beta} (\bar\gamma_{\mu\nu\lambda\rho\tau} C)^{\dot\gamma\dot\delta} = 1920 \left(C_\alpha{}^{\dot\gamma} C_\beta{}^{\dot\delta} + C_\alpha{}^{\dot\delta} C_\beta{}^{\dot\gamma} \right) - 240\, (\gamma^\mu C)_{\alpha\beta} (\bar\gamma_\mu C)^{\dot\gamma\dot\delta}. \tag{B.50d}$$

Hence, the three different tensors appearing on the right hand side of the previous equations are sufficient to express $\langle S_\alpha S_\beta S^{\dot\gamma} S^{\dot\delta} \rangle$.

Correlator $\langle \psi^\mu S_\alpha S_\beta S_\gamma S^{\dot\delta} \rangle$

The six tensors $C_\alpha{}^{\dot\delta} (\gamma^\mu C)_{\beta\gamma}$, $(\gamma^\nu \bar\gamma^\mu C)_\alpha{}^{\dot\delta} (\gamma_\nu C)_{\beta\gamma}$ and permutations in α, β, γ can be used to express this correlation function. However, (B.48) multiplied by $(\bar\gamma^\mu)^{\dot\delta\dot\delta}$ admits to eliminate one of them:

$$(\gamma^\nu \bar\gamma^\mu C)_\alpha{}^{\dot\delta} (\gamma_\nu C)_{\beta\gamma} + (\gamma^\nu \bar\gamma^\mu C)_\beta{}^{\dot\delta} (\gamma_\nu C)_{\gamma\alpha} + (\gamma^\nu \bar\gamma^\mu C)_\gamma{}^{\dot\delta} (\gamma_\nu C)_{\alpha\beta} = 0. \tag{B.51}$$

B.4 Relations for $D = 10$

Correlator $\langle \psi^\mu \psi^\nu S_\alpha S_\beta S_\gamma S_\delta \rangle$

Among the 15 index terms obtained from $\eta^{\mu\nu}(\gamma^\lambda C)_{\alpha\beta}(\gamma_\lambda C)_{\gamma\delta}$, $(\gamma^\mu C)_{\alpha\beta}(\gamma^\nu C)_{\gamma\delta}$ and $(\gamma^{\mu\nu\lambda}C)_{\alpha\beta}(\gamma_\lambda C)_{\gamma\delta}$ including permutations in $\alpha, \beta, \gamma, \delta$, there are four relations. We choose to work with antisymmetric γ-products because in this case the tensors involving $\eta^{\mu\nu}$ decouple from the others in the following relations. Equation (B.48) directly implies

$$\eta^{\mu\nu}(\gamma^\lambda C)_{\alpha\beta}(\gamma_\lambda C)_{\gamma\delta} + \eta^{\mu\nu}(\gamma^\lambda C)_{\alpha\gamma}(\gamma_\lambda C)_{\beta\delta} + \eta^{\mu\nu}(\gamma^\lambda C)_{\alpha\delta}(\gamma_\lambda C)_{\gamma\beta} = 0 \qquad (B.52)$$

and from (B.51) we derive:

$$(\gamma^{\mu\nu\lambda}C)_{\alpha\beta}(\gamma_\lambda C)_{\gamma\delta} + (\gamma^{\mu\nu\lambda}C)_{\alpha\gamma}(\gamma_\lambda C)_{\delta\beta} + (\gamma^{\mu\nu\lambda}C)_{\alpha\delta}(\gamma_\lambda C)_{\beta\gamma}$$
$$+ (\gamma^\mu C)_{\gamma\delta}(\gamma^\nu C)_{\alpha\beta} - (\gamma^\mu C)_{\alpha\beta}(\gamma^\nu C)_{\gamma\delta} + (\gamma^\mu C)_{\beta\delta}(\gamma^\nu C)_{\alpha\gamma}$$
$$- (\gamma^\mu C)_{\alpha\gamma}(\gamma^\nu C)_{\beta\delta} + (\gamma^\mu C)_{\beta\gamma}(\gamma^\nu C)_{\alpha\delta} - (\gamma^\mu C)_{\alpha\delta}(\gamma^\nu C)_{\beta\gamma} = 0. \qquad (B.53)$$

Further relations of the last type can be found by permuting the spinor indices in (B.53). In the result (5.65) for this correlator, we have used only two out of the three independent permutations to eliminate $(\gamma^{\mu\nu\lambda}C)_{\alpha\gamma}(\gamma_\lambda C)_{\beta\delta}$ and $(\gamma^{\mu\nu\lambda}C)_{\beta\delta}(\gamma_\lambda C)_{\alpha\gamma}$. The missing third identity can be written as

$$(\gamma^{\mu\nu\lambda}C)_{\alpha\beta}(\gamma_\lambda C)_{\gamma\delta} + (\gamma^{\mu\nu\lambda}C)_{\alpha\delta}(\gamma_\lambda C)_{\gamma\beta} + (\gamma^{\mu\nu\lambda}C)_{\gamma\beta}(\gamma_\lambda C)_{\alpha\delta}$$
$$+ (\gamma^{\mu\nu\lambda}C)_{\gamma\delta}(\gamma_\lambda C)_{\alpha\beta} - 4(\gamma^{[\mu}C)_{\alpha\gamma}(\gamma^{\nu]}C)_{\beta\delta} = 0. \qquad (B.54)$$

Correlator $\langle \psi^\mu \psi^\nu S_\alpha S_\beta S^{\dot\gamma} S^{\dot\delta} \rangle$

This correlator can be expressed in terms of the Clebsch–Gordan coefficients $\eta^{\mu\nu} C_\alpha{}^{\dot\gamma} C_\beta{}^{\dot\delta}$, $\eta^{\mu\nu}(\gamma^\lambda C)_{\alpha\beta}(\bar\gamma_\lambda C)^{\dot\gamma\dot\delta}$, $(\gamma^\mu C)_{\alpha\beta}(\bar\gamma^\nu C)^{\dot\gamma\dot\delta}$, $(\gamma^\nu C)_{\alpha\beta}(\bar\gamma^\mu C)^{\dot\gamma\dot\delta}$. In addition we have to consider $(\gamma^\mu \bar\gamma^\nu C)_\alpha{}^{\dot\gamma} C_\beta{}^{\dot\delta}$, $(\gamma^\mu \bar\gamma^\lambda C)_\alpha{}^{\dot\gamma}(\gamma^\nu \bar\gamma_\lambda C)_\beta{}^{\dot\delta}$, permutations of all these terms in $\alpha, \beta, \dot\gamma, \dot\delta$, as well as the three-$\gamma$-chains $(\gamma^\mu \bar\gamma^\nu \gamma^\lambda C)_{\alpha\beta}(\bar\gamma_\lambda C)^{\dot\gamma\dot\delta}$ and $(\bar\gamma^\mu \gamma^\nu \bar\gamma^\lambda C)^{\dot\gamma\dot\delta}(\gamma_\lambda C)_{\alpha\beta}$. However only twelve of these fifteen index terms are independent. One relation is found be replacing ν with λ and μ with ν in (B.51) and multiplying with $\bar\gamma^{\mu\dot\gamma\dot\gamma}$, another by treating the complex conjugate of (B.51) in the same manner:

$$(\gamma^\mu \bar\gamma^\nu \gamma^\lambda C)_{\alpha\beta}(\bar\gamma_\lambda C)^{\dot\gamma\dot\delta} = -2(\gamma^\mu C)_{\alpha\beta}(\bar\gamma^\nu C)^{\dot\gamma\dot\delta} - (\gamma^\mu \bar\gamma^\lambda C)_\alpha{}^{\dot\gamma}(\gamma^\nu \bar\gamma_\lambda C)_\beta{}^{\dot\delta}$$
$$- (\gamma^\mu \bar\gamma^\lambda C)_\alpha{}^{\dot\delta}(\gamma^\nu \bar\gamma_\lambda C)_\beta{}^{\dot\gamma}, \qquad (B.55a)$$

$$(\bar\gamma^\mu \gamma^\nu \bar\gamma^\lambda C)^{\dot\gamma\dot\delta}(\gamma_\lambda C)_{\alpha\beta} = -2(\gamma^\nu C)_{\alpha\beta}(\bar\gamma^\mu C)^{\dot\gamma\dot\delta} - 2(\gamma^\mu \bar\gamma^\nu C)_\alpha{}^{\dot\gamma} C_\beta{}^{\dot\delta}$$
$$+ 2(\gamma^\mu \bar\gamma^\nu C)_\alpha{}^{\dot\delta} C_\beta{}^{\dot\gamma} - 2(\gamma^\mu \bar\gamma^\nu C)_\beta{}^{\dot\gamma} C_\alpha{}^{\dot\delta} + 2(\gamma^\mu \bar\gamma^\nu C)_\beta{}^{\dot\delta} C_\alpha{}^{\dot\gamma}$$
$$- (\gamma^\mu \bar\gamma^\lambda C)_\alpha{}^{\dot\gamma}(\gamma^\nu \bar\gamma_\lambda C)_\beta{}^{\dot\delta} - (\gamma^\mu \bar\gamma^\lambda C)_\beta{}^{\dot\gamma}(\gamma^\nu \bar\gamma_\lambda C)_\alpha{}^{\dot\delta}. \qquad (B.55b)$$

A third equation is found by symmetrizing the previous result in the vector indices μ and ν:

$$(\gamma^\mu \bar{\gamma}^\lambda C)_\beta{}^{\dot{\delta}} (\gamma^\nu \bar{\gamma}_\lambda C)_\alpha{}^{\dot{\gamma}} = 2\eta^{\mu\nu} (\gamma^\lambda C)_{\alpha\beta} (\bar{\gamma}_\lambda C)^{\dot{\gamma}\dot{\delta}} - 2(\gamma^\mu C)_{\alpha\beta} (\bar{\gamma}^\nu C)^{\dot{\gamma}\dot{\delta}}$$
$$-2(\gamma^\nu C)_{\alpha\beta} (\bar{\gamma}^\mu C)^{\dot{\gamma}\dot{\delta}} - (\gamma^\mu \bar{\gamma}^\lambda C)_\alpha{}^{\dot{\gamma}} (\gamma^\nu \bar{\gamma}_\lambda C)_\beta{}^{\dot{\delta}}$$
$$-(\gamma^\mu \bar{\gamma}^\lambda C)_\alpha{}^{\dot{\delta}} (\gamma^\nu \bar{\gamma}_\lambda C)_\beta{}^{\dot{\gamma}} - (\gamma^\mu \bar{\gamma}^\lambda C)_\beta{}^{\dot{\gamma}} (\gamma^\nu \bar{\gamma}_\lambda C)_\alpha{}^{\dot{\delta}}. \tag{B.56}$$

Correlator $\langle S_\alpha S_\beta S_\gamma S_\delta S_\epsilon S^{\dot{\zeta}} \rangle$

The relevant index terms for this correlation function up to permutations in the spinor indices are $(\gamma^\mu C)_{\alpha\beta} (\gamma_\mu C)_{\gamma\delta} C_\epsilon{}^{\dot{\zeta}}$, $(\gamma_\mu C)_{\alpha\beta} (\gamma_\nu C)_{\gamma\delta} (\gamma_\mu \bar{\gamma}^\nu C)_\epsilon{}^{\dot{\zeta}}$. Using (B.48) one can eliminate six out of the fifteen tensors of the first type. Changing the index $\dot{\delta}$ in (B.51) to $\dot{\zeta}$ and multiplying with $(\gamma_\nu C)_{\delta\epsilon}$ gives rise to the relation:

$$(\gamma^\mu \bar{\gamma}^\nu C)_\alpha{}^{\dot{\zeta}} (\gamma_\mu C)_{\beta\gamma} (\gamma_\nu C)_{\delta\epsilon} + (\gamma^\mu \bar{\gamma}^\nu C)_\beta{}^{\dot{\zeta}} (\gamma_\mu C)_{\alpha\gamma} (\gamma_\nu C)_{\delta\epsilon}$$
$$+ (\gamma^\mu \bar{\gamma}^\nu C)_\gamma{}^{\dot{\zeta}} (\gamma_\mu C)_{\alpha\beta} (\gamma_\nu C)_{\delta\epsilon} = 0. \tag{B.57}$$

By permuting the spinor indices in this equation one obtains eight further independent relations that can be used to eliminate in total nine tensors of the second type.

Correlator $\langle S_\alpha S_\beta S_\gamma S^{\dot{\delta}} S^{\dot{\epsilon}} S^{\dot{\zeta}} \rangle$

For this correlations function we have to consider the 24 tensors built from $C_\alpha{}^{\dot{\delta}} C_\beta{}^{\dot{\epsilon}} C_\gamma{}^{\dot{\zeta}}$, $(\gamma^\mu C)_{\alpha\beta} (\bar{\gamma}_\mu C)^{\dot{\delta}\dot{\epsilon}} C_\gamma{}^{\dot{\zeta}}$, $(\gamma^\mu \bar{\gamma}^\nu C)_\alpha{}^{\dot{\delta}} (\gamma_\mu C)_{\beta\gamma} (\bar{\gamma}_\nu C)^{\dot{\epsilon}\dot{\zeta}}$ and permutations in the spinor indices. However only 19 of these are independent. By multiplying (B.48) with $\bar{\gamma}^{\nu\dot{\delta}\dot{\delta}} (\bar{\gamma}_\nu C)^{\dot{\epsilon}\dot{\zeta}}$ and proceeding in the same way with the complex conjugate one obtains the equations

$$(\gamma^\mu \bar{\gamma}^\nu C)_\alpha{}^{\dot{\delta}} (\gamma_\mu C)_{\beta\gamma} (\bar{\gamma}_\nu C)^{\dot{\epsilon}\dot{\zeta}} + (\gamma^\mu \bar{\gamma}^\nu C)_\beta{}^{\dot{\delta}} (\gamma_\mu C)_{\alpha\gamma} (\bar{\gamma}_\nu C)^{\dot{\epsilon}\dot{\zeta}}$$
$$+ (\gamma^\mu \bar{\gamma}^\nu C)_\gamma{}^{\dot{\delta}} (\gamma_\mu C)_{\alpha\beta} (\bar{\gamma}_\nu C)^{\dot{\epsilon}\dot{\zeta}} = 0, \tag{B.58a}$$

$$(\gamma^\mu \bar{\gamma}^\nu C)_\alpha{}^{\dot{\delta}} (\gamma_\mu C)_{\beta\gamma} (\bar{\gamma}_\nu C)^{\dot{\epsilon}\dot{\zeta}} + (\gamma^\mu \bar{\gamma}^\nu C)_\alpha{}^{\dot{\epsilon}} (\gamma_\mu C)_{\beta\gamma} (\bar{\gamma}_\nu C)^{\dot{\delta}\dot{\zeta}}$$
$$+ (\gamma^\mu \bar{\gamma}^\nu C)_\alpha{}^{\dot{\zeta}} (\gamma_\mu C)_{\beta\gamma} (\bar{\gamma}_\nu C)^{\dot{\delta}\dot{\epsilon}} = 0. \tag{B.58b}$$

Upon permutation in the spinor indices these identities yield in total five independent equations which are sufficient to reduce the number of index terms to 19.

APPENDIX C

Generalized Θ Functions

In this Appendix we demonstrate some techniques and our conventions regarding generalized Θ functions. These enter the calculation of RNS correlators at loop-level. For a scattering process of open strings with g loops the underlying CFT of the RNS fields has support on a genus g Riemann surface. The generalized Θ functions then ensure that the correlation functions satisfy certain periodicity properties under shifting the NS fermions and R spin fields along the homology cycles of this manifold. We present in the following the methods necessary to check the periodicity properties as well as a class of very useful identities going under the name of Fay's trisecant identity. We follow [155].

C.1 Periodicity Properties

Correlations functions of RNS fields at the g loop-level have to satisfy certain periodicity conditions if the fields are transported around the $2g$ homology cycles α_I, β_I of the genus g Riemann surface. The sign configuration are encoded in the spin structure $\vec{a}$, $\vec{b}$,

$$\langle \psi^\mu(z_1 + \alpha_I)\, \phi_2(z_2) \ldots \phi_N(z_N) \rangle^{\vec{a}}_{\vec{b}} = e^{-i\pi a_I} \langle \psi^\mu(z_1)\, \phi_2(z_2) \ldots \phi_N(z_N) \rangle^{\vec{a}}_{\vec{b}}, \qquad \text{(C.1a)}$$

$$\langle \psi^\mu(z_1 + \beta_I)\, \phi_2(z_2) \ldots \phi_N(z_N) \rangle^{\vec{a}}_{\vec{b}} = e^{+i\pi b_I} \langle \psi^\mu(z_1)\, \phi_2(z_2) \ldots \phi_N(z_N) \rangle^{\vec{a}}_{\vec{b}}, \qquad \text{(C.1b)}$$

where ϕ_i are some other RNS fields. The situation is different for R spin fields. These fields create branch cuts on the Riemann surface and the NS fermions change sign when going around these points. If the spin field is translated around a homology cycle, the branch cut is extend along all the way and therefore the whole spin structure of the correlator changes:

$$\langle S_\alpha(z_1 + \alpha_I)\, \phi_2(z_2) \ldots \phi_N(z_N) \rangle^{\vec{a}}_{\vec{b}} \sim \langle S_\gamma(z_1)\, \phi_2(z_2) \ldots \phi_N(z_N) \rangle^{\vec{a}}_{\vec{b}+\vec{e}_I} \qquad \text{(C.2a)}$$

$$\langle S_\alpha(z_1 + \beta_I)\, \phi_2(z_2) \ldots \phi_N(z_N) \rangle^{\vec{a}}_{\vec{b}} \sim \langle S_\gamma(z_1)\, \phi_2(z_2) \ldots \phi_N(z_N) \rangle^{\vec{a}+\vec{e}_I}_{\vec{b}} \qquad \text{(C.2b)}$$

For checking that the presented loop correlators satisfy these properties the Θ function identity

$$\Theta^{\vec{a}}_{\vec{b}}(\vec{x} + \vec{s} + \Omega \vec{t}|\Omega) = \exp\left[-i\pi\left(\vec{t}^t\,\Omega\,\vec{t} + \vec{t}^t\,(2\,\vec{x} + \vec{b} + 2\,\vec{s})\right)\right]\Theta^{\vec{a}+2\vec{t}}_{\vec{b}+2\vec{s}}(\vec{x}|\Omega) \tag{C.3}$$

is handy. The prime forms change under $z \to z + \alpha_I$ or $z \to z + \beta_I$ like

$$E(z + \alpha_I, w) = E(z, w)\,, \tag{C.4a}$$

$$E(z + \beta_I, w) = \exp\left[-i\pi\,\Omega_{II} - 2\pi i \int_w^z \omega_I\right] E(z, w)\,. \tag{C.4b}$$

C.2 Fay's Trisecant Identity

We now list some versions of Fay's trisecant identity. These are relations between generalized Θ functions $\Theta^{\vec{a}}_{\vec{b}}$ and the prime forms $E_{ij} \equiv E(z_i, z_j)$ defined in Chapter 5.1.1. From our perspective they can be understood as the loop generalization of the tree-level z-crossing identity

$$z_{13}\,z_{24} = z_{12}\,z_{34} + z_{14}\,z_{23}\,. \tag{C.5}$$

The most general form of Fay's trisecant identity is given in [131]:

$$\Theta^{\vec{a}}_{\vec{b}}\left(\sum_{k=1}^N \int_{y_k}^{x_k} \vec{\omega} - \vec{e}\right) \left[\Theta^{\vec{a}}_{\vec{b}}(\vec{e})\right]^{N-1} \frac{\prod_{i<j}^N E(x_i, x_j)\,E(y_i, y_j)}{\prod_{i,j=1}^N E(x_i, y_j)}$$

$$= (-1)^{N(N-1)/2} \det_{i,j}\left[E(x_i, y_j)^{-1}\,\Theta^{\vec{a}}_{\vec{b}}\left(\int_{y_j}^{x_i} \vec{\omega} - \vec{e}\right)\right]. \tag{C.6}$$

Here x_i, y_j, $i, j = 1, 2, ..., N$, denote arbitrary positions on the Riemann surface of genus g and $\vec{e} \in \mathbb{C}^g$ with $\Theta^{\vec{a}}_{\vec{b}}(\vec{e}) \neq 0$. The particular choice $\vec{e} = \frac{1}{2}\sum_{k=1}^N \int_{y_k}^{x_k} \vec{\omega} - \vec{\Delta}$ yields (C.6) in its most convenient form for the manipulation of loop correlators:

$$\Theta^{\vec{a}}_{\vec{b}}\left(\frac{1}{2}\sum_{k=1}^N \int_{y_k}^{x_k} \vec{\omega} + \vec{\Delta}\right) \left[\Theta^{\vec{a}}_{\vec{b}}\left(\frac{1}{2}\sum_{k=1}^N \int_{y_k}^{x_k} \vec{\omega} - \vec{\Delta}\right)\right]^{N-1} \frac{\prod_{i<j}^N E(x_i, x_j)\,E(y_i, y_j)}{\prod_{i,j=1}^N E(x_i, y_j)}$$

$$= (-1)^{N(N-1)/2} \det_{i,j}\left[E(x_i, y_j)^{-1}\,\Theta^{\vec{a}}_{\vec{b}}\left(-\frac{1}{2}\sum_{k=1}^N \int_{y_k}^{x_k} \vec{\omega} + \int_{y_j}^{x_i} \vec{\omega} + \vec{\Delta}\right)\right]. \tag{C.7}$$

These identities are of great use when determining and manipulating the coefficients of the different index terms in a loop correlator. First, they are needed when one changes the basis of Clebsch–Gordan coefficients and different z coefficients have to be summed up. Second, they are required in the opposite way for the derivation of loop correlators in Lorentz covariant form

C.2 Fay's Trisecant Identity

as presented in Chapter 5.2.2, especially when the contributions to different index terms have to be separated.

We now present various versions of (C.6) for different N. For $N = 2$ and $(z_1, z_2, z_3, z_4) = (x_1, y_1, x_2, y_2)$ we obtain the loop generalization of (C.5):

$$E_{13} E_{24} \Theta_{\vec{b}}^{\vec{a}}\left(\frac{1}{2}\int_{z_2}^{z_1}\vec{\omega} + \frac{1}{2}\int_{z_4}^{z_3}\vec{\omega} + \vec{\Delta}\right) \Theta_{\vec{b}}^{\vec{a}}\left(\frac{1}{2}\int_{z_2}^{z_1}\vec{\omega} + \frac{1}{2}\int_{z_4}^{z_3}\vec{\omega} - \vec{\Delta}\right)$$

$$= E_{12} E_{34} \Theta_{\vec{b}}^{\vec{a}}\left(\frac{1}{2}\int_{z_3}^{z_1}\vec{\omega} + \frac{1}{2}\int_{z_4}^{z_2}\vec{\omega} + \vec{\Delta}\right) \Theta_{\vec{b}}^{\vec{a}}\left(\frac{1}{2}\int_{z_3}^{z_1}\vec{\omega} + \frac{1}{2}\int_{z_4}^{z_2}\vec{\omega} - \vec{\Delta}\right)$$

$$+ E_{14} E_{23} \Theta_{\vec{b}}^{\vec{a}}\left(\frac{1}{2}\int_{z_2}^{z_1}\vec{\omega} + \frac{1}{2}\int_{z_3}^{z_4}\vec{\omega} + \vec{\Delta}\right) \Theta_{\vec{b}}^{\vec{a}}\left(\frac{1}{2}\int_{z_2}^{z_1}\vec{\omega} + \frac{1}{2}\int_{z_3}^{z_4}\vec{\omega} - \vec{\Delta}\right). \tag{C.8}$$

Indeed, this reduces to (C.5) for $g = 0$ because then $E_{ij} \to z_{ij}$ and $\Theta_{\vec{b}}^{\vec{a}} \to 1$. The next order version $N = 3$ becomes relevant for correlators including at least six fields:

$$\frac{E_{13} E_{15} E_{35} E_{24} E_{26} E_{46}}{E_{12} E_{14} E_{16} E_{23} E_{34} E_{36} E_{25} E_{45} E_{56}}$$

$$\times \Theta_{\vec{b}}^{\vec{a}}\left(\frac{1}{2}\int_{z_2}^{z_1}\vec{\omega} + \frac{1}{2}\int_{z_4}^{z_3}\vec{\omega} + \frac{1}{2}\int_{z_6}^{z_5}\vec{\omega} - \vec{\Delta}\right) \left[\Theta_{\vec{b}}^{\vec{a}}\left(\frac{1}{2}\int_{z_2}^{z_1}\vec{\omega} + \frac{1}{2}\int_{z_4}^{z_3}\vec{\omega} + \frac{1}{2}\int_{z_6}^{z_5}\vec{\omega} + \vec{\Delta}\right)\right]^2$$

$$= \frac{1}{E_{12} E_{34} E_{56}} \Theta_{\vec{b}}^{\vec{a}}\left(-\frac{1}{2}\int_{z_2}^{z_1}\vec{\omega} + \frac{1}{2}\int_{z_4}^{z_3}\vec{\omega} + \frac{1}{2}\int_{z_6}^{z_5}\vec{\omega} + \vec{\Delta}\right)$$

$$\times \Theta_{\vec{b}}^{\vec{a}}\left(\frac{1}{2}\int_{z_2}^{z_1}\vec{\omega} - \frac{1}{2}\int_{z_4}^{z_3}\vec{\omega} + \frac{1}{2}\int_{z_6}^{z_5}\vec{\omega} + \vec{\Delta}\right) \Theta_{\vec{b}}^{\vec{a}}\left(\frac{1}{2}\int_{z_2}^{z_1}\vec{\omega} + \frac{1}{2}\int_{z_4}^{z_3}\vec{\omega} - \frac{1}{2}\int_{z_6}^{z_5}\vec{\omega} + \vec{\Delta}\right)$$

$$- \frac{1}{E_{12} E_{36} E_{54}} \Theta_{\vec{b}}^{\vec{a}}\left(-\frac{1}{2}\int_{z_2}^{z_1}\vec{\omega} + \frac{1}{2}\int_{z_6}^{z_3}\vec{\omega} + \frac{1}{2}\int_{z_4}^{z_5}\vec{\omega} + \vec{\Delta}\right)$$

$$\times \Theta_{\vec{b}}^{\vec{a}}\left(\frac{1}{2}\int_{z_2}^{z_1}\vec{\omega} - \frac{1}{2}\int_{z_6}^{z_3}\vec{\omega} + \frac{1}{2}\int_{z_4}^{z_5}\vec{\omega} + \vec{\Delta}\right) \Theta_{\vec{b}}^{\vec{a}}\left(\frac{1}{2}\int_{z_2}^{z_1}\vec{\omega} + \frac{1}{2}\int_{z_6}^{z_3}\vec{\omega} - \frac{1}{2}\int_{z_4}^{z_5}\vec{\omega} + \vec{\Delta}\right)$$

$$+ \frac{1}{E_{14} E_{36} E_{52}} \Theta_{\vec{b}}^{\vec{a}}\left(-\frac{1}{2}\int_{z_4}^{z_1}\vec{\omega} + \frac{1}{2}\int_{z_6}^{z_3}\vec{\omega} + \frac{1}{2}\int_{z_2}^{z_5}\vec{\omega} + \vec{\Delta}\right)$$

$$\times \Theta_{\vec{b}}^{\vec{a}}\left(\frac{1}{2}\int_{z_4}^{z_1}\vec{\omega} - \frac{1}{2}\int_{z_6}^{z_3}\vec{\omega} + \frac{1}{2}\int_{z_2}^{z_5}\vec{\omega} + \vec{\Delta}\right) \Theta_{\vec{b}}^{\vec{a}}\left(\frac{1}{2}\int_{z_4}^{z_1}\vec{\omega} + \frac{1}{2}\int_{z_6}^{z_3}\vec{\omega} - \frac{1}{2}\int_{z_2}^{z_5}\vec{\omega} + \vec{\Delta}\right)$$

$$- \frac{1}{E_{14} E_{32} E_{56}} \Theta_{\vec{b}}^{\vec{a}}\left(-\frac{1}{2}\int_{z_4}^{z_1}\vec{\omega} + \frac{1}{2}\int_{z_2}^{z_3}\vec{\omega} + \frac{1}{2}\int_{z_6}^{z_5}\vec{\omega} + \vec{\Delta}\right)$$

$$\times \Theta_{\vec{b}}^{\vec{a}}\left(\frac{1}{2}\int_{z_4}^{z_1}\vec{\omega} - \frac{1}{2}\int_{z_2}^{z_3}\vec{\omega} + \frac{1}{2}\int_{z_6}^{z_5}\vec{\omega} + \vec{\Delta}\right) \Theta_{\vec{b}}^{\vec{a}}\left(\frac{1}{2}\int_{z_4}^{z_1}\vec{\omega} + \frac{1}{2}\int_{z_2}^{z_3}\vec{\omega} - \frac{1}{2}\int_{z_6}^{z_5}\vec{\omega} + \vec{\Delta}\right)$$

$$+ \frac{1}{E_{16} E_{32} E_{54}} \Theta_{\vec{b}}^{\vec{a}}\left(-\frac{1}{2}\int_{z_6}^{z_1}\vec{\omega} + \frac{1}{2}\int_{z_2}^{z_3}\vec{\omega} + \frac{1}{2}\int_{z_4}^{z_5}\vec{\omega} + \vec{\Delta}\right)$$

$$\times \Theta_{\vec{b}}^{\vec{a}}\left(\frac{1}{2}\int_{z_6}^{z_1}\vec{\omega} - \frac{1}{2}\int_{z_2}^{z_3}\vec{\omega} + \frac{1}{2}\int_{z_4}^{z_5}\vec{\omega} + \vec{\Delta}\right) \Theta_{\vec{b}}^{\vec{a}}\left(\frac{1}{2}\int_{z_6}^{z_1}\vec{\omega} + \frac{1}{2}\int_{z_2}^{z_3}\vec{\omega} - \frac{1}{2}\int_{z_4}^{z_5}\vec{\omega} + \vec{\Delta}\right)$$

$$- \frac{1}{E_{16} E_{34} E_{52}} \Theta_{\vec{b}}^{\vec{a}}\left(-\frac{1}{2}\int_{z_6}^{z_1}\vec{\omega} + \frac{1}{2}\int_{z_4}^{z_3}\vec{\omega} + \frac{1}{2}\int_{z_2}^{z_5}\vec{\omega} + \vec{\Delta}\right)$$

$$\times \Theta^{\vec{a}}_{\vec{b}}\left(\frac{1}{2}\int_{z_6}^{z_1}\vec{\omega} - \frac{1}{2}\int_{z_4}^{z_3}\vec{\omega} + \frac{1}{2}\int_{z_2}^{z_5}\vec{\omega} + \vec{\Delta}\right)\Theta^{\vec{a}}_{\vec{b}}\left(\frac{1}{2}\int_{z_6}^{z_1}\vec{\omega} + \frac{1}{2}\int_{z_4}^{z_3}\vec{\omega} - \frac{1}{2}\int_{z_2}^{z_5}\vec{\omega} + \vec{\Delta}\right). \tag{C.9}$$

We do not display the case $N = 4$ in full beauty as it contains 26 terms with four Θ functions each. It has the structure

$$\frac{E_{13}\,E_{15}\,E_{17}\,E_{35}\,E_{37}\,E_{57}\,E_{24}\,E_{26}\,E_{28}\,E_{46}\,E_{48}\,E_{68}}{E_{12}\,E_{14}\,E_{16}\,E_{18}\,E_{23}\,E_{25}\,E_{27}\,E_{34}\,E_{36}\,E_{38}\,E_{45}\,E_{47}\,E_{56}\,E_{58}\,E_{67}\,E_{78}}$$
$$\times \Theta^{\vec{a}}_{\vec{b}}\left(\frac{1}{2}\left[\int_{z_2}^{z_1}\vec{\omega} + \int_{z_4}^{z_3}\vec{\omega} + \int_{z_6}^{z_5}\vec{\omega} + \int_{z_8}^{z_7}\vec{\omega}\right] - \vec{\Delta}\right)\left[\Theta^{\vec{a}}_{\vec{b}}\left(\frac{1}{2}\left[\int_{z_2}^{z_1}\vec{\omega} + \int_{z_4}^{z_3}\vec{\omega} + \int_{z_6}^{z_5}\vec{\omega} + \int_{z_8}^{z_7}\vec{\omega}\right] + \vec{\Delta}\right)\right]^3$$
$$= \frac{1}{E_{12}\,E_{34}\,E_{56}\,E_{78}}$$
$$\times \Theta^{\vec{a}}_{\vec{b}}\left(\frac{1}{2}\left[\int_{z_2}^{z_1}\vec{\omega} + \int_{z_4}^{z_3}\vec{\omega} + \int_{z_6}^{z_5}\vec{\omega} - \int_{z_8}^{z_7}\vec{\omega}\right] + \vec{\Delta}\right)\Theta^{\vec{a}}_{\vec{b}}\left(\frac{1}{2}\left[\int_{z_2}^{z_1}\vec{\omega} + \int_{z_4}^{z_3}\vec{\omega} - \int_{z_6}^{z_5}\vec{\omega} + \int_{z_8}^{z_7}\vec{\omega}\right] + \vec{\Delta}\right)$$
$$\times \Theta^{\vec{a}}_{\vec{b}}\left(\frac{1}{2}\left[\int_{z_2}^{z_1}\vec{\omega} - \int_{z_4}^{z_3}\vec{\omega} + \int_{z_6}^{z_5}\vec{\omega} + \int_{z_8}^{z_7}\vec{\omega}\right] + \vec{\Delta}\right)\Theta^{\vec{a}}_{\vec{b}}\left(\frac{1}{2}\left[-\int_{z_2}^{z_1}\vec{\omega} + \int_{z_4}^{z_3}\vec{\omega} + \int_{z_6}^{z_5}\vec{\omega} + \int_{z_8}^{z_7}\vec{\omega}\right] + \vec{\Delta}\right)$$
$$- \frac{1}{E_{12}\,E_{34}\,E_{58}\,E_{76}}$$
$$\times \Theta^{\vec{a}}_{\vec{b}}\left(\frac{1}{2}\left[\int_{z_2}^{z_1}\vec{\omega} + \int_{z_4}^{z_3}\vec{\omega} + \int_{z_8}^{z_5}\vec{\omega} - \int_{z_6}^{z_7}\vec{\omega}\right] + \vec{\Delta}\right)\Theta^{\vec{a}}_{\vec{b}}\left(\frac{1}{2}\left[\int_{z_2}^{z_1}\vec{\omega} + \int_{z_4}^{z_3}\vec{\omega} - \int_{z_8}^{z_5}\vec{\omega} + \int_{z_6}^{z_7}\vec{\omega}\right] + \vec{\Delta}\right)$$
$$\times \Theta^{\vec{a}}_{\vec{b}}\left(\frac{1}{2}\left[\int_{z_2}^{z_1}\vec{\omega} - \int_{z_4}^{z_3}\vec{\omega} + \int_{z_8}^{z_5}\vec{\omega} + \int_{z_6}^{z_7}\vec{\omega}\right] + \vec{\Delta}\right)\Theta^{\vec{a}}_{\vec{b}}\left(\frac{1}{2}\left[-\int_{z_2}^{z_1}\vec{\omega} + \int_{z_4}^{z_3}\vec{\omega} + \int_{z_8}^{z_5}\vec{\omega} + \int_{z_6}^{z_7}\vec{\omega}\right] + \vec{\Delta}\right)$$
$$\pm 22 \text{ further permutations in } (z_2, z_4, z_6, z_8). \tag{C.10}$$

This relation is highly important for calculating the six-dimensional loop correlator consisting of four left- and right-handed spin fields each. Only with this version of Fay's trisecant identity we have been able to generalize this correlator to (5.70) with an arbitrary number of left- and right-handed spin fields.

APPENDIX D

Details of the Amplitude Calculation

In this Appendix we collect details for the calculation of the open string amplitude presented in Chapter 6. In particular we state the kinematical terms entering the expressions $\mathcal{K}^i$ and show the necessary steps for checking gauge invariance or giving results in spinor product notation.

D.1 The Kinematical Structure

The index terms appearing in the correlators (6.24)-(6.26) are multiplied in the calculation with the momenta of the gauge bosons, k_1, k_2, their polarization vectors ξ_1, ξ_2 and the polarization spinors of the gauginos, u_3, u_5 and $\bar{u}_4, \bar{u}_6$. Taking these expressions together, the terms entering $\mathcal{K}^i$ take the form:

$$\begin{aligned}
R^1 &= 4\,(k_1\,k_2)\,(\xi_1\,\xi_2)\,\langle 35\rangle\,[46]\,, & R^{10} &= 4\,(k_4\,\xi_1)\,(k_1\,\xi_2)\,\langle 35\rangle\,[46]\,, \\
R^2 &= 4\,(k_1\,\xi_2)\,(k_2\,\xi_1)\,\langle 35\rangle\,[46]\,, & R^{11} &= 4\,(k_4\,\xi_1)\,(k_3\,\xi_2)\,\langle 35\rangle\,[46]\,, \\
R^3 &= 4\,(k_2\,\xi_1)\,(k_3\,\xi_2)\,\langle 35\rangle\,[46]\,, & R^{12} &= 4\,(k_4\,\xi_1)\,(k_4\,\xi_2)\,\langle 35\rangle\,[46]\,, \\
R^4 &= 4\,(k_2\,\xi_1)\,(k_4\,\xi_2)\,\langle 35\rangle\,[46]\,, & R^{13} &= 4\,(k_4\,\xi_1)\,(k_5\,\xi_2)\,\langle 35\rangle\,[46]\,, \\
R^5 &= 4\,(k_2\,\xi_1)\,(k_5\,\xi_2)\,\langle 35\rangle\,[46]\,, & R^{14} &= 4\,(k_5\,\xi_1)\,(k_1\,\xi_2)\,\langle 35\rangle\,[46]\,, \\
R^6 &= 4\,(k_3\,\xi_1)\,(k_1\,\xi_2)\,\langle 35\rangle\,[46]\,, & R^{15} &= 4\,(k_5\,\xi_1)\,(k_3\,\xi_2)\,\langle 35\rangle\,[46]\,, \\
R^7 &= 4\,(k_3\,\xi_1)\,(k_3\,\xi_2)\,\langle 35\rangle\,[46]\,, & R^{16} &= 4\,(k_5\,\xi_1)\,(k_4\,\xi_2)\,\langle 35\rangle\,[46]\,, \\
R^8 &= 4\,(k_3\,\xi_1)\,(k_4\,\xi_2)\,\langle 35\rangle\,[46]\,, & R^{17} &= 4\,(k_5\,\xi_1)\,(k_5\,\xi_2)\,\langle 35\rangle\,[46]\,, \\
R^9 &= 4\,(k_3\,\xi_1)\,(k_5\,\xi_2)\,\langle 35\rangle\,[46]\,, &&
\end{aligned} \qquad (D.1)$$

$$\begin{aligned}
S^1 &= 2\,(k_1\,k_2)\,\xi_{1\mu}\,\xi_{2\nu}\,(\bar{u}_4\,\varepsilon\,\bar{\sigma}^\mu\,\sigma^\nu\,\bar{u}_6)\,\langle 35\rangle\,, & T^1 &= 2\,(k_1\,k_2)\,\xi_{1\mu}\,\xi_{2\nu}\,(u_3\,\sigma^\mu\,\bar{\sigma}^\nu\,\varepsilon\,u_5)\,[46]\,, \\
S^2 &= 2\,(\xi_1\,\xi_2)\,k_{1\mu}\,k_{2\nu}\,(\bar{u}_4\,\varepsilon\,\bar{\sigma}^\mu\,\sigma^\nu\,\bar{u}_6)\,\langle 35\rangle\,, & T^2 &= 2\,(\xi_1\,\xi_2)\,k_{1\mu}\,k_{2\nu}\,(u_3\,\sigma^\mu\,\bar{\sigma}^\nu\,\varepsilon\,u_5)\,[46]\,,
\end{aligned}$$

$$S^3 = 2\,(k_1\,\xi_2)\,\xi_{1\mu}\,k_{2\nu}\,(\bar{u}_4\,\varepsilon\,\bar{\sigma}^\mu\,\sigma^\nu\,\bar{u}_6)\,\langle 35 \rangle\,, \quad T^3 = 2\,(k_1\,\xi_2)\,\xi_{1\mu}\,k_{2\nu}\,(u_3\,\sigma^\mu\,\bar{\sigma}^\nu\,\varepsilon\,u_5)\,[46]\,,$$
$$S^4 = 2\,(\xi_1\,k_2)\,k_{1\mu}\,\xi_{2\nu}\,(\bar{u}_4\,\varepsilon\,\bar{\sigma}^\mu\,\sigma^\nu\,\bar{u}_6)\,\langle 35 \rangle\,, \quad T^4 = 2\,(\xi_1\,k_2)\,k_{1\mu}\,\xi_{2\nu}\,(u_3\,\sigma^\mu\,\bar{\sigma}^\nu\,\varepsilon\,u_5)\,[46]\,, \quad \text{(D.2)}$$

$$
\begin{aligned}
U^1 &= 2\,(k_1\,k_2)\,\xi_{1\mu}\,\xi_{2\nu}\,(u_3\,\sigma^\mu\,\bar{u}_4)\,(u_5\,\sigma^\nu\,\bar{u}_6)\,, & V^1 &= 2\,(k_1\,\xi_2)\,k_{1\mu}\,\xi_{1\nu}\,(u_3\,\sigma^\mu\,\bar{u}_6)\,(u_5\,\sigma^\nu\,\bar{u}_4)\,, \\
U^2 &= 2\,(\xi_1\,\xi_2)\,k_{1\mu}\,k_{2\nu}\,(u_3\,\sigma^\mu\,\bar{u}_4)\,(u_5\,\sigma^\nu\,\bar{u}_6)\,, & V^2 &= 2\,(k_3\,\xi_2)\,k_{1\mu}\,\xi_{1\nu}\,(u_3\,\sigma^\mu\,\bar{u}_6)\,(u_5\,\sigma^\nu\,\bar{u}_4)\,, \\
U^3 &= 2\,(k_1\,\xi_2)\,\xi_{1\mu}\,k_{2\nu}\,(u_3\,\sigma^\mu\,\bar{u}_4)\,(u_5\,\sigma^\nu\,\bar{u}_6)\,, & V^3 &= 2\,(k_4\,\xi_2)\,k_{1\mu}\,\xi_{1\nu}\,(u_3\,\sigma^\mu\,\bar{u}_6)\,(u_5\,\sigma^\nu\,\bar{u}_4)\,, \\
U^4 &= 2\,(k_2\,\xi_1)\,k_{1\mu}\,\xi_{2\nu}\,(u_3\,\sigma^\mu\,\bar{u}_4)\,(u_5\,\sigma^\nu\,\bar{u}_6)\,, & V^4 &= 2\,(k_5\,\xi_2)\,k_{1\mu}\,\xi_{1\nu}\,(u_3\,\sigma^\mu\,\bar{u}_6)\,(u_5\,\sigma^\nu\,\bar{u}_4)\,, \\
U^5 &= 2\,(k_1\,\xi_2)\,k_{1\mu}\,\xi_{1\nu}\,(u_3\,\sigma^\mu\,\bar{u}_4)\,(u_5\,\sigma^\nu\,\bar{u}_6)\,, & V^5 &= 2\,(k_2\,\xi_1)\,k_{2\mu}\,\xi_{2\nu}\,(u_3\,\sigma^\mu\,\bar{u}_6)\,(u_5\,\sigma^\nu\,\bar{u}_4)\,, \\
U^6 &= 2\,(k_3\,\xi_2)\,k_{1\mu}\,\xi_{1\nu}\,(u_3\,\sigma^\mu\,\bar{u}_4)\,(u_5\,\sigma^\nu\,\bar{u}_6)\,, & V^6 &= 2\,(k_3\,\xi_1)\,k_{2\mu}\,\xi_{2\nu}\,(u_3\,\sigma^\mu\,\bar{u}_6)\,(u_5\,\sigma^\nu\,\bar{u}_4)\,, \\
U^7 &= 2\,(k_4\,\xi_2)\,k_{1\mu}\,\xi_{1\nu}\,(u_3\,\sigma^\mu\,\bar{u}_4)\,(u_5\,\sigma^\nu\,\bar{u}_6)\,, & V^7 &= 2\,(k_4\,\xi_1)\,k_{2\mu}\,\xi_{2\nu}\,(u_3\,\sigma^\mu\,\bar{u}_6)\,(u_5\,\sigma^\nu\,\bar{u}_4)\,, \\
U^8 &= 2\,(k_5\,\xi_2)\,k_{1\mu}\,\xi_{1\nu}\,(u_3\,\sigma^\mu\,\bar{u}_4)\,(u_5\,\sigma^\nu\,\bar{u}_6)\,, & V^8 &= 2\,(k_5\,\xi_1)\,k_{2\mu}\,\xi_{2\nu}\,(u_3\,\sigma^\mu\,\bar{u}_6)\,(u_5\,\sigma^\nu\,\bar{u}_4)\,, \\
U^9 &= 2\,(k_2\,\xi_1)\,k_{2\mu}\,\xi_{2\nu}\,(u_3\,\sigma^\mu\,\bar{u}_4)\,(u_5\,\sigma^\nu\,\bar{u}_6)\,, \\
U^{10} &= 2\,(k_3\,\xi_1)\,k_{2\mu}\,\xi_{2\nu}\,(u_3\,\sigma^\mu\,\bar{u}_4)\,(u_5\,\sigma^\nu\,\bar{u}_6)\,, \\
U^{11} &= 2\,(k_4\,\xi_1)\,k_{2\mu}\,\xi_{2\nu}\,(u_3\,\sigma^\mu\,\bar{u}_4)\,(u_5\,\sigma^\nu\,\bar{u}_6)\,, \\
U^{12} &= 2\,(k_5\,\xi_1)\,k_{2\mu}\,\xi_{2\nu}\,(u_3\,\sigma^\mu\,\bar{u}_4)\,(u_5\,\sigma^\nu\,\bar{u}_6)\,, & & \text{(D.3)}
\end{aligned}
$$

$$
\begin{aligned}
W^1 &= 2\,(k_1\,\xi_2)\,k_{1\mu}\,\xi_{1\nu}\,(u_5\,\sigma^\mu\,\bar{u}_4)\,(u_3\,\sigma^\nu\,\bar{u}_6)\,, & X^1 &= 2\,(k_1\,\xi_2)\,k_{1\mu}\,\xi_{1\nu}\,(u_5\,\sigma^\mu\,\bar{u}_6)\,(u_3\,\sigma^\nu\,\bar{u}_4)\,, \\
W^2 &= 2\,(k_3\,\xi_2)\,k_{1\mu}\,\xi_{1\nu}\,(u_5\,\sigma^\mu\,\bar{u}_4)\,(u_3\,\sigma^\nu\,\bar{u}_6)\,, & X^2 &= 2\,(k_3\,\xi_2)\,k_{1\mu}\,\xi_{1\nu}\,(u_5\,\sigma^\mu\,\bar{u}_6)\,(u_3\,\sigma^\nu\,\bar{u}_4)\,, \\
W^3 &= 2\,(k_4\,\xi_2)\,k_{1\mu}\,\xi_{1\nu}\,(u_5\,\sigma^\mu\,\bar{u}_4)\,(u_3\,\sigma^\nu\,\bar{u}_6)\,, & X^3 &= 2\,(k_4\,\xi_2)\,k_{1\mu}\,\xi_{1\nu}\,(u_5\,\sigma^\mu\,\bar{u}_6)\,(u_3\,\sigma^\nu\,\bar{u}_4)\,, \\
W^4 &= 2\,(k_5\,\xi_2)\,k_{1\mu}\,\xi_{1\nu}\,(u_5\,\sigma^\mu\,\bar{u}_4)\,(u_3\,\sigma^\nu\,\bar{u}_6)\,, & X^4 &= 2\,(k_5\,\xi_2)\,k_{1\mu}\,\xi_{1\nu}\,(u_5\,\sigma^\mu\,\bar{u}_6)\,(u_3\,\sigma^\nu\,\bar{u}_4)\,, \\
W^5 &= 2\,(k_2\,\xi_1)\,k_{2\mu}\,\xi_{2\nu}\,(u_5\,\sigma^\mu\,\bar{u}_4)\,(u_3\,\sigma^\nu\,\bar{u}_6)\,, & X^5 &= 2\,(k_2\,\xi_1)\,k_{2\mu}\,\xi_{2\nu}\,(u_5\,\sigma^\mu\,\bar{u}_6)\,(u_3\,\sigma^\nu\,\bar{u}_4)\,, \\
W^6 &= 2\,(k_3\,\xi_1)\,k_{2\mu}\,\xi_{2\nu}\,(u_5\,\sigma^\mu\,\bar{u}_4)\,(u_3\,\sigma^\nu\,\bar{u}_6)\,, & X^6 &= 2\,(k_3\,\xi_1)\,k_{2\mu}\,\xi_{2\nu}\,(u_5\,\sigma^\mu\,\bar{u}_6)\,(u_3\,\sigma^\nu\,\bar{u}_4)\,, \\
W^7 &= 2\,(k_4\,\xi_1)\,k_{2\mu}\,\xi_{2\nu}\,(u_5\,\sigma^\mu\,\bar{u}_4)\,(u_3\,\sigma^\nu\,\bar{u}_6)\,, & X^7 &= 2\,(k_4\,\xi_1)\,k_{2\mu}\,\xi_{2\nu}\,(u_5\,\sigma^\mu\,\bar{u}_6)\,(u_3\,\sigma^\nu\,\bar{u}_4)\,, \\
W^8 &= 2\,(k_5\,\xi_1)\,k_{2\mu}\,\xi_{2\nu}\,(u_5\,\sigma^\mu\,\bar{u}_4)\,(u_3\,\sigma^\nu\,\bar{u}_6)\,, & X^8 &= 2\,(k_5\,\xi_1)\,k_{2\mu}\,\xi_{2\nu}\,(u_5\,\sigma^\mu\,\bar{u}_6)\,(u_3\,\sigma^\nu\,\bar{u}_4)\,, \quad \text{(D.4)}
\end{aligned}
$$

$$
\begin{aligned}
Y_1 &= k_{1\mu}\,\xi_{1\nu}\,k_{2\lambda}\,\xi_{2\rho}\,(u_3\,\sigma^\mu\,\bar{\sigma}^\nu\,\sigma^\lambda\,\bar{u}_4)\,(u_5\,\sigma^\rho\,\bar{u}_6)\,, \\
Y_2 &= k_{1\mu}\,\xi_{1\nu}\,k_{2\lambda}\,\xi_{2\rho}\,(u_3\,\sigma^\lambda\,\bar{\sigma}^\rho\,\sigma^\mu\,\bar{u}_4)\,(u_5\,\sigma^\nu\,\bar{u}_6)\,, \\
Y_3 &= k_{1\mu}\,\xi_{1\nu}\,k_{2\lambda}\,\xi_{2\rho}\,(u_3\,\sigma^\nu\,\bar{\sigma}^\lambda\,\sigma^\rho\,\bar{u}_6)\,(u_5\,\sigma^\mu\,\bar{u}_4)\,, \\
Y_4 &= k_{1\mu}\,\xi_{1\nu}\,k_{2\lambda}\,\xi_{2\rho}\,(u_3\,\sigma^\mu\,\bar{\sigma}^\nu\,\sigma^\rho\,\bar{u}_6)\,(u_5\,\sigma^\lambda\,\bar{u}_4)\,, \\
Y_5 &= k_{1\mu}\,\xi_{1\nu}\,k_{2\lambda}\,\xi_{2\rho}\,(u_5\,\sigma^\nu\,\bar{\sigma}^\lambda\,\sigma^\rho\,\bar{u}_6)\,(u_3\,\sigma^\mu\,\bar{u}_4)\,, \\
Y_6 &= k_{1\mu}\,\xi_{1\nu}\,k_{2\lambda}\,\xi_{2\rho}\,(u_3\,\sigma^\mu\,\bar{\sigma}^\nu\,\sigma^\lambda\,\bar{u}_6)\,(u_5\,\sigma^\rho\,\bar{u}_4)\,, \quad \text{(D.5)}
\end{aligned}
$$

D.2 Gauge Invariance

$$Z_1 = \langle 35 \rangle k_{1\mu} \xi_{1\nu} k_{2\lambda} \xi_{2\rho} (\bar{u}_4 \, \varepsilon \, \bar{\sigma}^\mu \, \sigma^\nu \, \bar{\sigma}^\lambda \, \sigma^\rho \, \bar{u}_6),$$
$$Z_2 = [46] k_{1\mu} \xi_{1\nu} k_{2\lambda} \xi_{2\rho} (u_3 \, \sigma^\mu \, \bar{\sigma}^\nu \, \sigma^\lambda \, \bar{\sigma}^\rho \, \varepsilon \, u_5). \tag{D.6}$$

These terms combine as follows into the kinematical factors $\mathcal{K}^i$:

$$\begin{aligned}
\mathcal{K}^1 &= R^1/s_1 - R^2, \\
\mathcal{K}^2 &= R^3 + R^7 - U^6 - U^9 - U^{10} + V^2 + V^5, \\
\mathcal{K}^3 &= R^4 + R^{12} + S^2 - S^4 + U^1 - U^3 - V^3 - V^5 \\
&\quad + X^3 + X^5 - Y^2 + Y^4 + Y^5 - Y^6 - Z^1, \\
\mathcal{K}^4 &= R^5 + R^{17} + W^4 + W^5 + W^8 - X^4 - X^5, \\
\mathcal{K}^5 &= -R^6 - R^7 + U^5 + U^6 + U^{10} - V^1 - V^2, \\
\mathcal{K}^6 &= -R^{10} - R^{12} + V^1 + V^3 - X^1 - X^3 + Z^1, \\
\mathcal{K}^7 &= -R^{14} - R^{17} - W^1 - W^4 - W^8 + X^1 + X^4, \\
\mathcal{K}^8 &= -R^1 + R^2, \\
\mathcal{K}^9 &= S^1 - S^3 - U^1 + U^3 + Y^2 - Y^4 - Y^5 + Y^6, \\
\mathcal{K}^{10} &= -T^1 - T^2 + T^3 + T^4, \\
\mathcal{K}^{11} &= S^2 - S^4 - U^2 + U^4 - Y^2 + Y^3, \\
\mathcal{K}^{12} &= R^{11} - U^{11} - V^2 + X^2, \\
\mathcal{K}^{13} &= R^{15} - U^{12} + W^2 - X^2, \\
\mathcal{K}^{14} &= R^8 - U^7 + V^3 - Y^1 + Y^4, \\
\mathcal{K}^{15} &= R^{16} + W^3 - X^3 + Y^1 - Y^4, \\
\mathcal{K}^{16} &= R^9 - U^8 + V^4 + W^6, \\
\mathcal{K}^{17} &= R^{13} - V^4 + W^7 + X^4. \tag{D.7}
\end{aligned}$$

For a homogeneous notation in (6.30) we have introduced

$$\begin{aligned}
\mathcal{K}^{18} &= V^6, & \mathcal{K}^{19} &= V^7, & \mathcal{K}^{20} &= V^8, & \mathcal{K}^{21} &= Z^2, \\
\mathcal{K}^{22} &= X^6, & \mathcal{K}^{23} &= X^7. & \mathcal{K}^{24} &= X^8. \tag{D.8}
\end{aligned}$$

D.2 Gauge Invariance

Checking gauge invariance of the amplitude is achieved by setting the polarization vector of one of the gluons equal to its momentum, $\xi^\mu = k^\mu$. This corresponds to a pure gauge configuration and the amplitude must then vanish. In the following we demonstrate the steps for checking gauge invariance in case of the first gluon, i.e. $\xi_1^\mu = k_1^\mu$, while ξ_2 is arbitrary. The kinematical terms (D.1)-(D.6) then become

$$\begin{aligned}
(R^2, R^6, R^{10}, R^{14}) &= L^1 \left(s_{12}, s_{13}, s_{14}\, s_{15} \right), & (U^9, U^{10}, U^{11}, U^{12}) &= P^1 \left(s_{12}, s_{13}, s_{14}\, s_{15} \right), \\
(R^3, R^7, R^{11}, R^{15}) &= L^2 \left(s_{12}, s_{13}, s_{14}\, s_{15} \right), & (V^5, V^6, V^7, V^8) &= P^2 \left(s_{12}, s_{13}, s_{14}\, s_{15} \right), \\
(R^4, R^8, R^{12}, R^{16}) &= L^3 \left(s_{12}, s_{13}, s_{14}\, s_{15} \right), & (W^5, W^6, W^7, W^8) &= P^3 \left(s_{12}, s_{13}, s_{14}\, s_{15} \right), \\
(R^5, R^9, R^{13}, R^{17}) &= L^4 \left(s_{12}, s_{13}, s_{14}\, s_{15} \right), & (X^5, X^6, X^7, X^8) &= P^4 \left(s_{12}, s_{13}, s_{14}\, s_{15} \right),
\end{aligned} \quad \text{(D.9)}$$

where we have defined

$$\begin{aligned}
L^1 &\equiv \tfrac{2}{\alpha'} (k_1\, \xi_2) \braket{35} [46], & P^1 &\equiv \tfrac{1}{\alpha'} k_{2\mu}\, \xi_{2\nu} (u_3\, \sigma^\mu\, \bar{u}_4)(u_5\, \sigma^\mu\, \bar{u}_6), \\
L^2 &\equiv \tfrac{2}{\alpha'} (k_3\, \xi_2) \braket{35} [46], & P^2 &\equiv \tfrac{1}{\alpha'} k_{2\mu}\, \xi_{2\nu} (u_3\, \sigma^\mu\, \bar{u}_6)(u_5\, \sigma^\mu\, \bar{u}_4), \\
L^3 &\equiv \tfrac{2}{\alpha'} (k_4\, \xi_2) \braket{35} [46], & P^3 &\equiv \tfrac{1}{\alpha'} k_{2\mu}\, \xi_{2\nu} (u_5\, \sigma^\mu\, \bar{u}_4)(u_3\, \sigma^\mu\, \bar{u}_6), \\
L^4 &\equiv \tfrac{2}{\alpha'} (k_5\, \xi_2) \braket{35} [46], & P^4 &\equiv \tfrac{1}{\alpha'} k_{2\mu}\, \xi_{2\nu} (u_5\, \sigma^\mu\, \bar{u}_6)(u_3\, \sigma^\mu\, \bar{u}_4).
\end{aligned} \quad \text{(D.10)}$$

Additional simplifications occur as the following terms coincide:

$$\begin{aligned}
R^1 &= R^2, & T^1 &= T^4, & U^5 &= X^1, & V^1 &= W^1, \\
S^1 &= S^4, & T^2 &= T^3, & U^6 &= X^2, & V^2 &= W^2, \\
S^2 &= S^3, & U^1 &= U^4, & U^7 &= X^3, & V^3 &= W^3, \\
& & U^2 &= U^3, & U^8 &= X^4, & V^4 &= W^4.
\end{aligned} \quad \text{(D.11)}$$

Further concordance is found when we imply the formula (B.6). One obtains:

$$\begin{aligned}
U^5 &= X^1 = V^1 = W^1, & U^9 + X^5 &= V^5 + W^5, \\
U^6 &= X^2 = V^2 = W^2, & U^{10} + X^6 &= V^6 + W^6, \\
U^7 &= X^3 = V^3 = W^3, & U^{11} + X^7 &= V^7 + W^7, \\
U^8 &= X^4 = V^4 = W^4, & U^{12} + X^8 &= V^8 + W^8.
\end{aligned} \quad \text{(D.12)}$$

Let us now discuss the three- and four-σ-chains (D.5) and (D.6). The terms Y^1, Y^4, Y^6, Z^1 and Z^2 vanish as a result of

$$k_\mu k_\nu (\sigma^\mu\, \bar{\sigma}^\nu \varepsilon)_{\alpha\dot{\beta}} = k_\mu k_\nu \left[(\sigma^{\mu\nu} \varepsilon)_{\alpha\dot{\beta}} - \eta^{\mu\nu} \right]. \quad \text{(D.13)}$$

The first term disappears because $\sigma^{\mu\nu}$ is antisymmetric in the vector indices, but $k_\mu k_\nu$ is symmetric. The second term simply vanishes due to $k^2 = 0$. The remaining expressions Y^i are not

D.3 Results ins Spinor Product Notation

independent. Upon using the formula (B.16) in the form of

$$(\sigma^\lambda \bar{\sigma}^\rho \varepsilon)_{\alpha\gamma} (\varepsilon \bar{\sigma}^\mu \sigma^\nu)_{\dot{\beta}\dot{\delta}} = (\sigma^\lambda \bar{\sigma}^\rho \sigma^\mu)_{\alpha\dot{\beta}} \sigma^\nu_{\gamma\dot{\delta}} - (\sigma^\lambda \bar{\sigma}^\rho \sigma^\nu)_{\alpha\dot{\delta}} \sigma^\mu_{\gamma\dot{\beta}},$$
$$(\sigma^\mu \bar{\sigma}^\nu \varepsilon)_{\alpha\gamma} (\varepsilon \bar{\sigma}^\lambda \sigma^\rho)_{\dot{\beta}\dot{\delta}} = (\sigma^\nu \bar{\sigma}^\lambda \sigma^\rho)_{\gamma\dot{\beta}} \sigma^\mu_{\alpha\dot{\delta}} - (\sigma^\mu \bar{\sigma}^\lambda \sigma^\rho)_{\alpha\dot{\delta}} \sigma^\nu_{\gamma\dot{\beta}} \quad (D.14)$$

we find for the leftover kinematic terms

$$Y^2 = Y^3 + U_1 - U_2 - S_1 + S_2, \qquad Y^3 = Y^5. \quad (D.15)$$

These results taken together lead to tremendous simplifications for the $\mathcal{K}$'s:

$$\begin{aligned}
\mathcal{K}^1 &= (1 - s_{12}) L^1, & \mathcal{K}^{15} &= s_{15} L^3, \\
\mathcal{K}^2 &= (s_{12} + s_{13}) (L^2 - P^3 + P^4) - s_{13} P^2, & \mathcal{K}^{16} &= s_{13} (L^4 + P^3), \\
\mathcal{K}^3 &= (s_{12} + s_{14}) L^3 - s_{12} (P^2 - P^4), & \mathcal{K}^{17} &= s_{14} (L^4 + P^3), \\
\mathcal{K}^4 &= (s_{12} + s_{15}) (L^4 + P^3) - s_{12} P^4, & \mathcal{K}^{18} &= s_{13} P^2, \\
\mathcal{K}^5 &= -s_{13} (L^1 + L^2 - P^2 - P^3 + P^4), & \mathcal{K}^{19} &= s_{14} P^2, \\
\mathcal{K}^6 &= -s_{14} (L^1 + L^3), & \mathcal{K}^{20} &= s_{15} P^2, \\
\mathcal{K}^7 &= -s_{15} (L^1 + L^4 - P^3), & \mathcal{K}^{22} &= s_{13} P^4, \\
\mathcal{K}^{12} &= s_{14} (L^2 - P^2 - P^3 + P^4), & \mathcal{K}^{23} &= s_{14} P^4, \\
\mathcal{K}^{13} &= s_{15} (L^2 - P^2 - P^3 + P^4), & \mathcal{K}^{24} &= s_{15} P^4. \\
\mathcal{K}^{14} &= s_{13} L^3, & & \quad (D.16)
\end{aligned}$$

In terms of these kinematical structures the amplitude must vanish on account of the pure gauge configuration of the first gluon. The seven terms above L^1, L^2, L^3, L^4 and P^2, P^3, P^4, are independent. The vanishing of the amplitude must therefore be due to cancellations between the integral terms H_i. This circumstance yields precisely the equations in (6.33), while similar calculations for inspecting gauge invariance of the second gluon lead to (6.34) and (6.35).

D.3 Results ins Spinor Product Notation

The first gluon is chosen to have negative helicity and reference momentum k_4. The second has positive helicity and we take k_5 as reference momentum. The polarization vectors become for this choice:

$$\xi_1^{-\mu} = -\frac{1}{\sqrt{2}} \frac{\bar{\sigma}^{\mu\dot{\alpha}\alpha} k_{1\alpha} \bar{k}_{4\dot{\alpha}}}{[14]}, \qquad \xi_2^{+\mu} = -\frac{1}{\sqrt{2}} \frac{\bar{\sigma}^{\mu\dot{\alpha}\alpha} k_{5\alpha} \bar{k}_{2\dot{\alpha}}}{\langle 52 \rangle}. \quad (D.17)$$

Multiplication of the polarization vectors with another four-momentum results in

$$k_{i\mu}\xi_1^{-\mu} = -\frac{1}{\sqrt{2}}\frac{\langle 1i\rangle [i4]}{[14]}, \qquad k_{i\mu}\xi_2^{+\mu} = -\frac{1}{\sqrt{2}}\frac{\langle 5i\rangle [i2]}{\langle 52\rangle}. \tag{D.18}$$

Together with $k_\mu \sigma^\mu_{\alpha\dot\alpha} = k_\alpha \bar{k}_{\dot\alpha}$ the kinematic terms (D.1)-(D.6) can then entirely be written in terms of spinor products. Due to $\langle ii\rangle = [jj] = 0$ many of these terms vanish. The remaining ones are:

$$\left(R^2, R^3, R^4\right) = 2\frac{\langle 12\rangle[24]\langle 35\rangle[46]}{[14]\langle 52\rangle}\left(\langle 51\rangle[12], \langle 53\rangle[32], \langle 54\rangle[42]\right),$$

$$\left(R^6, R^7, R^8\right) = 2\frac{\langle 13\rangle[34]\langle 35\rangle[46]}{[14]\langle 52\rangle}\left(\langle 51\rangle[12], \langle 53\rangle[32], \langle 54\rangle[42]\right),$$

$$\left(R^{14}, R^{15}, R^{16}\right) = 2\frac{\langle 15\rangle[54]\langle 35\rangle[46]}{[14]\langle 52\rangle}\left(\langle 51\rangle[12], \langle 53\rangle[32], \langle 54\rangle[42]\right),$$

$$\left(U^5, U^6, U^7\right) = 2\frac{\langle 31\rangle[14]\langle 15\rangle[46]}{[14]\langle 52\rangle}\left(\langle 51\rangle[12], \langle 53\rangle[32], \langle 54\rangle[42]\right),$$

$$\left(X^5, X^6, X^8\right) = 2\frac{\langle 25\rangle[24]\langle 35\rangle[26]}{[14]\langle 52\rangle}\left(\langle 12\rangle[24], \langle 13\rangle[34], \langle 15\rangle[54]\right),$$

$$\left(S^2, T^2, U^2\right) = 2\frac{\langle 15\rangle[24]}{[14]\langle 52\rangle}\left(\langle 12\rangle[26]\langle 35\rangle[41], \langle 13\rangle[12]\langle 25\rangle[46], \langle 13\rangle[14]\langle 52\rangle[26]\right). \tag{D.19}$$

Additionally, it turns out that the following terms coincide,

$$\begin{aligned}
R^1 &= R^2, & W^1 &= U^5, & W^5 &= X^5, & Y^3 &= -Y^4 = Y^5 = U^2, \\
S^4 &= S^2, & W^2 &= U^6, & W^6 &= X^6, & & \\
T^3 &= T^2, & W^3 &= U^7, & W^7 &= X^7, & &
\end{aligned} \tag{D.20}$$

while all other expressions in (D.1)-(D.6) vanish. With these findings the kinematics $\mathcal{K}^i$ simplify considerably. The results can further be reduced with the help of the relations found from gauge invariance. Ultimately, only eight factors contribute to the amplitude:

$$\begin{aligned}
\mathcal{K}^1 &= (1/s_1 - 1)R^1, & \mathcal{K}^7 &= -R^{14} - U^5, \\
\mathcal{K}^2 &= R^3 + R^7 - U^6, & \mathcal{K}^{13} &= R^{15} + U^6, \\
\mathcal{K}^3 &= R^4 + X^5, & \mathcal{K}^{14} &= R^8 - U^2 - U^7 + X^6, \\
\mathcal{K}^5 &= -R^6 - R^7 + U^5 + U^6, & \mathcal{K}^{15} &= R^{16} + U^2 + U^7 + X^8.
\end{aligned} \tag{D.21}$$

Inserting the respective expressions above yields (6.43).

APPENDIX E

Spinor Helicity Formalism

In this Appendix we present further details regarding the spinor helicity formalism in four space-time dimensions [162–164]. The following presentation is inspired by [165, 173].

E.1 Clifford Algebra

First, let us present our conventions which follow [21]. The metric in our discussion takes the form $\eta_{\mu\nu} = \text{diag}(-1, +1, +1, +1)$. Gamma matrices, representations of the Clifford algebra

$$\{\Gamma^\mu, \Gamma^\nu\} = -2\eta^{\mu\nu}, \tag{E.1}$$

are for example given by[1]

$$\Gamma^\mu = \begin{pmatrix} 0 & \sigma^\mu \\ \bar{\sigma}^\mu & 0 \end{pmatrix}, \qquad \Gamma_5 \equiv -i\,\Gamma^0\,\Gamma^1\,\Gamma^2\,\Gamma^3 = \begin{pmatrix} -1 & 0 \\ 0 & 1 \end{pmatrix}, \tag{E.2}$$

with $\sigma^\mu = (-1, -\sigma^i)$, $\bar{\sigma}^\mu = (-1, \sigma^i)$ and σ^i the standard Pauli matrices

$$\sigma^1 = \begin{pmatrix} 0 & 1 \\ 1 & 0 \end{pmatrix}, \qquad \sigma^2 = \begin{pmatrix} 0 & -i \\ i & 0 \end{pmatrix}, \qquad \sigma^3 = \begin{pmatrix} 1 & 0 \\ 0 & -1 \end{pmatrix}. \tag{E.3}$$

In component form the matrices have the index structure $\sigma^\mu_{\alpha\dot\beta}$, $\bar{\sigma}^{\mu\,\dot\alpha\beta}$ and behave under complex conjugation as

$$(\sigma^\mu_{\alpha\dot\beta})^* = \sigma^\mu_{\beta\dot\alpha}, \qquad (\bar{\sigma}^{\mu\,\dot\alpha\beta})^* = \bar{\sigma}^{\mu\,\dot\beta\alpha}. \tag{E.4}$$

Spinor indices can be raised and lowered using the diagonal blocks of the charge conjugation matrix $\varepsilon_{\alpha\beta}$, $\varepsilon^{\dot\alpha\dot\beta}$. These are antisymmetric 2×2 matrices, whose inverse is denoted by $\varepsilon^{\alpha\beta}$ and

---

[1] This representation is very convenient to use in the following, but does not coincide with the representation constructed in Appendix A.4.

$\varepsilon_{\dot\alpha\dot\beta}$ respectively. We use $\varepsilon_{12} = \varepsilon^{21} = -1$. With the help of these matrices σ^μ and $\bar\sigma^\mu$ are related via

$$\sigma^\mu_{\alpha\dot\alpha} = \varepsilon_{\alpha\beta}\,\varepsilon_{\dot\alpha\dot\beta}\,\bar\sigma^{\mu\,\dot\beta\beta}\,. \tag{E.5}$$

E.2 Momentum Spinors

Any four-momentum k_μ of a particle with mass m can be contracted with $\sigma^\mu_{\alpha\dot\alpha}$,

$$k \equiv k_\mu\,\sigma^\mu = \begin{pmatrix} k^0 + k^3 & k^1 - ik^2 \\ k^1 + ik^2 & k^0 - k^3 \end{pmatrix}. \tag{E.6}$$

In index notation this reads $k_{\alpha\dot\alpha} = k_\mu\,\sigma^\mu_{\alpha\dot\alpha}$ and $k^{\dot\alpha\alpha} = k_\mu\,\bar\sigma^{\mu\,\dot\alpha\alpha}$. Applying the identity $\mathrm{Tr}(\sigma^\mu\,\bar\sigma^\nu) = -2\,\eta^{\mu\nu}$ these relations can be inverted

$$k_\mu = -\frac{1}{2}\,k_{\alpha\dot\alpha}\,\bar\sigma^{\mu\,\dot\alpha\alpha} = -\frac{1}{2}\,k^{\dot\alpha\alpha}\,\sigma^\mu_{\alpha\dot\alpha}\,. \tag{E.7}$$

For real momenta the matrix k is hermitian and can therefore be written as

$$k = \sum_{i=1}^{2} \kappa_i\,\mathbf{v}_i\,\mathbf{v}_i^\dagger\,, \tag{E.8}$$

where $\mathbf{v}_i$ are the eigenvectors and κ_i the corresponding eigenvalues [174]. The determinant of k yields

$$\det k = -k_\mu\,k^\mu = -m^2\,. \tag{E.9}$$

In our work we are dealing with massless particles. Therefore the determinant vanishes and hence k has only one eigenvector $\mathbf{v}$ and eigenvalue λ. Following (E.8) the matrix in component notation can be decomposed as

$$k_{\alpha\dot\alpha} = k_\alpha\,\bar k_{\dot\alpha}\,, \tag{E.10}$$

where we have defined $k_\alpha \equiv \sqrt{\kappa}\,\mathbf{v}_\alpha$. The quantities k_α, $\bar k_{\dot\alpha}$ are obviously related by complex conjugation, $k_\alpha = (\bar k_{\dot\alpha})^*$. The same decomposition holds for

$$k^{\dot\alpha\alpha} = \bar k^{\dot\alpha}\,k^\alpha \tag{E.11}$$

with $k^\alpha = (\bar k^{\dot\alpha})^*$. Due to (E.5) these are related to the previously introduced k_α, $\bar k_{\dot\alpha}$ by

$$k_\alpha = \varepsilon_{\alpha\beta}\,k^\beta\,, \qquad \bar k_{\dot\alpha} = \varepsilon_{\dot\alpha\dot\beta}\,\bar k^{\dot\beta}\,. \tag{E.12}$$

E.2 Momentum Spinors

The commuting quantities k_α and $\bar{k}_{\dot\alpha}$ carry spinor indices and are therefore referred to as *momentum spinors*. Note that we have not made use of the Dirac equation and hence, the introduction of momentum spinors is possible for massless bosons and fermions. In the latter case k_α and $\bar{k}_{\dot\alpha}$ are chiral solutions of the Dirac equation, i.e. solutions with definite helicity $\pm$. This fact lends its name to the formalism.

We prove this by directly constructing chiral, plane-wave solutions of the massless Dirac equation

$$\slashed{k}\, u(k) = \begin{pmatrix} k_{\alpha\dot\alpha}\, \bar\chi^{\dot\alpha} \\ k^{\dot\alpha\alpha}\, \lambda_\alpha \end{pmatrix} = 0\,, \tag{E.13}$$

where $u(k) = (\lambda_\alpha, \bar\chi^{\dot\alpha})^t$ and λ, χ are two-component Weyl spinors. Due to the block-diagonal form of Γ_5 in (E.2)

$$u_-(k) = \begin{pmatrix} \lambda_\alpha \\ 0 \end{pmatrix}, \qquad u_+(k) = \begin{pmatrix} 0 \\ \bar\lambda^{\dot\alpha} \end{pmatrix} \tag{E.14}$$

satisfy the identities

$$P_\pm\, u_\pm(k) = u_\pm(k)\,, \qquad P_\pm\, u_\mp(k) = 0 \tag{E.15}$$

with the projector $P_\pm \equiv \frac{1}{2}(1\pm\Gamma_5)$. Hence, for appropriate λ_α and $\bar\lambda^{\dot\alpha}$ the spinors $u_\pm$ are solutions to (E.13) with definite helicity $\pm$. We define the Dirac conjugate of $u(k)$ by $\bar u(k) \equiv -u(k)^\dagger\, \Gamma^0$. Then

$$\bar u_-(k) = (0, \bar\lambda_{\dot\alpha})\,, \qquad \bar u_+(k) = (\lambda^\alpha, 0)\,, \tag{E.16}$$

where $(\lambda_\alpha)^* = \bar\lambda_{\dot\alpha}$, $(\lambda^\alpha)^* = \bar\lambda^{\dot\alpha}$. Indices can be raised and lowered in the same fashion as in (E.12). The four-momentum entering (E.13) can be parameterized in polar coordinates as[2]

$$k^\mu = (E, E\sin\theta\cos\phi, E\sin\theta\sin\phi, E\cos\phi)\,. \tag{E.17}$$

Then (E.6) becomes

$$k_{\alpha\dot\alpha} = 2E\begin{pmatrix} c^2 & s\,c\,e^{-i\phi} \\ s\,c\,e^{i\phi} & s^2 \end{pmatrix}, \qquad k^{\dot\alpha\alpha} = 2E\begin{pmatrix} s^2 & -s\,c\,e^{-i\phi} \\ -s\,c\,e^{i\phi} & c^2 \end{pmatrix} \tag{E.18}$$

with the abbreviations $s \equiv \sin(\theta/2)$ and $c \equiv \cos(\theta/2)$. The eigenvalue of $k_{\alpha\dot\alpha}$ is $\kappa = 2E$ and the corresponding eigenvector $\mathbf{v} = (c\,e^{-i\phi}, s)^t$. As before we set $k_\alpha = \sqrt{2E}\,\mathbf{v}_\alpha$ and find

$$k_\alpha = \sqrt{2E}\begin{pmatrix} c\,e^{-i\phi} \\ s \end{pmatrix}, \qquad \bar k_{\dot\alpha} = \sqrt{2E}\,(c\,e^{i\phi}, s)\,,$$

---

[2] In the following we attach vector and spinor indices to the quantities in order to distinguish them from each other.

$$\bar{k}^{\dot{\alpha}} = \sqrt{2E} \begin{pmatrix} s \\ -c\, e^{i\phi} \end{pmatrix}, \qquad k^{\alpha} = \sqrt{2E}\,(s, -c\, e^{-i\phi}). \tag{E.19}$$

With the explicit expressions (E.18) and (E.19) one can now check that indeed

$$k_{\alpha\dot{\alpha}} = k_{\alpha}\,\bar{k}_{\dot{\alpha}}, \qquad k^{\dot{\alpha}\alpha} = \bar{k}^{\dot{\alpha}}\,k^{\alpha}. \tag{E.20}$$

Additionally these satisfy

$$k_{\alpha\dot{\alpha}}\,\bar{k}^{\dot{\alpha}} = k^{\dot{\alpha}\alpha}\,k_{\alpha} = 0, \qquad \bar{k}_{\dot{\alpha}}\,k^{\dot{\alpha}\alpha} = k^{\alpha}\,k_{\alpha\dot{\alpha}} = 0. \tag{E.21}$$

By comparison with (E.13) this proves that $u_{\pm}(k)$ satisfy the Dirac equation if one identifies $\lambda_{\alpha} \equiv k_{\alpha}$ and $\bar{\lambda}^{\dot{\alpha}} \equiv \bar{k}^{\dot{\alpha}}$.

E.3   Spinor Products

Using $\sigma^{\mu}_{\alpha\dot{\alpha}}\,\sigma_{\mu\beta\dot{\beta}} = -2\,\varepsilon_{\alpha\beta}\,\varepsilon_{\dot{\alpha}\dot{\beta}}$ and the fact that k_{α}, $\bar{k}_{\dot{\alpha}}$ are commuting quantities we can write the Mandelstam variable s_{ij} in terms of momentum spinors:

$$s_{ij} \equiv (k_i + k_j)^2 = 2\,k_i\,k_j = \frac{1}{2}\,k_i^{\dot{\alpha}\alpha}\,k_j^{\dot{\beta}\beta}\,\sigma^{\mu}_{\alpha\dot{\alpha}}\,\sigma_{\mu\beta\dot{\beta}} = \langle ij \rangle\,[ij]. \tag{E.22}$$

Here we have introduced the bracket notation for spinor products

$$\langle p\,q \rangle \equiv p^{\alpha}\,q_{\alpha}, \qquad [p\,q] \equiv \bar{p}_{\dot{\alpha}}\,\bar{q}^{\dot{\alpha}} \tag{E.23}$$

as in [108] and the short-hand notation $\langle k_i\,k_j \rangle \equiv \langle i\,j \rangle$ and $[k_i k_j] \equiv [i\,j]$. In the case of fermions the bras and kets can be associated to the Weyl spinors (E.14) and (E.16):

$$\langle k| = \bar{u}_{+}(k), \qquad |k\rangle = u_{-}(k), \qquad [k| = \bar{u}_{-}(k), \qquad |k] = u_{+}(k). \tag{E.24}$$

The antisymmetry of $\varepsilon_{\alpha\beta}$ establishes that the spinor products are also antisymmetric,

$$\langle i\,j \rangle = -\langle j\,i \rangle, \qquad [i\,j] = -[j\,i] \quad \Rightarrow \quad \langle i\,i \rangle = [i\,i] = 0. \tag{E.25}$$

For real momenta the different brackets are related by complex conjugation, $\langle i\,j \rangle^{*} = -[j\,i]$.

In the manipulations of spinor products the following relations are helpful. The Fierz identity

$$\varepsilon_{\alpha\gamma}\,\varepsilon_{\beta\delta} = \varepsilon_{\alpha\beta}\,\varepsilon_{\gamma\delta} + \varepsilon_{\alpha\delta}\,\varepsilon_{\beta\gamma} \tag{E.26}$$

E.4 Polarization Vectors

yields upon multiplication with momentum spinors the Schouten identity

$$\langle i\,j\rangle\,\langle k\,l\rangle = \langle i\,k\rangle\,\langle j\,l\rangle + \langle i\,l\rangle\,\langle k\,j\rangle. \tag{E.27}$$

For n-particle scattering momentum conservation $\sum_{i=1}^n k_i^\mu = 0$ can be formulated in terms of momentum spinors. For this purpose we apply (E.6) and (E.10) and multiply with two further spinors in order to find

$$\sum_{\substack{i=1\\i\neq j,k}}^n [j\,i]\,\langle i\,k\rangle = 0. \tag{E.28}$$

E.4 Polarization Vectors

The polarization vectors entering the calculation of amplitudes through the gluon vertex operators with momentum k^μ and reference momentum r^μ become in the helicity formalism

$$\xi^{+\,\mu}(k,r) = -\frac{1}{\sqrt{2}}\frac{\bar{\sigma}^{\mu\,\dot{\alpha}\alpha}\,r_\alpha\,\bar{k}_{\dot{\alpha}}}{\langle r\,k\rangle}, \qquad \xi^{-\,\mu}(k,r) = -\frac{1}{\sqrt{2}}\frac{\bar{\sigma}^{\mu\,\dot{\alpha}\alpha}\,k_\alpha\,\bar{r}_{\dot{\alpha}}}{[k\,r]}, \tag{E.29}$$

where ξ^+ comes with a right-handed, ξ^- with a left-handed gluon state. Upon contraction with $\sigma^\mu_{\alpha\dot\alpha}$ these yield

$$\xi^+_{\alpha\dot\alpha}(k,r) = \sqrt{2}\,\frac{r_\alpha\,\bar{k}_{\dot\alpha}}{\langle r\,k\rangle}, \qquad \xi^-_{\alpha\dot\alpha}(k,r) = \sqrt{2}\,\frac{k_\alpha\,\bar{r}_{\dot\alpha}}{[k\,r]}. \tag{E.30}$$

It is easy to show that

$$\xi^{+\,\mu}\xi^-_\mu = 1, \qquad \xi^{+\,\mu}\xi^+_\mu = \xi^{-\,\mu}\xi^-_\mu = 0 \tag{E.31}$$

if the polarizations have the same momentum and reference vector. Multiplying a polarization vector with the momentum of another particle gives

$$k_{i\mu}\,\xi^{+\,\mu}(k,r) = \frac{1}{\sqrt{2}}\frac{\langle r\,i\rangle\,[k\,i]}{\langle r\,k\rangle}, \qquad k_{i\mu}\,\xi^{-\,\mu}(k,r) = \frac{1}{\sqrt{2}}\frac{\langle k\,i\rangle\,[r\,i]}{[k\,r]}. \tag{E.32}$$

Then it is clear that these products vanish for $k_{i\mu} = k_\mu$ or $k_{i\mu} = r_\mu$.

The reference momentum r entering (E.29) can be freely chosen. It shifts under the change $r \to s$ by an amount proportional to the momentum of the gluon,

$$\begin{aligned}\xi^{+\,\mu}(k,r) - \xi^{+\,\mu}(k,s) &= \frac{1}{\sqrt{2}}\frac{r_\alpha\,s_\beta - r_\beta\,s_\alpha}{\langle r\,k\rangle\,\langle s\,k\rangle}\,(\varepsilon\,\sigma^\nu\,\bar\sigma^\mu)^{\beta\alpha}\,k_\nu \\ &= \frac{1}{\sqrt{2}}\frac{r_\alpha\,s_\beta}{\langle r\,k\rangle\,\langle s\,k\rangle}\,(\varepsilon\,\sigma^\nu\,\bar\sigma^\mu + \varepsilon\,\sigma^\mu\,\bar\sigma^\nu)^{\beta\alpha}\,k_\nu \\ &= \sqrt{2}\,\frac{\langle r\,s\rangle}{\langle r\,k\rangle\,\langle s\,k\rangle}\,k^\mu. \end{aligned} \tag{E.33}$$

In the last step we have used the Clifford algebra (E.1). In the same way one can show that

$$\xi^{-\mu}(k,r) - \xi^{-\mu}(k,s) = -\sqrt{2}\,\frac{[r\,s]}{[k\,r]\,[k\,s]}\,k^\mu\,. \tag{E.34}$$

Shifting ξ by an amount proportional to its momentum corresponds to an on-shell gauge transformation acting on the gluon. The choice of the reference momentum is therefore a gauge choice.

Bibliography

[1] D. Härtl, O. Schlotterer and S. Stieberger, *Higher point spin field correlators in D=4 superstring theory*, Nucl.Phys. **B834** (2010) 163–221 [arXiv:0911.5168].

[2] D. Härtl and O. Schlotterer, *Higher loop spin field correlators in various dimensions*, Nucl.Phys. **B849** (2011) 364–409 [arXiv:1011.1249].

[3] D. Härtl, *Spin field correlators in various dimensions*, Nucl.Phys.Proc.Suppl. **216** (2011) 231–233.

[4] D. Härtl, O. Schlotterer and S. Stieberger, *Couplings of brane & bulk string states vs. pure brane couplings*, to appear.

[5] S. Glashow, *Partial Symmetries of Weak Interactions*, Nucl.Phys. **22** (1961) 579–588.

[6] S. Weinberg, *A Model of Leptons*, Phys.Rev.Lett. **19** (1967) 1264–1266.

[7] A. Salam, *Weak and Electromagnetic Interactions*, Proceedings Of The Nobel Symposium (1968) 367–377.

[8] M. E. Peskin and D. V. Schroeder, *An Introduction to quantum field theory*, Addison-Wesley (1995).

[9] P. W. Higgs, *Broken symmetries, massless particles and gauge fields*, Phys.Lett. **12** (1964) 132–133.

[10] F. Englert and R. Brout, *Broken Symmetry and the Mass of Gauge Vector Mesons*, Phys.Rev.Lett. **13** (1964) 321–322.

[11] G. Guralnik, C. Hagen and T. Kibble, *Global Conservation Laws and Massless Particles*, Phys.Rev.Lett. **13** (1964) 585–587.

[12] P. Langacker, *Introduction to the Standard Model and Electroweak Physics*, [arXiv:0901.0241].

[13] S. Weinberg, *Implications of Dynamical Symmetry Breaking*, Phys.Rev. **D13** (1976) 974–996.

[14] E. Gildener, *Gauge Symmetry Hierarchies*, Phys.Rev. **D14** (1976) 1667.

[15] L. Susskind, *Dynamics of Spontaneous Symmetry Breaking in the Weinberg-Salam Theory*, Phys.Rev. **D20** (1979) 2619–2625.

[16] **Particle Data Group** Collaboration, K. Nakamura et. al., *Review of particle physics*, J.Phys.G **G37** (2010) 075021.

[17] J. E. Kim and G. Carosi, *Axions and the Strong CP Problem*, Rev.Mod.Phys. **82** (2010) 557–602 [arXiv:0807.3125].

[18] **WMAP Collaboration** Collaboration, E. Komatsu et. al., *Seven-Year Wilkinson Microwave Anisotropy Probe (WMAP) Observations: Cosmological Interpretation*, Astrophys.J.Suppl. **192** (2011) 18 [arXiv:1001.4538].

[19] R. Peccei and H. R. Quinn, *CP Conservation in the Presence of Instantons*, Phys.Rev.Lett. **38** (1977) 1440–1443.

[20] R. Peccei and H. R. Quinn, *Constraints Imposed by CP Conservation in the Presence of Instantons*, Phys.Rev. **D16** (1977) 1791–1797.

[21] J. Wess and J. Bagger, *Supersymmetry and supergravity*, Princeton University Press (1992).

[22] R. Haag, J. T. Lopuszanski and M. Sohnius, *All Possible Generators of Supersymmetries of the S Matrix*, Nucl.Phys. **B88** (1975) 257.

[23] S. R. Coleman and J. Mandula, *All Possible Symmetries Of The S Matrix*, Phys.Rev. **159** (1967) 1251–1256.

[24] S. P. Martin, *A Supersymmetry primer*, [hep-ph/9709356].

[25] H. Georgi and S. Glashow, *Unity of All Elementary Particle Forces*, Phys.Rev.Lett. **32** (1974) 438–441.

[26] H. Fritzsch and P. Minkowski, *Unified Interactions of Leptons and Hadrons*, Annals Phys. **93** (1975) 193–266.

[27] A. Einstein, *On the General Theory of Relativity*, Sitzungsber.Preuss.Akad.Wiss.Berlin (Math.Phys.) **1915** (1915) 778–786.

Bibliography

[28] R. M. Wald, *General Relativity*, The University of Chicago Press (1984).

[29] S. M. Carroll, *Spacetime and geometry: An introduction to general relativity*, Addison-Wesley (2004).

[30] S. Weinberg, *The Cosmological Constant Problem*, Rev.Mod.Phys. **61** (1989) 1–23.

[31] J. D. Bekenstein, *Black holes and entropy*, Phys.Rev. **D7** (1973) 2333–2346.

[32] S. Hawking, *Particle Creation by Black Holes*, Commun.Math.Phys. **43** (1975) 199–220.

[33] M. B. Green, J. Schwarz and E. Witten, *Superstring Theory. Vol. 1: Introduction*, Cambridge University Press (1987).

[34] M. B. Green, J. Schwarz and E. Witten, *Superstring Theory. Vol. 2: Loop Amplitudes, Anomalies and Phenomenology*, Cambridge University Press (1987).

[35] J. Polchinski, *String theory. Vol. 1: An introduction to the bosonic string*, Cambridge University Press (1998).

[36] J. Polchinski, *String theory. Vol. 2: Superstring theory and beyond*, Cambridge University Press (1998).

[37] G. Veneziano, *Construction of a crossing - symmetric, Regge behaved amplitude for linearly rising trajectories*, Nuovo Cim. **A57** (1968) 190–197.

[38] T. Thiemann, *Lectures on loop quantum gravity*, Lect.Notes Phys. **631** (2003) 41–135 [`gr-qc/0210094`].

[39] E. Witten, *String theory dynamics in various dimensions*, Nucl.Phys. **B443** (1995) 85–126 [`hep-th/9503124`].

[40] D. Tong, *String Theory*, [`arXiv:0908.0333`].

[41] F. Gliozzi, J. Scherk and D. I. Olive, *Supersymmetry, Supergravity Theories and the Dual Spinor Model*, Nucl.Phys. **B122** (1977) 253–290.

[42] M. B. Green and J. H. Schwarz, *Anomaly Cancellation in Supersymmetric D=10 Gauge Theory and Superstring Theory*, Phys.Lett. **B149** (1984) 117–122.

[43] J. Polchinski, *Dirichlet branes and Ramond-Ramond charges*, Phys.Rev.Lett. **75** (1995) 4724–4727 [`hep-th/9510017`].

[44] R. Blumenhagen, B. Kors, D. Lüst and S. Stieberger, *Four-dimensional string compactifications with D-branes, orientifolds and fluxes*, Phys.Rept. **445** (2007) 1–193 [hep-th/0610327].

[45] A. Strominger and C. Vafa, *Microscopic origin of the Bekenstein-Hawking entropy*, Phys.Lett. **B379** (1996) 99–104 [hep-th/9601029].

[46] e. Erdmenger, Johanna, *String cosmology: Modern string theory concepts from the Big Bang to cosmic structure*, Wiley-VCH (2009).

[47] T. Weigand, *Lectures on F-theory compactifications and model building*, Class.Quant.Grav. **27** (2010) 214004 [arXiv:1009.3497].

[48] S. Raby, *Searching for the Standard Model in the String Landscape: SUSY GUTs*, Rept.Prog.Phys. **74** (2011) 036901 [arXiv:1101.2457].

[49] N. Seiberg and E. Witten, *String theory and noncommutative geometry*, JHEP **9909** (1999) 032 [hep-th/9908142].

[50] D. Lüst, *T-duality and closed string non-commutative (doubled) geometry*, JHEP **1012** (2010) 084 [arXiv:1010.1361].

[51] R. Blumenhagen and E. Plauschinn, *Nonassociative Gravity in String Theory?*, J.Phys.A **A44** (2011) 015401 [arXiv:1010.1263].

[52] J. M. Maldacena, *The Large N limit of superconformal field theories and supergravity*, Adv.Theor.Math.Phys. **2** (1998) 231–252 [hep-th/9711200].

[53] S. Gubser, I. R. Klebanov and A. M. Polyakov, *Gauge theory correlators from noncritical string theory*, Phys.Lett. **B428** (1998) 105–114 [hep-th/9802109].

[54] E. Witten, *Anti-de Sitter space and holography*, Adv.Theor.Math.Phys. **2** (1998) 253–291 [hep-th/9802150].

[55] J. Erdmenger, N. Evans, I. Kirsch and E. Threlfall, *Mesons in Gauge/Gravity Duals - A Review*, Eur.Phys.J. **A35** (2008) 81–133 [arXiv:0711.4467].

[56] S. A. Hartnoll, C. P. Herzog and G. T. Horowitz, *Building a Holographic Superconductor*, Phys.Rev.Lett. **101** (2008) 031601 [arXiv:0803.3295].

[57] S. A. Hartnoll, C. P. Herzog and G. T. Horowitz, *Holographic Superconductors*, JHEP **0812** (2008) 015 [arXiv:0810.1563].

[58] M. Ammon, J. Erdmenger, M. Kaminski and P. Kerner, *Superconductivity from gauge/gravity duality with flavor*, Phys.Lett. **B680** (2009) 516–520 [arXiv:0810.2316].

[59] M. Ammon, J. Erdmenger, M. Kaminski and P. Kerner, *Flavor Superconductivity from Gauge/Gravity Duality*, JHEP **0910** (2009) 067 [arXiv:0903.1864].

[60] P. Candelas, G. T. Horowitz, A. Strominger and E. Witten, *Vacuum Configurations for Superstrings*, Nucl.Phys. **B258** (1985) 46–74.

[61] M. Graña, *Flux compactifications in string theory: A Comprehensive review*, Phys.Rept. **423** (2006) 91–158 [hep-th/0509003].

[62] S. Kachru, R. Kallosh, A. D. Linde and S. P. Trivedi, *De Sitter vacua in string theory*, Phys.Rev. **D68** (2003) 046005 [hep-th/0301240].

[63] O. Lebedev, H. P. Nilles and M. Ratz, *De Sitter vacua from matter superpotentials*, Phys.Lett. **B636** (2006) 126–131 [hep-th/0603047].

[64] R. Bousso and J. Polchinski, *Quantization of four form fluxes and dynamical neutralization of the cosmological constant*, JHEP **0006** (2000) 006 [hep-th/0004134].

[65] S. Cullen, M. Perelstein and M. E. Peskin, *TeV strings and collider probes of large extra dimensions*, Phys.Rev. **D62** (2000) 055012 [hep-ph/0001166].

[66] D. Lüst, S. Stieberger and T. R. Taylor, *The LHC string hunter's companion*, Nucl.Phys. **B808** (2009) 1–52 [arXiv:0807.3333].

[67] D. Lüst, O. Schlotterer, S. Stieberger and T. Taylor, *The LHC string hunter's companion (II): five-particle amplitudes and universal properties*, Nucl.Phys. **B828** (2010) 139–200 [arXiv:0908.0409].

[68] W.-Z. Feng, D. Lüst, O. Schlotterer, S. Stieberger and T. R. Taylor, *Direct production of lightest Regge resonances*, Nucl.Phys. **B843** (2011) 570–601 [arXiv:1007.5254].

[69] N. Arkani-Hamed, S. Dimopoulos and G. Dvali, *The Hierarchy problem and new dimensions at a millimeter*, Phys.Lett. **B429** (1998) 263–272 [hep-ph/9803315].

[70] I. Antoniadis, N. Arkani-Hamed, S. Dimopoulos and G. Dvali, *New dimensions at a millimeter to a Fermi and superstrings at a TeV*, Phys.Lett. **B436** (1998) 257–263 [hep-ph/9804398].

[71] N. Arkani-Hamed, S. Dimopoulos and G. Dvali, *Phenomenology, astrophysics and cosmology of theories with submillimeter dimensions and TeV scale quantum gravity*, Phys.Rev. **D59** (1999) 086004 [hep-ph/9807344].

[72] **CDF** Collaboration, T. Aaltonen et. al., *Invariant Mass Distribution of Jet Pairs Produced in Association with a W boson in $p\bar{p}$ Collisions at $\sqrt{s} = 1.96$ TeV*, Phys.Rev.Lett. **106** (2011) 171801 [arXiv:1104.0699].

[73] L. A. Anchordoqui, H. Goldberg, X. Huang, D. Lüst and T. R. Taylor, *Stringy origin of Tevatron Wjj anomaly*, [arXiv:1104.2302].

[74] L. A. Anchordoqui, H. Goldberg, S. Nawata and T. R. Taylor, *Direct photons as probes of low mass strings at the CERN LHC*, Phys.Rev. **D78** (2008) 016005 [arXiv:0804.2013].

[75] L. A. Anchordoqui, H. Goldberg, D. Lüst, S. Stieberger and T. R. Taylor, *String Phenomenology at the LHC*, Mod.Phys.Lett. **A24** (2009) 2481–2490 [arXiv:0909.2216].

[76] L. A. Anchordoqui, H. Goldberg, D. Lüst, S. Nawata, S. Stieberger et. al., *Dijet signals for low mass strings at the LHC*, Phys.Rev.Lett. **101** (2008) 241803 [arXiv:0808.0497].

[77] L. A. Anchordoqui, H. Goldberg, D. Lüst, S. Nawata, S. Stieberger et. al., *LHC phenomenology for string hunters*, Nucl.Phys. **B821** (2009) 181–196 [arXiv:0904.3547].

[78] **CMS** Collaboration, V. Khachatryan et. al., *Search for Dijet Resonances in 7 TeV pp Collisions at CMS*, Phys.Rev.Lett. **105** (2010) 211801 [arXiv:1010.0203].

[79] D. Lüst, P. Mayr, R. Richter and S. Stieberger, *Scattering of gauge, matter, and moduli fields from intersecting branes*, Nucl.Phys. **B696** (2004) 205–250 [hep-th/0404134].

[80] K. Becker, G. Guo and D. Robbins, *Higher derivative brane couplings from T-duality*, JHEP **1009** (2010) 029 [arXiv:1007.0441].

[81] A. A. Tseytlin, *On nonAbelian generalization of Born-Infeld action in string theory*, Nucl.Phys. **B501** (1997) 41–52 [hep-th/9701125].

[82] H. Kawai, D. Lewellen and S. Tye, *A relation between tree amplitudes of closed and open strings*, Nucl.Phys. **B269** (1986) 1.

[83] S. Stieberger, *Open & closed vs. pure open string disk amplitudes*, [arXiv:0907.2211].

[84] Z. Bern, J. Carrasco and H. Johansson, *New relations for gauge-theory amplitudes*, Phys.Rev. **D78** (2008) 085011 [arXiv:0805.3993].

[85] Z. Bern, T. Dennen, Y.-t. Huang and M. Kiermaier, *Gravity as the square of gauge theory*, Phys.Rev. **D82** (2010) 065003 [arXiv:1004.0693].

Bibliography

[86] Z. Bern, J. J. M. Carrasco and H. Johansson, *Perturbative quantum gravity as a double copy of gauge theory*, Phys.Rev.Lett. **105** (2010) 061602 [arXiv:1004.0476].

[87] C. R. Mafra, O. Schlotterer, S. Stieberger and D. Tsimpis, *Six open string disk amplitude in pure spinor superspace*, [arXiv:1011.0994].

[88] C. R. Mafra, O. Schlotterer and S. Stieberger, *Explicit BCJ Numerators from Pure Spinors*, [arXiv:1104.5224].

[89] Z. Bern, J. J. Carrasco, L. J. Dixon, H. Johansson and R. Roiban, *Amplitudes and Ultraviolet Behavior of N=8 Supergravity*, [arXiv:1103.1848].

[90] S. Stieberger, *Constraints on Tree-Level Higher Order Gravitational Couplings in Superstring Theory*, Phys.Rev.Lett. **106** (2011) 111601 [arXiv:0910.0180].

[91] N. Beisert, H. Elvang, D. Z. Freedman, M. Kiermaier, A. Morales and S. Stieberger, *E7(7) constraints on counterterms in N=8 supergravity*, Phys.Lett. **B694** (2010) 265–271 [arXiv:1009.1643].

[92] P. Ramond, *Dual Theory for Free Fermions*, Phys.Rev. **D3** (1971) 2415–2418.

[93] A. Neveu and J. Schwarz, *Factorizable dual model of pions*, Nucl.Phys. **B31** (1971) 86–112.

[94] A. Neveu and J. Schwarz, *Quark Model of Dual Pions*, Phys.Rev. **D4** (1971) 1109–1111.

[95] D. Lüst and S. Theisen, *Lectures on string theory*, Lect.Notes Phys. **346** (1989) 1–346.

[96] R. Blumenhagen and E. Plauschinn, *Introduction to conformal field theory*, Lect.Notes Phys. **779** (2009) 1–256.

[97] P. Di Francesco, P. Mathieu and D. Senechal, *Conformal field theory*, Springer (1997).

[98] D. Friedan, E. J. Martinec and S. H. Shenker, *Conformal invariance, supersymmetry and string theory*, Nucl.Phys. **B271** (1986) 93.

[99] J. Cohn, D. Friedan, Z.-a. Qiu and S. H. Shenker, *Covariant quantization of supersymmetric string theories: The spinor field of the Ramond-Neveu-Schwarz model*, Nucl. Phys. **B278** (1986) 577.

[100] N. Berkovits, *A new description of the superstring*, [hep-th/9604123].

[101] C. R. Mafra, *Four-point one-loop amplitude computation in the pure spinor formalism*, JHEP **0601** (2006) 075 [hep-th/0512052].

[102] N. Berkovits and C. R. Mafra, *Some Superstring Amplitude Computations with the Non-Minimal Pure Spinor Formalism*, JHEP **0611** (2006) 079 [hep-th/0607187].

[103] C. R. Mafra, O. Schlotterer, S. Stieberger and D. Tsimpis, *A recursive formula for N-point SYM tree amplitudes*, [arXiv:1012.3981].

[104] V. Kostelecky, O. Lechtenfeld, W. Lerche, S. Samuel and S. Watamura, *Conformal techniques, bosonization and tree level string amplitudes*, Nucl.Phys. **B288** (1987) 173.

[105] V. Knizhnik, *Covariant Fermionic Vertex in Superstrings*, Phys.Lett. **B160** (1985) 403–407.

[106] C. Johnson, *D-branes*, Cambridge University Press (2003).

[107] J. Polchinski, S. Chaudhuri and C. V. Johnson, *Notes on D-branes*, [hep-th/9602052].

[108] S. Stieberger and T. R. Taylor, *Supersymmetry relations and MHV amplitudes in superstring theory*, Nucl.Phys. **B793** (2008) 83–113 [arXiv:0708.0574].

[109] I. Antoniadis, K. Benakli and A. Laugier, *Contact interactions in D-brane models*, JHEP **0105** (2001) 044 [hep-th/0011281].

[110] M. Cvetic and I. Papadimitriou, *Conformal field theory couplings for intersecting D-branes on orientifolds*, Phys.Rev. **D68** (2003) 046001 [hep-th/0303083].

[111] S. Abel and A. Owen, *Interactions in intersecting brane models*, Nucl.Phys. **B663** (2003) 197–214 [hep-th/0303124].

[112] S. Abel and A. Owen, *N point amplitudes in intersecting brane models*, Nucl.Phys. **B682** (2004) 183–216 [hep-th/0310257].

[113] S. Stieberger and T. R. Taylor, *Amplitude for N-Gluon superstring scattering*, Phys.Rev.Lett. **97** (2006) 211601 [hep-th/0607184].

[114] S. Stieberger and T. R. Taylor, *Multi-gluon scattering in open superstring theory*, Phys.Rev. **D74** (2006) 126007 [hep-th/0609175].

[115] S. Stieberger and T. R. Taylor, *Complete six-gluon disk amplitude in superstring theory*, Nucl.Phys. **B801** (2008) 128–152 [arXiv:0711.4354].

[116] D. Lüst, *Seeing through the String Landscape - a String Hunter's Companion in Particle Physics and Cosmology*, JHEP **0903** (2009) 149 [arXiv:0904.4601].

[117] O. Schlotterer, *Higher spin scattering in superstring theory*, [arXiv:1011.1235].

Bibliography

[118] D. Oprisa and S. Stieberger, *Six gluon open superstring disk amplitude, multiple hypergeometric series and Euler-Zagier sums*, [hep-th/0509042].

[119] S. J. Parke and T. Taylor, *An amplitude for n gluon scattering*, Phys.Rev.Lett. **56** (1986) 2459.

[120] E. P. Verlinde and H. L. Verlinde, *Multiloop calculations in covariant superstring theory*, Phys. Lett. **B192** (1987) 95.

[121] O. Lechtenfeld and A. Parkes, *On covariant multiloop superstring amplitudes*, Nucl. Phys. **B332** (1990) 39.

[122] E. D'Hoker and D. H. Phong, *The geometry of string perturbation theory*, Rev. Mod. Phys. **60** (1988) 917.

[123] O. Lechtenfeld, *Superconformal ghost correlations on Riemann surfaces*, Phys. Lett. **B232** (1989) 193.

[124] E. Witten, *Elliptic genera and quantum field theory*, Commun. Math. Phys. **109** (1987) 525.

[125] A. Schellekens and N. Warner, *Anomalies, characters and strings*, Nucl.Phys. **B287** (1987) 317.

[126] W. Lerche, B. Nilsson, A. Schellekens and N. Warner, *Anomaly cancelling terms from the elliptic genus*, Nucl.Phys. **B299** (1988) 91.

[127] W. Lerche, A. Schellekens and N. Warner, *Lattices and strings*, Phys.Rept. **177** (1989) 1.

[128] O. Lechtenfeld and W. Lerche, *On nonrenormalization theorems for four-dimensional superstrings*, Phys. Lett. **B227** (1989) 373.

[129] D. Lüst and S. Theisen, *Superstring partition functions and the characters of exceptional groups*, Phys. Lett. **B227** (1989) 367.

[130] J. J. Atick and A. Sen, *Correlation functions of spin operators on a torus*, Nucl.Phys. **B286** (1987) 189.

[131] J. Fay, *Theta functions on Riemann surfaces*, Springer (1973).

[132] D. Mumford, *Tata Lectures on Theta I, II*, Springer (1983/84).

[133] J. Igusa, *Theta functions*, Springer (1972).

[134] S. Stieberger and T. Taylor, *NonAbelian Born-Infeld action and type I. Heterotic duality (1): Heterotic F^6 terms at two loops*, Nucl.Phys. **B647** (2002) 49–68 [hep-th/0207026].

[135] S. Stieberger and T. Taylor, *NonAbelian Born-Infeld action and type I. Heterotic duality (2): Nonrenormalization theorems*, Nucl.Phys. **B648** (2003) 3–34 [hep-th/0209064].

[136] T. Ortin, *Gravity and strings*, Cambridge University Press (2007).

[137] P. C. West, *Supergravity, brane dynamics and string duality*, [hep-th/9811101].

[138] J. Strathdee, *Extended Poincare supersymmetry*, Int.J.Mod.Phys. **A2** (1987) 273.

[139] G. Wick, *The Evaluation of the Collision Matrix*, Phys.Rev. **80** (1950) 268–272.

[140] V. Kostelecky, O. Lechtenfeld, W. Lerche, S. Samuel and S. Watamura, *A four point amplitude for the O(16) x O(16) heterotic string*, Phys.Lett. **B182** (1986) 331.

[141] V. Kostelecky, O. Lechtenfeld, S. Samuel, D. Verstegen, S. Watamura et. al., *The six fermion amplitude in the superstring*, Phys.Lett. **B183** (1987) 299.

[142] V. Knizhnik and A. Zamolodchikov, *Current Algebra and Wess-Zumino Model in Two-Dimensions*, Nucl.Phys. **B247** (1984) 83–103.

[143] R. Cahn, *Semi-simple Lie algebras and their representations*, Dover Publications (2006).

[144] M. van Leeuwen, *LiE, A software package for Lie group computations*, Euromath Bulletin 1 **No 2** (1994).

[145] M. Fierz, *Zur Fermischen Theorie des β-Zerfalls*, Zeitschrift für Physik **104** (1937) 553–565.

[146] D. Tsimpis, *Introduction to superspace*, unpublished (2009).

[147] S. Wolfram, *The Mathematica Book*, Cambridge University Press (1996).

[148] M. Bousquet-Melou, *Counting walks in the quarter plane*, Trends Math., Birkhäuser, Basel (2002) 49–67.

[149] M. Bousquet-Melou, *Walks with small steps in the quarter plane*, [arXiv:0810.4387].

[150] J. J. Atick and A. Sen, *Spin field correlators on an arbitrary genus riemann surface and nonrenormalization theorems in string Theories*, Phys.Lett. **B186** (1987) 339.

[151] J. J. Atick and A. Sen, *Covariant one loop fermion emission amplitudes in closed string theories*, Nucl.Phys. **B293** (1987) 317.

[152] L. Alvarez-Gaume, G. W. Moore, P. C. Nelson, C. Vafa and J. Bost, *Bosonization in arbitrary genus*, Phys.Lett. **B178** (1986) 41–47.

[153] L. Alvarez-Gaume, J. Bost, G. W. Moore, P. C. Nelson and C. Vafa, *Bosonization on higher genus Riemann surfaces*, Commun.Math.Phys. **112** (1987) 503.

[154] L. Alvarez-Gaume, G. W. Moore and C. Vafa, *Theta Functions, Modular Invariance and Strings*, Commun.Math.Phys. **106** (1986) 1–40.

[155] O. Schlotterer, *Higher loop spin field correlators in D=4 superstring theory*, JHEP **1009** (2010) 050 [arXiv:1001.3158].

[156] R. Medina and L. A. Barreiro, *Higher N-point amplitudes in open superstring theory*, PoS **IC2006** (2006) 038 [hep-th/0611349].

[157] L. A. Barreiro and R. Medina, *5-field terms in the open superstring effective action*, JHEP **0503** (2005) 055 [hep-th/0503182].

[158] L. Brink, J. H. Schwarz and J. Scherk, *Supersymmetric Yang-Mills Theories*, Nucl.Phys. **B121** (1977) 77.

[159] T. Banks and L. J. Dixon, *Constraints on string vacua with space-time supersymmetry*, Nucl.Phys. **B307** (1988) 93–108.

[160] Z. Koba and H. B. Nielsen, *Reaction amplitude for n mesons: A generalization of the Veneziano-Bardakci-Ruegg-Virasora model*, Nucl.Phys. **B10** (1969) 633–655.

[161] F. A. Berends and W. Giele, *Recursive calculations for processes with n gluons*, Nucl.Phys. **B306** (1988) 759.

[162] Z. Xu, D.-H. Zhang and L. Chang, *Helicity amplitudes for multiple Bremsstrahlung in massless nonabelian gauge theories*, Nucl.Phys. **B291** (1987) 392.

[163] M. L. Mangano and S. J. Parke, *Multiparton amplitudes in gauge theories*, Phys.Rept. **200** (1991) 301–367 [hep-th/0509223].

[164] L. J. Dixon, *Calculating scattering amplitudes efficiently*, [hep-ph/9601359].

[165] M. Bianchi, H. Elvang and D. Z. Freedman, *Generating tree amplitudes in N=4 SYM and N=8 SG*, JHEP **0809** (2008) 063 [arXiv:0805.0757].

[166] J. Brödel and L. J. Dixon, R^4 *counterterm and* $E(7)(7)$ *symmetry in maximal supergravity*, JHEP **1005** (2010) 003 [arXiv:0911.5704].

[167] L. J. Slater, *Generalized Hypergeometric Functions*, Cambridge University Press (1966).

[168] H. Exton, *Multiple Hypergeometric Functions and Applications*, Ellis Horwood (1985).

[169] I. S. Gradshteyn and I. M. Ryzhik, *Table of Integrals, Series and Products*, Academic Press (1994).

[170] D. Zagier, *Values of zeta functions and their applications*, Birkhäuser (1994) 497–512.

[171] R. Kleiss and H. Kuijf, *Multi-gluon cross-sections and five jet production at hadron colliders*, Nucl.Phys. **B312** (1989) 616.

[172] V. Gates, E. Kangaroo, M. Roachcock and W. Gall, *Stuperspace*, Physica **15D** (1985) 289–293.

[173] S. Dittmaier, *Weyl-van der Waerden formalism for helicity amplitudes of massive particles*, Phys.Rev. **D59** (1999) 016007 [`hep-ph/9805445`].

[174] S. Bosch, *Lineare Algebra*, Springer (2009).

Die VDM Verlagsservicegesellschaft sucht für wissenschaftliche Verlage abgeschlossene und herausragende

Dissertationen, Habilitationen, Diplomarbeiten, Master Theses, Magisterarbeiten usw.

für die kostenlose Publikation als Fachbuch.

Sie verfügen über eine Arbeit, die hohen inhaltlichen und formalen Ansprüchen genügt, und haben Interesse an einer honorarvergüteten Publikation?

Dann senden Sie bitte erste Informationen über sich und Ihre Arbeit per Email an *info@vdm-vsg.de*.

Sie erhalten kurzfristig unser Feedback!

VDM Verlagsservicegesellschaft mbH
Dudweiler Landstr. 99
D - 66123 Saarbrücken
www.vdm-vsg.de

Telefon +49 681 3720 174
Fax +49 681 3720 1749

Die VDM Verlagsservicegesellschaft mbH vertritt

Printed by Books on Demand GmbH, Norderstedt / Germany